14th Pacific Rim Conference on Ceramic and Technology with 2021 Glass & Optical Materials Division Annual Meeting

Online
13 – 16 December 2021

ISBN: 978-1-7138-6324-3

Printed from e-media with permission by:

Curran Associates, Inc.
57 Morehouse Lane
Red Hook, NY 12571

Some format issues inherent in the e-media version may also appear in this print version.

Printed with permission by Curran Associates, Inc. (2023)

The American Ceramic Society
550 Polaris Parkway, Suite 510
Westerville, Ohio 43082

Phone: 1-866-721-3322 (toll free U.S. only)
1-614-890-4700 (outside U.S.)
Fax: 1-614-899-6109

ceramics.org

Additional copies of this publication are available from:

Curran Associates, Inc.
57 Morehouse Lane
Red Hook, NY 12571 USA
Phone: 845-758-0400
Fax: 845-758-2634
Email: curran@proceedings.com
Web: www.proceedings.com

The American Ceramic Society

14th Pacific Rim Conference on Ceramic and Glass Technology

with

2021 Glass & Optical Materials Division Annual Meeting

ABSTRACT BOOK

December 13–16, 2021
Virtual Conference

The American Ceramic Society
www.ceramics.org

Introduction

This volume contains abstracts for over 700 presentations during the Virtual 14th Pacific Rim Conference on Ceramic and Glass Technology with 2021 Glass & Optical Materials Division Annual Meeting. The abstracts are reproduced as submitted by authors, a format that provides for longer, more detailed descriptions of papers. The American Ceramic Society accepts no responsibility for the content or quality of the abstract content. Abstracts are arranged by day, then by symposium and session title. An Author Index appears at the back of this book. The Meeting Guide contains locations of sessions with times, titles and authors of papers, but not presentation abstracts.

How to Use the Abstract Book

Refer to the Table of Contents to determine page numbers on which specific session abstracts begin. At the beginning of each session are headings that list session title, location and session chair. Starting times for presentations and paper numbers precede each paper title. The Author Index lists each author and the page number on which their abstract can be found.

The PACRIM program abstracts are found on pages 17–130.

The GOMD program abstracts are found on pages 131-177.

MEETING REGULATIONS

The American Ceramic Society is a nonprofit scientific organization that facilitates the exchange of knowledge meetings and publication of papers for future reference. The Society owns and retains full right to control its publications and its meetings. The Society has an obligation to protect its members and meetings from intrusion by others who may wish to use the meetings for their own private promotion purpose. Literature found not to be in agreement with the Society's goals, in competition with Society services or of an offensive nature will not be displayed anywhere in the vicinity of the meeting. Promotional literature of any kind may not be displayed without the Society's permission and unless the Society provides tables for this purpose. Literature not conforming to this policy or displayed in other than designated areas will be disposed. The Society will not permit unauthorized scheduling of activities during its meeting by any person or group when those activities are conducted at its meeting place in interference with its programs and scheduled activities. The Society does not object to appropriate activities by others during its meetings if it is consulted with regard to time, place, and suitability. Any person or group wishing to conduct any activity at the time and location of the Society meeting must obtain permission from the Executive Director or Director of Meetings, giving full details regarding desired time, place and nature of activity.

Diversity Statement: The American Ceramic Society values diverse and inclusive participation within the field of ceramic science and engineering. ACerS strives to promote involvement and access to leadership opportunity regardless of race, ethnicity, gender, religion, age, sexual orientation, nationality, disability, appearance, geographic location, career path or academic level.

The American Ceramic Society will record all virtual sessions and make them available for viewing by registered attendees until February 28, 2022 through the Bravura Connect O portal.

By participating in the conference, you grant The American Ceramic the right to use your name, whether in print, electronic or other media, including The American Ceramic Society's website.

Registration Requirements: Attendance at any meeting of the Society shall be limited to duly registered persons.

Disclaimer: Statements of fact and opinion are the responsibility of the authors alone and do not imply an opinion on the part of the officers, staff or members of The American Ceramic Society. The American Ceramic Society assumes no responsibility for the statements and opinions advanced by the contributors to its publications or by the speakers at its programs; nor does The American Ceramic Society assume any liability for losses or injuries suffered by attendees at its meetings. Registered names and trademarks, etc. used in its publications, even without specific indications thereof, are not to be considered unprotected by the law. Mention of trade names of commercial products does not constitute endorsement or recommendations for use by the publishers, editors or authors.

Final determination of the suitability of any information, procedure or products for use contemplated by any user, and the manner of that use, is the sole responsibility of the user. Expert advice should be obtained at all times when implementation is being considered, particularly where hazardous materials or processes are encountered.

Table of Contents

PACRIM Symposium 31: Advanced Functional Materials, Devices, and Systems for Environmental Conservation, Pollution Control, and Critical Materials

PACRIM Symposium 35: Advanced Additive Manufacturing Technologies for Bio-applications; Materials, Processes, and Systems

PACRIM Symposium 38: Nanotechnology in Medicine

PACRIM Symposium 39: Biomimetics and Bioinspired Processing of Advanced Materials

PACRIM Symposium 40: 6th International Richard M. Fulrath Symposium, "Frontiers of Ceramics for a Sustainable Society"

PACRIM Symposium 42: Young Investigator Forum - Next-Generation Materials for Multifunctional Applications and Sustainable Development, and Concurrent Societal Challenges in the New Millennium

PACRIM Symposium 4: Novel, Green, and Strategic Processing and Manufacturing Technologies

PACRIM Symposium 10: Sol-Gel Processing and Related Liquid-Phase Synthesis of Ceramics

PACRIM Symposium 12: Specific Reaction Field and Material Fabrication Design

PACRIM Symposium 14: Functional Nanomaterials for Energy Harvesting and Solar Fuels

PACRIM Symposium 17: Multifunctional Coatings for Structural, Energy, and Environmental Applications

PACRIM Symposium 19: Geopolymers: Low Energy and Environmentally Friendly Ceramics

PACRIM Symposium 21: Crystalline Materials for Electrical, Optical, and Medical Applications

PACRIM Symposium 24: Solid Oxide Fuel Cells and Hydrogen Technologies

PACRIM Symposium 25: Direct Thermal to Electrical Energy Conversion Materials, Applications, and Thermal Energy Harnessing Challenges

PACRIM Symposium 26: Materials for Solar Thermal Energy Conversion and Storage

PACRIM Symposium 27: Advanced Materials and Technologies for Electrochemical Energy Storage Systems

PACRIM Symposium 34: Glass and Ceramics for Nuclear Waste Treatment and Sequestration

PACRIM Symposium 1: Characterization and Modeling of Ceramic Interfaces: Structure, Bonding, and Grain Growth

PACRIM Symposium 3: Advanced Structure Analysis and Characterization of Ceramics

PACRIM Symposium 5: Polymer Derived Ceramics (PDCs) and Composites

PACRIM Symposium 8: Porous Ceramics: Innovative Processing and Advanced Applications

PACRIM Symposium 9: Additive Manufacturing and 3D Printing Technologies

PACRIM Symposium 13: Novel Nanocrystal Technologies for Advanced Ceramic Materials & Devices

PACRIM Symposium 5: Polymer Derived Ceramics (PDCs) and Composites

PACRIM Symposium 8: Porous Ceramics: Innovative Processing and Advanced Applications

PACRIM Symposium 9: Additive Manufacturing and 3D Printing Technologies

PACRIM Symposium 13: Novel Nanocrystal Technologies for Advanced Ceramic Materials & Devices

PACRIM Symposium 18: Advanced Wear Resistant Materials: Tribology and Reliability

PACRIM Symposium 23: Transparent Ceramic Materials and Devices

PACRIM Symposium 24: Solid Oxide Fuel Cells and Hydrogen Technologies

PACRIM Symposium 28: Atomic Structure and Electrochemical Property Diagnosis Toward Full Crystal Rechargeable Batteries

PACRIM Symposium 33: Photocatalysts for Energy and Environmental Applications

Monday, December 13, 2021

PACRIM Plenary Lectures

Plenary Session

Room: Regency A/B/C
Session Chair: Michael Halbig, NASA Glenn Research Center

9:00 AM

(PACRIM-001-2021) Thermal Properties of Advanced Ceramics

M. White*[1]

1. Dalhousie University, Department of Chemistry, Canada

One of the many attributes of ceramics is their ability to withstand high temperatures. Their range of compositions and microstructures also allows significant applications that make use of the exceptional thermal properties of ceramics. In this presentation, the fundamentals of thermal properties, including thermal expansion and thermal conductivity, will be reviewed, with an eye to future applications. A running theme throughout the presentation will be consideration of sustainable aspects, including elemental availability and embodied energy.

9:40 AM

(PACRIM-002-2021) Superconducting quantum computer and its future issues

J. Tsai*[1]

1. Tokyo University of Science, Department of Physics, Japan

Superconducting qubit technology is arguably the most advanced platform for the quantum computer. In recent years, commercial superconducting quantum computers and quantum annealers have already started to serve the general public. However, to further scale up the performance, there still exist many important issues for superconducting quantum system. It is crucial to improve the accuracy of the gate operations, and in this direction, further increase in the coherence time of the qubit is important. Coherence time is strongly influenced by the environments surrounding the superconducting qubit such as surface of substrate and qubit itself, and optimizations of these surface material/processing issues would be of vital importance. The further integration of the number of the qubits on a chip is required for practical quantum computers. The wiring and packaging of the integrated qubit chip is one of the challenging problems in the scaling of the quantum system. Currently, the difficulty is dealt with sophisticated 3D wring technologies. The recently proposed micro architecture that utilizes a seudo-2D wiring network might be able to help circumvent the hurdle.

10:40 AM

(PACRIM-003-2021) Recent Advances in the Disruptive Technology of Additive Manufacturing at Boeing

M. Orme*[1]

1. The Boeing Company, Boeing Additive Manufacturing, USA

Additive Manufacturing offers numerous benefits in the aerospace sector including: light-weighting, part consolidation, on-demand manufacturing, faster time to market, and addressing pain points in supply chain, all of which leads to cost/fuel savings. More important than part-specific value added, additive manufacturing enables the opportunity to optimize the end product, not just the part within the product. This presentation discusses the challenges of full scale production of air-worthy additively manufactured hardware, and how Boeing is addressing these challenges, including the creation of the digital twin at every instance along our value stream, and the incorporation of this data into our digital thread to enable scale and quality. Finally, this talk will describe two case studies that illustrate the value added of additive manufacturing insertions in current Boeing platforms.

11:20 AM

(PACRIM-004-2021) Halide perovskite-based photovoltaics - from materials to devices

S. Seok*[1]

1. Ulsan National Institute of Science and Technology, School of Energy and Chemical Engineering, Republic of Korea

Oxide perovskite materials have occupied a very important position in functional materials such as ferroelectric, piezoelectric, thermoelectric, and even superconducting. They are represented by the general formula ABX_3 and have the same crystal structure as calcium titanate ($CaTiO_3$), where A and B sites accept inorganic cations of various valence and ionic radius. Likewise, halide perovskites in which halogen is substituted at the site of an oxygen anion has been applied to solar cells, showing very excellent properties, and has emerged as a promising material in various fields. Specifically, from a photovoltaic point of view, halide perovskites exhibit beneficial properties for high-performance photovoltaic systems such as a suitable band gap (1.5 - 1.4 eV), high absorption coefficient (10^4 - 10^5 cm^{-1}), low exciton binding energy (< 50 meV), and long charge-carrier diffusion length (~175 μm). These excellent properties allowed the power conversion efficiency (PCE) of perovskite solar cells (PSC) to reach values above 25%, which is comparable to that of silicon solar cells studied over a long period of time. However, such rapid increase in PCE has been achieved by optimizing the device structure, the uniform thin film deposition process, and the material composition of the halide perovskite, in addition to its excellent material properties.

PACRIM Symposium 4: Novel, Green, and Strategic Processing and Manufacturing Technologies

Novel, Green, and Strategic Processing I

Room: Georgia A
Session Chair: Tatsuki Ohji, National Institute of Advanced Industrial Science and Technology (AIST)

3:00 PM

(PACRIM-006-2021) Effects of sintering additive and parameters on the solid-state single-crystal growth (Invited)

I. Milisavljevic[1]; Y. Wu*[1]

1. Alfred University, Kazuo Inamori School of Engineering, New York State College of Ceramics, USA

Solid-state single-crystal growth (SSCG) has been recognized in recent years as a promising alternative method for the growth of single-crystals. The simplicity and cost-effectiveness of the SSCG technique are reflected in the application of the conventional sintering equipment and the processing at the temperatures below the melting point of the materials. However, despite its novelty and numerous advantages over the conventional techniques, the knowledge about the SSCG technique is still very limited. In this work, $Y_3Al_5O_{12}$ (YAG) is used to investigate the effects of different sintering additives, such as MgO, CaO, and SiO_2 on the single-crystal conversion phenomenon during the spark plasma sintering. Polished YAG single-crystal seed, buried inside the powder samples before sintering, was used to instigate the conversion of the polycrystalline matrix. The effects of sintering temperature, time, and the post-sintering treatment atmosphere on the conversion kinetics and mechanisms were also discussed. It was observed that the presence of CaO as a sintering additive had a significant effect on the conversion kinetics by increasing the single-crystal growth rate almost ten-fold. The XRD and SEM characterization techniques were used to investigate the phase compositions and the microstructures of the area around the converted single-crystal, respectively.

3:30 PM

(PACRIM-007-2021) Effects of Sintering Additives on Densification and Thermal Conductivity of Sintered Reaction-Bonded Silicon Nitride Ceramics (Invited)

Y. Zhou*[1]; H. Hyuga[1]; K. Hirao[1]

1. National Institute of Advanced Industrial Science and Technology (AIST), Japan

Silicon nitride ceramics with both high thermal conductivity and good mechanical properties are promising substrate materials for packaging of the next-generation power devices. Our recent studies have revealed that silicon nitride ceramics with high thermal conductivity could be fabricated by a sintering of reaction-bonded silicon nitride (SRBSN) method by employing a high purity silicon powder as a starting material and 2 mol% of Y_2O_3 and 5 mol% of MgO as sintering aids. In the present study, various amounts and combinations of Y_2O_3 and MgO were used as sintering additives for the SRBSN process. Nitridation was carried out at 1400 °C for 4 h under a nitrogen pressure of 0.1 MPa, followed by post-sintering at 1900 °C for 6 or 24 h under a nitrogen pressure of 1 MPa. It was found that the amount and composition of the sintering additives exerted great influences on the densification behavior of the nitrided compacts and thermal conductivity of the sintered silicon nitride ceramics. It indicates that thermal conductivity of sintered reaction-bonded silicon nitride ceramics can be improved by judiciously choosing the composition and amount of sintering additives.

4:00 PM

(PACRIM-008-2021) Effect of TiB_2 particle size on the microstructures and mechanical properties of hot-pressed B_4C-TiB_2 composites

W. Guo*[1]; Q. He[1]; A. Wang[1]; T. Tian[1]; L. Hu[1]; W. Wang[1]; Z. Fu[1]

1. Wuhan University of Technology, China

B_4C-13vol%TiB_2 composites were fabricated via hot pressing using B_4C and ball milled TiB_2 powder mixtures. The effects of TiB_2 with different particle sizes on the microstructure and mechanical properties were investigated. The results showed that TiB_2 particle sizes played an important role on the properties of B_4C-TiB_2 composites. In addition, the SiO_2 introduced by ball milling also affects the densification and properties of B_4C-TiB_2 composites. The typical values of relative density, hardness, flexural strength, and fracture toughness of the composites reached 99.30%, 30.81GPa, 554 MPa, and 7.69 MPa•m$^{1/2}$, respectively. The toughening mechanisms of B_4C-TiB_2 composites could be explained by crack deflection and crack branching.

4:20 PM

(PACRIM-009-2021) Enhanced toughness and strength of boron carbide ceramics with reduced graphene oxide fabricated by hot pressing

A. Wang*[1]; Q. He[1]; L. Hu[1]; Z. Zhang[2]; Z. Fu[1]; W. Wang[1]; Y. Xiong[3]

1. Wuhan University of Technology, China
2. Hebei University of Engineering, China
3. Hubei University of Technology, China

Boron carbide (B_4C) hybrids with different contents of graphene oxide (GO) were prepared by a heterogeneous co-precipitation method using cetyltrimethyl ammonium bromide (CTAB) as the cationic surfactant. The as-obtained mixtures were further hot-pressed at 1950 °C under 30 MPa for 60 min, by which B_4C–reduced GO (rGO) composites were fabricated. It was found that the addition of only 0.5 wt% rGO could alter the predominance of trans-granular fracture in monolithic B_4C ceramic material to mixed trans-granular and inter-granular modes in B_4C–rGO composites. The flexural strength and fracture toughness of the B_4C–2 wt% rGO were increased by 31% (from 350 to 455 MPa) and 83% (from 3.20 to 5.85 MPa•m$^{1/2}$), respectively, compared with those of pure B_4C. The improved mechanical properties are attributed to the mechanisms of pull-out and bridging of rGO and crack deflection, as evidenced by microstructural observations. The energy dissipation in the present B_4C–rGO composites was further verified using two micromechanical models.

4:40 PM

(PACRIM-010-2021) Microstructural characterization and mechanical properties of Hot-Pressed boron-rich boron carbides

T. Tian*[1]; W. Wang[1]; Z. Fu[1]

1. Wuhan University of Technology, China

Boron carbide has a very wide range of solubility due to the substitution of B and C atoms in the crystle. In this study, boron carbides with different stoichiometric ratios were prepared using hot pressing to explore the influences of B/C atomic ratio on its microstructure and properties. XRD analysis shows that excessive boron atoms lead to lattice expansion. It also can be seen that boron atoms has disordered substitution on the chain and icosahedron by Raman spectroscopy. The microstructure analysis reveals that the addition of boron could improve the densification, more stacking faults and twins were observed in boron-rich boron carbide. Moreover, mechanical property testing were investigated that the Vicker hardness, modulus and sound velocity all decrease with the increase of boron content. Meantime, the flexural strength, fracture toughness and electrical conductivity of the samples achieve the optimal value while the stoichiometric ratio of B/C reaches 6.5.

5:00 PM

(PACRIM-011-2021) Novel, Green, and Strategic Processing of TiO_2 Nanostructures via Cold Crystallization (Invited)

Y. Masuda*[1]

1. National Institute of Advanced Industrial Science and Technology (AIST), Japan

TiO_2 nanostructures have been developed via cold crystallization. Nuleation and crystal growth were controlled in aqueous solutions to form characteristic nanostructures. Nanostructured TiO_2 coatings were successfully formed on polyethylene terephthalate (PET) films and fluorine-doped tin oxide (FTO) films in aqueous solutions. They contained an assembly of nanoneedles that grow perpendicular to the films. The surface area of the coatings on PET films reached around 284 times that of a bare PET film. Micro-, nano-, or subnanosized surface roughness and inside pores contributed to the high nitrogen adsorption. The coatings on FTO films showed an acetaldehyde removal rate of 2.80 μmol/h; this value is similar to those of commercial products certified by the Photocatalysis Industry Association of Japan. The rate increased greatly to 10.16 μmol/h upon annealing in air at 500 °C for 4 h; this value exceeded those of commercial products. Further, the coatings showed a NO_x removal rate of 1.04 μmol/h; this value is similar to those of commercial products. The rate decreased to 0.42 μmol/h upon annealing. NO_x removal was affected by the photocatalyst's surface area rather than its crystallinity.

PACRIM Symposium 6: Advanced Powder Processing and Manufacturing Technologies

Advanced Recycling Technology and Energy-saving Processes

Room: Oxford

1:30 PM

(PACRIM-012-2021) Advanced powder processing for recycling of silicon-based photovoltaic panel (Invited)

C. Tokoro*[1]

1. Waseda University, Japan

Photovoltaic (PV) power generation is one of the most promising renewable energy technologies. Silicon-based PV panels, which are currently the mainstream, consist of glass, and copper/silver wire and silicon sealed with resin. Since glass accounts for more than 70% by weight, it is necessary to collect it with high purity to realize horizontal recycling as much as possible. To accomplish this, a recycling process for recovering high-purity glass by a hot-knife method or a shaving method has been proposed. In these methods, the metal wire and silicon sealed with resin remain for further treatment. In this presentation, I will introduce the selective separation of metal wires from the resin by the electrical pulsed discharge method. Although it is difficult to pulverize PV panel cells because metal wires are sealed in a hard resin, this method is also effective as a pretreatment for pulverization for the liberation of them. Also, there is another recycling process that shreds the entire PV panel without separating the glass and the cell. In this case, locked glass and resin particles are generated in the pulverized product. To recycle it into glass wool and tile additives, it is necessary to reprocess it to improve the purity of glass as much as possible. In this presentation, I will introduce that the eccentric stirring type mill was effective for selective grinding of glass and liberation from resin.

1:50 PM

(PACRIM-013-2021) Numerical investigation of gravity separation mechanism on a shaking table

Y. Tsunazawa*[1]; Y. Kon[1]

1. National Institute of Advanced Industrial Science and Technology (AIST), Geological Survey of Japan, Japan

A shaking table is one of the physical separation methods utilizing the difference in particle density. Because of the lower initial and operating costs, it has been widely employed in mineral processing and recycling processes. For the effective separation, it is important to control particle behavior on a shaking table. Since particle behavior is strongly affected by instrument and operating parameters, knowledge about the segregation and separation mechanism of the shaking table remains limited. To investigate the separation mechanism on a shaking table, the discrete element method is applied to gravity separation of heavy and light particles. To simplify the system, two types of spherical particles which have the same diameter and different density are simulated. Focusing on the macroscopic particle behavior, the effects of shaking amplitude and frequency are evaluated. Simulation results show that a larger amplitude and longer frequency of the shaking table promote density segregation. They also show that the progress of density segregation at different amplitudes and frequencies of the shaking table has a periodicity, which explains the time variation of the average velocity difference. Consequently, the discrete element simulation shows density segregation due to the velocity difference can explain the gravity separation mechanism on a shaking table.

2:10 PM

(PACRIM-014-2021) High-Quality Recovery of Cathode Active Material from Spent Li-ion Battery by Pulsed Power Discharge

K. Teruya*[1]; S. Lim[1]; K. Mochizuki[2]; T. Namihira[3]; T. Koita[1]; C. Tokoro[1]

1. Waseda University, Department of Resources and Environmental Engineering, Faculty of Science and Engineering, Japan
2. Retoca Laboratory LLC, Japan
3. Institute of Industrial Nanomaterials, Kumamoto University, Japan

Advanced recycling is required for Li-ion batteries for proper treatment after their use. In this study, we investigated the separation of the cathode active material ($LiNi_xCo_yMn_zO_2$) from the aluminum foil current collector by the pulsed power discharge. Electrodes of the pulsed power discharge were connected at each end of 80 mm × 210 mm of the cathode sample and set in a 5 L of water filled in the chamber. When applying 4.5 kJ using 40 μF capacitors with a pulse width of 75 μs and current peak of 32.6 kA, 93.9% of the active material was separated from the Al foil and recovered as particles. Rapid Joule heating of the Al foil played an important role in separation process, since the Al foil could be heated enough to melt adhesives on its surface. Based on the results of X-ray fluorescence and X-ray diffraction analysis, the chemical constitution and crystal form of the recovered particles were almost the same as that of untreated cathode active material. However, a small amount of Al particles were contaminated into the recovered particles due to plasmalization of Al foil around the electrodes. Besides, a part of elements such as Li, F, Mn, and P dissolved into the water in the chamber after the pulsed power discharge. These elements were originated from the electrolyte ($LiPF_6$) but some part of cathode active material, especially Mn, was also dissolved into the water.

2:30 PM

(PACRIM-015-2021) Microwave and mmWave Dielectrics for 6th Generation Devices

J. Varghese*[1]

1. Fraunhofer IKTS, Hybrid Microsystems, Germany

The rapid growth in wireless data traffic, microwave and mm-wave communications has become significant issues in fifth-generation communication systems. The propagation of commercial wireless technologies, such as cellular phones, global positioning systems (GPS), automobile radar systems, intelligent transport system (ITS), satellite broadcasting, Internet of things (IoT) and 5G networks have placed growing demands on the realization of dielectric materials in the millimetre wave frequency range. Fifth-generation systems operating in millimetre-wave fields have a principal concern about materials property stability. For the last two decades, HTCC and LTCC technologies are ruling the RF and microwave wireless products in microelectronics due to their reduced size and cost-effectiveness. Among these technologies LTCC offers, (i) low cost, (ii) design flexibility, (iii) miniaturization and (iv) integration of passive and active components. Herein, we address the emerging considerations for the 6th generation devices for microwave and mm-wave applications together with materials property concerns based on recent development and progress in ultra-low temperature co-fired ceramic (ULTCC) technology.

Novel Forming and Sintering Technology

Room: Oxford

3:20 PM

(PACRIM-016-2021) Effect of Al contained in polymer derived SiC crystals on creating stable crystal grain boundaries (Invited)

T. Ishikawa*[1]; H. Oda[2]

1. Tokyo University of Science, Yamaguchi (Sanyo-Onoda City University), Japan
2. Ube Industries, Ltd., Japan

Si-Al-C-O materials are well-known as a good precursor material for obtaining a dense, excellent heat-resistant SiC polycrystalline materials. Although it has been found that the aluminum contained in Si-Al-C-O materials plays an important role in obtaining dense SiC materials, its actual behavior during heat-treatment processes has not been investigated. In this study, we investigated the behavior of aluminum contained in the SiC crystals, during the densification of SiC crystals produced from the Si-Al-C-O material. A part of the aluminum contained in the SiC crystals plays an important role in creating a thermodynamically stable grain boundary by the diffusional transportation of aluminum to the SiC grain boundaries at temperatures above 1700°C. Subsequently, most of the aluminum disappeared during the heat treatment at higher temperatures (~1900°C); as a consequence, a dense SiC solid solution with uniformly distributed residual aluminum (0.15wt%) was formed without a secondary phase at the grain boundary.

3:40 PM

(PACRIM-017-2021) Properties of Si_3N_4 ceramics from tape casting and gas pressure sintering (Invited)

J. Zhang*[1]

1. Shanghai Institute of Ceramics, Chinese Academy of Sciences, China

Si_3N_4 ceramics exhibited both excellent mechanical properties and high theoretical thermal conductivity, is a prospective substrate candidate for power electronic devices. However, the development of high thermal conductivity Si_3N_4 is quite difficult, because ceramic sintering is conducted at high temperature with longer holding time under N_2 pressure etc., the resulting cost of the obtained Si_3N_4 ceramics was quite high and cannot be accepted by the market. In this paper, a relative fast processing was proposed using Y_2O_3-MgO-C as the sintering additives and silicon powder as the starting material. High thermal conductivity Si_3N_4 ceramic was prepared through tape casting, nitridation and gas pressure sintering at 1900°C, 0.6MPa with 2h holding. The influence of carbon addition on the sintering behavior, phase, microstructure, mechanical properties and thermal conductivity of Si_3N_4 was studied. Si_3N_4 ceramics with the relative density and thermal conductivity as 99% and 98 W/m×K respectively can be obtained.

4:00 PM

(PACRIM-018-2021) Microstructure evolution and grain growth mechanisms of pure h-BN ceramic and h-BN composite ceramics during hot-pressing (Invited)

X. Duan*[1]; D. Jia[1]; Z. Zhang[1]; B. Qiu[1]; Z. Yang[1]; Y. Zhou[1]

1. Harbin Institute of Technology, China

Pure h-BN ceramic specimens were prepared by hot-pressing under different sintering temperatures and pressures using ball milled h-BN powders composed of amorphous and nanocrystalline BN. Microstructures and thermal conductivities of these h-BN ceramic specimens were characterized and measured. Higher sintering pressure is more favorable to the preferred orientation growth of plate-like h-BN grains along the pressure direction, forming microstructures where the c-axes of h-BN grains are preferentially oriented perpendicular to the pressure direction. There is multi-area co-growth phenomenon around the grain boundaries composed of the basal planes of h-BN grains, which results in the formation of stacking faults in the as-grown h-BN grains. To investigate the liquid phase sintering mechanisms, h-BN composite ceramics with La_2O_3-Al_2O_3-SiO_2 addition were fabricated by hot pressing. The results indicated that liquid phase generated by the reaction of La_2O_3, Al_2O_3 and SiO_2 exhibited a good wettability with h-BN grains, contributing to fill the pores and improve the densification. The appropriate ratio and content of La-Al-Si-O glass phase can promote the densification of composite ceramics and the growth of BN grains.

4:20 PM

(PACRIM-019-2021) Room-temperture densification of MgO bulk ceramics with dispersed nitride phosphor particles

J. Tatami*[1]; E. Takahashi[2]; T. Takahashi[2]

1. Yokohama National University, Japan
2. Kanagawa Institute of Industrial Science and Technology, Japan

Wave conversion materials with high thermal conductivity are necessary for high-power semiconductor lighting. Ceramics have higher thermal conductivity than existing matrices such as resin or glass in which phosphor particles are dispersed. However, the high densification of ceramics generally requires high-temperature sintering, which degrades and alters the phosphor particles. In this study, we aimed to achieve the high densification of MgO ceramics at room temperature. Applying high hydrostatic pressure with water addition improved the sample packing ratio and promoted the formation of $Mg(OH)_2$. As a result, the relative density was ≥95%. Additionally, various nitride phosphor particles ($CaAlSiN_3:Eu^{2+}$, β-SiAlON:Eu_{2+}, and α-SiAlON:Eu^{2+}) were dispersed in the MgO matrix at room temperature without degrading the luminescence property. The thermal conductivity of the obtained sample was about 8 W m-1K-1, 40 times higher than that of the epoxy matrix.

4:40 PM

(PACRIM-020-2021) Design of Interparticle Photo-cross-linkable Suspensions for Hybrid Manufacturing of Transparent Silica Glass Components

M. Iijima*[1]; R. Sato[1]; J. Tatami[1]

1. Yokohama National University, Japan

Hybrid manufacturing of ceramics and glass has attracted wide attention due to its possibility for designing complex structured components. Typical photo-curable slurries were prepared by dispersing reagent powders into photo-curable monomers; therefore, large fraction of flammable polymer remained in the printed green components which lead to the requirement of slow firing process to avoid structural collapse. To overcome such problem, we recently reported a new series of suspension which can be photo-cured using interparticle photo-cross-linking reactions by small amounts of multifunctional acrylates (MAs). However, the mechanical properties of the photo-cured compacts were not reliable enough to be applied to the hybrid manufacturing. Here, using silica particle system as a model, we introduce a strategy to improve the mechanical properties of the green compacts photo-cured by interparticle photo-cross-linking reactions. The key was to co-polymerize silane alkoxides having acryloyl groups with MAs among the silica particles. The silane alkoxides in the cross-linking networks successfully improved the mechanical properties of the photo-cured green compacts which enabled the subtractive milling. Further, the resulted green compacts were sinter-able into transparent glass components using rapid heating profiles without forming any structural collapses.

5:00 PM

(PACRIM-021-2021) A Novel Room-Temperature Synthesis Technique for Producing High-Density Electroceramic Composites

E. C. Smith*[1]; R. Ubic[1]

1. Boise State University, Micron School of Materials Science and Engineering, USA

Room-temperature fabrication (RTF) represents a major technological leap in ceramic processing. This procedure has the potential to save a significant amount of energy because it avoids the use of the high temperatures needed for conventional ceramic sintering techniques. In this work, a novel approach is taken to achieve high densities (>95%) in ceramic compacts at room temperature. The electroceramic matrix phase ($Ba_xSr_{1-x}TiO_3$ or $PbZr_{1-y}Ti_yO_3$) is mixed with an aqueous solution of a dielectric binder material (Li_2MoO_4) and pressed under high pressure and high vacuum using a cylindrical, vacuum-assisted pellet die in a uniaxial press. The die is then repeatedly immersed in a bath sonicator for various times, after which it is pressed again. A major advantage of this technique is that it uses a relatively small amount of binder material (~12vol%); thus, the electroceramic filler material is the dominant phase in the resultant composite pellets. Finally, the permittivity of these composites is measured as a function of temperature. In the case of $Ba_xSr_{1-x}TiO_3$, the Curie temperatures are measured for several RTF pellets containing particles of different compositions.

PACRIM Symposium 20: Multiferroic Materials, Devices, and Applications

Multiferroic Materials, Devices, and Applications I

Room: Cypress
Session Chairs: Laurent Bellaiche, University of Arkansas; Liuyan Zhao, University of Michigan

1:30 PM

(PACRIM-022-2021) Observation of room temperature polar skyrmions (Invited)

R. Ramesh*[1]

1. UCBerkeley, MSE/Physics, USA

Complex topological configurations are a fertile playground to explore novel emergent phenomena and exotic phases in condensed-matter physics. For example, the recent discovery of polarization vortices and the associated complex-phase coexistence and response under applied field in superlattices of (PbTiO3)n/(SrTiO3)n suggests the presence of a complex, multi-dimensional system capable of exotic physical responses. I will describe the discovery of polar skyrmions in a lead-titanate layer confined by strontium-titanate layers by atomic-resolution scanning transmission electron microscopy (STEM). Phase-field modeling and second-principles calculations reveal that the polar skyrmions have a skyrmion number of +1 and resonant soft X-ray diffraction experiments show circular dichroism confirming chirality. Such nanometer-scale polar skyrmions exhibit a strong signature of negative permittivity at the surface of the skyrmion, which is furthermore highly tunable with an electric field. They are a new state of matter and electric analogs of magnetic skyrmions, and may be envisaged for potential applications in information technologies. I will attempt to describe the exciting observations we have made through many collaborations.

2:00 PM

(PACRIM-023-2021) Controlling local conductivity in multiferroic hexagonal manganites (Invited)

D. Meier*[1]

1. Norwegian University of Science and Technology (NTNU), Materials Science and Engineering, Norway

The control of conductivity is critical to any electronic device. In this context, oxides are interesting as they permit the continuous changes in conductivity required for synaptic and multi-configurational devices. I will discuss three strategies for controlling conductivity on demand and with nanoscale spatial precision. For this study, we choose multiferroic $ErMnO_3$ as it offers outstanding chemical flexibility. A first strategy utilizes the intrinsic properties of domain walls. At head-to-head walls, for example, an electronic inversion layer arises, enabling switching between resistive and conducting behavior. Going beyond domain walls, I will show how dislocations can be induced by electric fields for controlling the resistance. To resolve the dislocations, we apply high-resolution imaging techniques, whereas their impact on the electric transport is analysed by local conductance measurements. In a third approach, we control conductivity via point defects, generating charge-neutral interstitial–vacancy pairs. These so-called anti-Frenkel defects enhance the conductivity by up to four orders of magnitude; they are non-volatile and keep structural/ferroelectric integrity, providing a pathway for minimally invasive conductivity control. With this, multifunctional oxides become possible in which conductivity control may be used in parallel with other phenomena such as ferroelectricity and magnetism.

2:30 PM

(PACRIM-024-2021) Enhanced magnetostriction in FeGa epitaxial thin films (Invited)

J. Heron*[1]

1. University of Michigan, USA

Composite multiferroics, composed of a magnetostrictive ferromagnet and a piezoelectric ferroelectric, hold promise for magnetic field sensors and energy efficient beyond-CMOS logic by harnessing magnetoelectric transduction. Enhancing device performance requires highly magnetostrictive materials, however, relatively little attention has been given to engineering magnetostriction in thin films. One promising magnetostrictor system are the $Fe_{1-x}Ga_x$ alloys. Yet, in bulk, the magnetostriction coefficient of $Fe_{1-x}Ga_x$ alloys versus gallium composition peaks near ~18% Ga due to a phase change from the disordered A2 phase to an ordered BCC like phase ($D0_3$). Here we discuss a means to boost (by 200-300% relative to bulk) the magnetostriction of $Fe_{1-x}Ga_x$ thin film alloys by extending the phase stability of the A2 phase to higher Ga compositions. Transport-based magnetoelectric characterization of a $Fe_{1-x}Ga_x$ - $[Pb(Mg_{1/3}Nb_{2/3})O_3]_{0.7}$-$[PbTiO_3]_{0.3}$ (PMN-PT) composite multiferroic heterostructure shows a reversible 90 electrical switch of magnetic anisotropy and a room temperature converse magnetoelectric coefficient of 5.5×10^{-6} s m^{-1}.

Multiferroic Materials, Devices, and Applications II

Room: Cypress
Session Chairs: Laurent Bellaiche, University of Arkansas; Liuyan Zhao, University of Michigan

3:20 PM

(PACRIM-025-2021) Visualization of domain structures in a ferroaxial crystal (Invited)

T. Kimura*[1]

1. University of Tokyo, Japan

Ferroaxial materials are defined as materials having spontaneous ordering of a rotational structural distortion with an axial vector symmetry. As in conventional ferroics, domain states will be present in the ferroaxial materials. However, the observation of ferroaxial domains is non-trivial due to a nature of its order parameter invariant under both time-reversal and space-inversion operations. In ferroelectric materials, phase transitions are known to be classified into mainly two types: displacive type and order-disorder type. In the same manner, ferroaxial transitions will also be classified into these two types. However, only the displacive type of phase transition has been reported for ferroaxial materials. We propose that $NiTiO_3$ undergoes an order-disorder type ferroaxial transition and is recognized as an order-disorder type ferroaxial material. Furthermore, we spatially resolve its ferroaxial domains by using linear electrogyration effect: optical rotation in proportion to an applied electric field. To detect small signals of electrogyration, we adopt a difference image-sensing technique. Moreover, the ferroaxial domains are confirmed on nano-scale spatial resolution with a combined use of scanning transmission electron microscopy and convergent-beam electron diffraction. This work was done in collaboration with T. Hayashida, Y. Uemura, K. Kimura, S. Matsuoka, D. Morikawa, S. Hirose, K. Tsuda, and T. Hasegawa.

3:50 PM

(PACRIM-026-2021) Optical magnetoelectric responses in multiferroics (Invited)

Y. Takahashi*[1]

1. University of Tokyo, Quantum-Phase electronics Center, Japan

The magnetoelectric coupling, which associates the electric properties with the magnetism, emerges in particular classes of crystalline solids including the multiferroics and topological materials. In multiferroics, the long-range spin orders break the space-inversion symmetry, resulting in the spin-driven ferroelectricity. These spin orders always exhibit the electrically active collective spin excitations, termed electromagnon, in terahertz region. Accordingly, the electric field of light has the direct interaction with spins on the resonance of the electromagnon. As a consequence, the dynamical magnetoelectric coupling is resonantly enhanced on the electromagnon resonance, giving rise to the various nonreciprocal optical effects. We investigated the terahertz optical responses of the spin-spiral multiferroics with spin-driven ferroelectricity or chirality by using the time-domain terahertz spectroscopy. In the spin-spiral phases, the strong directional dichroism, in which the optical absorption changes with the reversal of propagation vector of light, is observed on the electromagnon resonance. In addition, two distinct nonreciprocal optical rotations, gyrotropic birefringence and natural optical activity, are found in the spin-spiral phases. We demonstrate the electric field control of these nonreciprocal optical effects.

4:20 PM

(PACRIM-027-2021) Measurement of multiple order parameters and their correlation at the atomic scale (Invited)

Z. Jing*[1]; Y. Zhang[1]; K. Xu[1]

1. Tsinghua University, School of Materials Science and Engineering, China

The correlation between phase transition with symmetry broken and the concept of order parameter were proposed by Landau in 1937. For a long time, the mechanical property which closely related to the lattice in materials was most concerned, and then the term "phase transition" has been much popular in the academic and engineering fields of. With the development of functional materials and "Quantum Materials", many intriguing physical phenomena present, the lattice, charge, spin, orbital, topology come frequently. Thus, the term "order parameter" has been received more attention. In this report by use of a suite of advanced electron microscopy techniques applied synergistically at the atomic scale we are able to reveal how variations in the microstructure are coupled to modulations in the lattice, charge, orbital and spin structure, as well as related order parameters, leading to the modulation of properties in two kinds of quantum materials, thereby further guiding the research and development for a wide range of complex functional materials.

4:50 PM

(PACRIM-028-2021) Stabilization and properties of ferroelastic charged domain walls in self-assembled $BiFeO_3$ nanoislands (Invited)

M. Chen[1]; J. Wang[2]; J. Ma[3]; J. Ma*[1]; C. Nan[1]

1. Tsinghua University, China
2. Beijing Institute of Technology, China
3. Kunming University of Science and Technology, China

Topological states with low-dimensional textures are fundamentally important and technologically attractive in terms of thermodynamic stability at reduced dimensions, low threshold of controllability and exotic physical behaviors. In this talk, we firstly present the growth mechanism of square-shaped $BiFeO_3$ nanoislands array with the low-dimensional ferroelectric topological textures via pulsed laser deposition. By a delicate design of the wetting layer and growth temperature, self-assembled ferroelectric nanoislands were achieved with the $BiFeO_3$/(La,Sr)MnO_3/$LaAlO_3$ heterostructure. By combining STEM, GPA, and EDS, we explored the microscopic mechanism of stabilizing the charged domain walls with upward center-divergent quad-domain structures. The comprehensive factors include the fully relaxed rhombohedral phase in nanoislands, negative surface screening charges, and upward built-in bias at the $BiFeO_3$-(La,Sr)MnO_3 interface, which are respectively induced by interfacial periodic dislocations, surface reconstructions, and interfacial terminations. The mechanisms for the growth of self-assembled $BiFeO_3$ nanoislands and stabilizing the exotic domain/domain walls provide practical strategies to prepare nanostructures and stabilize the exotic domain structures for applications in the future nano-electronics with high density and low power consumption.

PACRIM Symposium 22: Microwave Dielectric Materials and Their Applications

Millimeter-wave Materials for 5G Applications

Room: Plaza C
Session Chair: Michael Lanagan, Penn State University

1:30 PM

(PACRIM-029-2021) Accurate and Repeatable 5G and Automotive mmWave Materials Measurement with Latest Commercially Available Fixtures (Invited)

S. Phommakesone*[1]

1. Keysight Technologies Inc., USA

Keysight Technologies, Inc. provides electronic design and test solutions to commercial communications, networking, aerospace, defense and government, automotive, energy, semiconductor, electronic, and education industries in the Americas, Europe, and the Asia Pacific regions. It produces electronic design automation (EDA) software; DC, RF and microwave test instruments such as oscilloscopes, logic and serial protocol analyzers, arbitrary waveform generators, bit error rate testers; optical modulation and component analyzers, DMM, function generators, frequency counters, audio analyzers, LCR meters; and repair, calibration, and consulting services, as well as resells refurbished used Keysight equipment. Part of my job at Keysight is to reach out to customers about their measurement science problems and learn how Keysight can provide a measurement solution. The "PaCRim 22: Microwave Dielectric Materials and Their Applications" is an opportunity for me to share Keysight's latest solutions for 5G, Automotive Radar material test and measurement. In this talk, I will present our suite of measurement solutions, discuss how each instrument works, and how to use them. I'll show some example data and the results. I will also discuss my role in developing new standard reference materials and answer any questions you have about how to make measurements with Keysight.

2:00 PM

(PACRIM-030-2021) Some Fundamental Issues for Millimeter Wave Dielectric Ceramics (Invited)

X. Chen*[1]

1. Zhejiang University, School of Materials Science and Engineering, China

Microwave dielectric ceramics have received the increasing attention as the key materials in microwave communication system. Recently, because of the development trend of mobile communication towards millimeter wave frequencies, low dielectric constant and ultra-low loss millimeter wave dielectric ceramics have become the great challenge. To face up to this challenge, we may need to search completely new materials systems or effectively enhance the dielectric performance to meet the requirements of millimeter wave applications. In the present talk, the following issues would be discussed: 1) Possible candidates for millimetter dielectric ceramics with R-P structure; 2) Parameters dominating dielectric loss and temperature coefficient of resonant frequency in millimeter wave dielectric ceramics; 3) Perspective on ultra-low dielectric constant dielectric ceramics.

2:30 PM

(PACRIM-031-2021) Low dielectric constant and low loss LTCCs for microwave and millimeter wave applications (Invited)

H. Kim*[1]; S. Arun[1]; H. Shin[1]

1. Korea Institute of Ceramic Engineering and Technology (KICET), Nanomaterials and nanotechnology center Div, Republic of Korea

Dielectric materials with low dielectric constant and low loss at the microwave and millimeter wave range are expected of great interest in the 5G and post 5G applications. Among them, low-temperature co-fired ceramics, LTCC technology has been widely used in the wireless communication devices due to its low loss at high frequency, stable temperature characteristics, and multilayer integration using co-firing process with highly conductive electrodes. The experimental work for the development of LTCCs, with the targeting dielectric properties of $\varepsilon r \leq 4$ and $\tan\delta \leq 0.003$ and with decent thermal properties at the measuring frequencies over 30GHz, to provide better signal transmission rate and to minimize power dissipation, will be presented. The compositional and microstructural approaches of lowering the dielectric constants using low-dielectric constant fillers in the glass or LTCC matrix, improving dielectric loss, influence of high thermal conductivity fillers and matrix on the dielectric properties will be discussed.

3:20 PM

(PACRIM-032-2021) Ceramic Magnetic and Dielctric Oxide Materials for Sub-6 GHz 5G Telecommunications Applications (Invited)

M. D. Hill*[1]; D. B. Cruickshank[1]; I. MacFarlane[2]

1. Trans-Tech, Inc., Research and Development, USA
2. Skyworks Ireland, Research and Development, Ireland

5G Communications will use two distinct frequency bands: 3-6 GHz and mm-waves (20-100 GHz). The technology infrastructure for the 3-6 GHz band is becoming well established while that for the mm-wave band remains largely undetermined. Circulators using insulating magnetic oxide materials remain a critical technology for isolating the transmit and receive component chains in base stations. The use of a high dielectric constant bismuth substituted yttrium iron garnets are a key enabling technology over a range of frequencies. The frequency of use will determine the saturation magnetization and therefore, the composition of the garnet material. Dielectric materials are expected to play a significant role as well, both as substrates for integrated devices involving circulators as well for filters. Materials to be discussed include ultra-low loss microwave dielectric materials and temperature stable low dielectric constant (< 15) high frequency dielectrics. Current results on component integration are presented as well.

3:50 PM

(PACRIM-033-2021) Dielectric and Materials Engineering of Graphene for High Frequency Applications (Invited)

D. Carey*[1]

1. University of Surrey, Electrical and Electronic Engineering, United Kingdom

Future communication technologies will require new materials and in the case of THz technologies new materials are required to overcome the so-called 'THz gap' between electronics and photonics. In this presentation I will discuss how graphene could be one material solution to THz communications and the THz gap. Specifically, I will discuss how graphene's electrical and dielectric properties can be controlled tuned and how graphene can be used for THz wave generation; comparison will be made with other material systems. I will also discuss some recent developments on liquid phase exfoliation of graphene designed to optimize flake size. Potential high frequency applications of graphene will also be discussed.

4:20 PM

(PACRIM-034-2021) Targeted Chemical Pressure yields Tunable Millimeter-Wave Dielectric (Invited)

N. Dawley[1]; E. J. Marksz[2]; A. Hagerstrom[3]; G. Olsen[6]; M. Holtz[1]; V. Goian[4]; C. Kadlec[4]; J. Zhang[1]; X. Lu[3]; J. Drisko[3]; R. Uecker[5]; S. Ganschow[5]; C. Long[3]; J. Booth[3]; S. Kamba[4]; C. Fennie[6]; D. Muller[6]; N. Orloff[3]; D. Schlom*[1]

1. Cornell University, Department of Materials Science and Engineering, USA
2. University of Maryland, Department of Materials Science & Engineering, USA
3. NIST, USA
4. Institute of Physics ASCR, Czechia
5. Leibniz-Institut für Kristallzüchtung, Germany
6. Cornell University, School of Applied and Engineering Physics, USA

Epitaxial strain can unlock enhanced properties in oxide materials, but restricts substrate choice and maximum film thickness, above which lattice relaxation and property degradation occur. In this talk I will describe how we employ a chemical alternative to epitaxial strain by providing targeted chemical pressure, distinct from random doping, to induce a ferroelectric instability with the strategic introduction of barium into today's best millimeter-wave tuneble dielectric, the epitaxially strained 50-nm-thick $n = 6$ $(SrTiO_3)_nSrO$ Ruddlesden–Popper dielectric grown on (110) $DyScO_3$. The defect mitigating nature of $(SrTiO_3)_nSrO$ results in unprecedented low loss at frequencies up to 125 GHz. No barium-containing Ruddlesden–Popper titanates are known, but the resulting atomically engineered superlattice material, $(SrTiO3)_{n-m}(BaTiO_3)_mSrO$, enables low-loss, tuneble dielectric properties to be achieved with lower epitaxial strain and a 200% improvement in the figure of merit at commercially relevant millimeter-wave frequencies. As tuneble dielectrics are key constituents of emerging millimeter-wave high-frequency devices in telecommunications, our findings could lead to higher performance adaptive and reconfigurable electronics at these frequencies.

4:50 PM

(PACRIM-035-2021) Low dielectric permittivity silicate nanoparticles: Synthesis and applications in mm-wave communications (Invited)

D. Kwon*[1]; W. Sung[1]; J. Kim[1]

1. Korea Aerospace University, Materials Science & Engineering, Republic of Korea

Rapid progression in the development of mm-wave based communication and radar systems drives the demand for novel ceramic dielectric components in the form of dielectric resonators, filters, antennas, waveguides and substrates. The prospective dielectric ceramics for the mm-wave applications are required to possess low dielectric constant (ε_r), high quality factor (Qf), and near zero temperature coefficient of resonance frequency (τ_f). Two kinds of silicate-based dielectric ceramics, namely cordierite and forsterite, are selected as the major composition of this study. They both show very attractive dielectric properties for mm-wave applications, but have a narrow densification temperature gaps near to melting point. Therefore, it is necessary to reduce sintering temperature to the practical range. The low temperature sintering of the cordierite and forsterite should be initiated from the nanopowder preparations; however, obtaining pure-phase and well-dispersed nanopowders remains a challenge in certain advanced applications. we proposed a novel hydrothermally assisted nanoparticle synthesis methods for both cordierite and forsterite. The nanoparticles could be synthesized in either crystalline or amorphous form. Extensive study on phase formation, densification at reduced temperatures, microstructure evolution, high frequency dielectric characterizations were carried out.

PACRIM Symposium 24: Solid Oxide Fuel Cells and Hydrogen Technologies

Proton Conducting SOFC I

Room: Georgia B

Session Chairs: Fatih Dogan, Missouri University of Science and Technology; Hiroyuki Shimada, National Institute of Advanced Industrial Science and Technology (AIST)

1:30 PM

(PACRIM-036-2021) Materials Issues for the Next Generation PCFC Systems (Invited)

S. Yamaguchi*[1]

1. University of Tokyo, Materials Engineering, Japan

Various approaches for developing novel proton-conducting fuel cell (PCFC) are underway to utilize its advantages over Solid Oxide Fuel Cell (SOFC), such as superior efficiency, higher fuel usage, and long lifetime with reduced cost. Since the first high-temperature proton conducting perovskite ceramics were discovered in 1981 by Prof. Iwahara, seeking new electrolyte materials has been a focused issue. However, we are now aware that the cathode performance is growing to be a major issue of the cell performance: Whether or not we can enjoy an extremely low cathode resistance with solid proton-conducting ceramic systems observed in the Bacon cell. In this presentation, the reviews on the challenge to develop both materials systems for electrodes and electrolytes and its performance of integrated PCFC systems will be presented to emphasize the importance of the cathode reaction and proton transference number of the electrolyte. Further discussion will be made on how to promote a proton-assisted reaction at the electrode surface.

2:00 PM

(PACRIM-037-2021) Progress and challenges on ultra-high efficiency protonic ceramic fuel cells in NEDO project (Invited)

Y. Mizutani*[1]; K. Amezawa[2]; M. MORI[3]

1. National Institute of Advanced Industrial Science and Technology (AIST), Innovative Functional Materials Research Institute, Japan
2. Tohoku University, IMRAM, Japan
3. CRIEPI, Materials Science Research Laboratory, Japan

Protonic ceramic fuel cells (PCFCs) is a candidate technology to realize stationary fuel cell system which have ultra-high efficiency of >70% (LHV, AC). A leading research of PCFC was conducted from FY2017-FY2019 for understanding technical callenges of PCFCs, and a new NEDO research project have been started from FY2020 cooperation with three working packages. The WP1 develop innovative cathode and component materials, WP2 develop high-efficiency and high-power density cells, and WP3 evaluate performance and durability of cells from aspect of system and future market. In this presentation, current status and future development of PCFC in NEDO project are introduced.

2:30 PM

(PACRIM-038-2021) Development of protonic ceramic fuel cell with BZYb electrolyte (Invited)

Y. Mikami*[1]; K. Yamauchi[1]; T. Kuroha[1]

1. Panasonic corporation, Japan

Protonic ceramic fuel cells (PCFC) are expected to be next-generation fuel cell with high power generation efficiency. For PCFC electrolyte, $BaZrO_3$ based oxides are potential material because they have high proton conductivity and chemical stability against CO_2. Recently, we have developed PCFC using $BaZr_{0.8}Yb_{0.2}O_{3-\delta}$ (BZYb) as the electrolyte. We confirmed that BZYb has high proton conductivity (1.1×10^{-2} S/cm at 600°C) which could lead to reduce the internal resistance. In addition, BZYb did not form the complex oxide with NiO after sintering at 1500°C. It enabled to fabricate PCFC with no secondary phase formation inside the electrolyte layer during co-sintering process of anode and electrolyte. By developed the anode supported cell with generally used electrode material (NiO-BZYb|BZYb|LSC), the maximum power density reached 0.5 W/cm^2 at 600°C. Acknowledgements A part of this work was supported by the New Energy and Industrial Technology Development Organization (NEDO).

Proton Conducting SOFC II

Room: Georgia B

Session Chairs: Koji Amezawa, Tohoku University; Yasunobu Mizutani, National Institute of Advanced Industrial Science and Technology (AIST)

3:20 PM

(PACRIM-039-2021) Effect of added transition element on proton transport properties of ytterbium-doped barium zirconate (Invited)

Y. Okuyama*[1]; Y. Sekitani[1]; T. Kuroha[2]; K. Yamauchi[2]; Y. Mikami[2]

1. University of Miyazaki, Japan
2. Panasonic corporation, Japan

Protonic ceramic fuel cells (PCFCs) is a fuel cell using a proton-conducting oxide as electrolyte. This system is eco-friendly because they do not emit carbon dioxide. PCFC has many advantages of power generation. One is intermediate operation temperature, which reduces the material cost and start-up time of the system. The other one is the high energy conversion efficiency. Because the fuel was not diluted with steam for PCFC, the high energy efficiency and the high utilization of fuel might be expected. Barium zironeate has high proton conductivity and high chemical stability against carbon dioxide. Therefore, this material has attracted the attention as electrolyte of PCFC. To clarify the dissolution effect of transition element on proton transport properties such as partial conductivity, hydration, migration for barium zirconate doped with ytterbium, the conductivity and proton concertation of BaZr0.8Yb0.2O3-δ added Ni or Fe, Co were measured by impedance analysis and thermogravimetric analysis. The partial conductivity of proton and hole were determined form the oxygen partial pressure dependence of conductivity based on defect equilibrium. Proton conductivity and proton concertation decrease with increase in content of transition element. We will discuss the influence of transition element on the proton transport number and the energy efficiency of PCFC.

3:50 PM

(PACRIM-040-2021) Development of protonic ceramic fuel cells with Yb doped $BaZrO_3$ electrolyte toward higher power density (Invited)

H. Shimada*[1]; Y. Yamaguchi[1]; H. Sumi[1]; K. Nomura[1]; W. Shin[1]; Y. Mizutani[1]

1. National Institute of Advanced Industrial Science and Technology (AIST), Japan

Solid oxide fuel cells (SOFCs) using proton-conducting ceramics as electrolyte material, called protonic ceramic fuel cells (PCFCs), have attracted much attention due to their high energy-conversion efficiency (from fuel to electricity). Besides high efficiency, high power density is essential to widespread application. In the present study, we have developed PCFC components such as anode/electrolyte assembly, interlayer, electrolyte, and cathode toward higher power density. We will report effects of these components on electrochemical performance in PCFCs.

4:20 PM

(PACRIM-041-2021) Engineering Ceramics for Advanced Solid Oxide Cells (Invited)

V. Thangadurai*[1]

1. University of Calgary, Chemistry, Canada

The high operating temperature of solid oxide cells (SOCs) can be lowered by employing proton conducting electrolytes and electrodes exhibiting high electrochemical activity at reduced temperatures. Perovskite-type $Ba_{0.5}Sr_{0.5}Ce_{0.6}Zr_{0.2}Gd_{0.1}Y_{0.1}O_{3-\delta}$ (BSCZGY) electrolyte developed in our lab exhibited excellent chemical stability under CO_2 and H_2O along with high proton conductivity. Grain-boundary space charge effect, electrical conductivity, and dielectric loss were correlated for BSCZGY. Full cell studies on anode supported BSCZGY electrolyte showed maximum power density of 1 W/cm^2 at 850 °C under air/H_2 + 3 vol. % H_2O. Synchrotron X-ray and neutron diffraction studies on double perovskite-type $Ba_3Ca_{1+x}Nb_{2-x}O_{9-\delta}$ (x = 0–0.45) showed that the B-site ordering increases with increasing Ca/Nb mixing, where highest electrical conductivity was observed for the x = 0.18 composition. The oxygen reduction reaction mechanism was elucidated for perovskite-type $Ba_{0.5}Sr_{0.5}Fe_{1-x}Cu_xO_{3-\delta}$ and garnet-type $Y_{3-x}Ca_xFe_5O_{12-\delta}$ using symmetrical cells with the help of impedance spectroscopy genetic programming. In this talk, an overview of our recent progress in SOC materials will be presented.

4:50 PM

(PACRIM-042-2021) Preparation of BZY and BZY-BCY solid-solutions by Solid-State Reaction Technique (Invited)

T. Moriga*[1]; K. HATAI[1]; Y. OTANI[1]; K. MURAI[1]; R. M. Matsuda[2]; M. MORI[2]

1. Tokushima University, Japan
2. Central Research Institute of Electric Power Industry, Japan

Perovskite-type Ba(Zr, Y)O_3 (BZY) and Ba(Ce,Y)O_3 (BCY) are known as good proton conductors for solid oxide fuel cells with high chemical stability and high electrical conductivity. BZY possesses a more excellent stability against steams and carbon dioxide than BCY whereas BZY shows a poorer sinterability than BCY. To overcome these weaknesses, we have been conducting research on synthesizing solid solutions between BZY and BCY by solid-state reaction technique. However, these materials causes evaporation of barium component at high sintering temperature of 1600°C. In this study, we will show effects of addition of ZnO as a sintering agent on formation of BZY and the BZY-BCY solid-solutions. Addition of a few wt% of ZnO inhibited a formation of secondary phases in both BZY and the BZY-BCY solid-solutions and increased the relative density from 70% without addition to over 90% with addition. However we also observed the formation of Y_2O_3 small particles on the surface of the grains in case of BZY sintering. Another trial to prepare BZY was made by solid-state reaction between $BaCO_3$ and yttria stabilized zirconia (YSZ). In spite of a reaction between the equimolar mixture, it was confirmed that Y_2O_3 was exhaled from YSZ, and BZY and $BaCO_3$ were produced above 1300°C. Formation mechanism of the BZY and BZY-BCY solid-solutions will be discussed.

5:20 PM

(PACRIM-043-2021) Low-temperature sintering of the $BaZrO_3$ based electrolyte material using reactive sintering method without sintering aids

Y. Yamaguchi*[1]; H. Shimada[1]; H. Sumi[1]; K. Nomura[1]; W. Shin[1]; Y. Mizutani[1]

1. National Institute of Advanced Industrial Science and Technology (AIST), Japan

Metal substituted $BaZaO_3$ materials attract much attentions as proton conducting electrolyte materials in protonic ceramic fuel cells (PCFCs). $BaZrO_3$ based materials have superior stability against carbon dioxide compared with other proton conducting materials. $BaZrO_3$, however, requires quite high temperature for sintering. In order to obtain the dense sintered body of pure $BaZrO_3$, the sample were heated more than 1700 °C. Therefore, it is difficult to fabricate the PCFCs with $BaZrO_3$ based electrolytes. Although sintering aids are widely used to decrease the sintering temperature, addition of sintering aids causes deterioration of PCFC's performance. In this study, we developed the novel reactive sintering process for $BaZrO_3$ based electrolyte materials without the use of sintering aids.

5:40 PM

(PACRIM-044-2021) Effect of ZnO sintering agent on microstructure of $BaCe_{0.8}Zr_{0.1}Y_{0.1}O_3$ perovskite

R. M. Matsuda*[1]; K. Nakamura[1]; M. MORI[1]; J. Dailly[2]

1. Central Research Institute of Electric Power Industry, Energy Transformation Research Laboratory, Japan
2. European Institute for Energy Research, Germany

Ba(Ce,Zr)O_3-based perovskites are considered as one of the most promising materials for PCC electrolyte although these materials are well known to be quite poorly sinterable. ZnO is known as effective sintering aid for the manufacture of dense Ba(Ce,Zr)O_3-based electrolyte layers. However, the higher Zn-doping lowers the proton conductivity of the perovskites. The objective of this study is an investigation of sintering property of $BaCe_{0.8}Zr_{0.1}Y_{0.1}O_3$ (BCZY) perovskite with ZnO. The powder is a commercial product and was prepared by spray pyrolysis. The relative density of BCZY pellets without sintering aid was approx. 43% at 1000°C and increased to 60% at 1400°C. In contrast, the relative densities of the samples with 0.1 – 3 wt. % ZnO were over 91% at 1400°C. Additionally, it is found that the shrinkage in the temperature range from 1100°C to 1400°C can be controlled by a small amount of ZnO addition. This is important for co-firing process with the hydrogen electrode layer. X-ray diffraction analysis showed that ZnO peaks were detected as second phase in the samples with high ZnO addition. Scanning electron microscopic investigations showed a densified microstructure, but also Zn and Y segregations in the sample. A further discussion with first-principles calculation about the effect of ZnO addition on sinterabilty of BCZY is also provided in the conference.

PACRIM Symposium 27: Advanced Materials and Technologies for Electrochemical Energy Storage Systems

Materials Design, Screening and Electrode Architecture

Room: Balmoral
Session Chair: Mickael Dollé, Université de Montreal

1:30 PM

(PACRIM-045-2021) Combining machine learning with physical modeling to predict the impact of manufacturing on lithium ion battery electrode properties (Invited)

A. A. Franco*[1]

1. Universite de Picardie Jules Verne, LRCS, France

In both academia and industry the optimization process of battery electrodes is mainly driven by empirical trial-and-error approaches. Such approaches are costly and time-consuming, and a paradigm shift in battery R&D is needed in order to overcome the challenges imposed by the climate change and the massive electrification of the automotive sector. I present here a digital platform able to predict the impact of manufacturing parameters on lithium ion battery electrode mesostructure and electrochemical performance. Such platform combines supervised and supervised machine learning techniques with a multiscale physical modeling workflow simulating the different steps along the manufacturing (mixing, slurry, coating, drying, calendering, electrolyte filling) and the resulting electrode electrochemical response. I illustrate the prediction and optimization acceleration capabilities of this platform for NMC, graphite and silicon-graphite LIB electrodes, in strong connection with experimental validation. I present a web app being released by us allowing to explore the produced data and to launch online LIB electrode manufacturing and electrochemical simulations. Finally, i present novel human/machine interfaces being developed by us allowing to interact in an intuitive way with these models to assess experimental data in the most efficient way.

2:00 PM

(PACRIM-046-2021) Site-selective doping strategy for high-voltage spinel cathode for lithium-ion batteries (Invited)

G. Liang[1]; W. Pang*[1]

1. University of Wollongong, Institute for Superconducting and Electronic Materials, Australia

The commercialisation of lithium-ion batteries (LIBs) has gained huge success and LIBs are taking an important part of our modern life, as confirmed by the prestigious award of the 2019 Nobel Prize in Chemistry. Spinel LiNi0.5Mn1.5O4 (LNMO) is a promising cathode for the next-generation high-energy-density LIB, where the inferior cycle stability and short service life, limit its wider applications. Here, we demonstrate a successful and sophisticated site-selective doping strategy, to facilitate an exceptionally-durable and high-performance LNMO. Through introducing Mg element as a dopant at specific sites in the spinel LNMO structure, the LNMO structure can be significantly stabilized against deformation during delithiation and lithiation processes. In addition to the mitigation of Mn dissolution, the mechanistic behavior of LNMO is alternated and only the unwanted two-phase reaction behavior at highly-delithiated states is suppressed. All of these contribute to the extraordinarily stable cycle and excellent high-rate performance. Notably, the doped LNMO maintains ~86% and ~87% of initial capacity after 1500 cycles at 1C rate and 2200 cycles at 10 C rate, respectively. This pioneering atomic-doping-engineering strategy offers a new insight for battery researchers and could be analogically extended to other electrode materials for high-performance LIBs.

2:30 PM

(PACRIM-047-2021) 2D Materials for Next Generation Li ion Batteries (Invited)

R. Shahbazian Yassar*[1]

1. University of Illinois at Chicago, USA

Two dimensional (2D) materials are emerging materials for innovative design of Li-ion batteries that are safe and high energy density. This presentation encompasses recent progress in the PI's research team on addressing the Li-ion battery challenges via 2D materials design and integration. I first showcase a Li-metal case where graphene oxide (GO) materials were used to control the deposition of Li-metal ions during charge and discharge reactions. We demonstrated high cycling performance of Li-metal cell modified with GO in comparison to typical Li-metal cells. In another work, we studied the electrochemical cycling of Li storage in phosphorene 2D materials and showed interesting structural ordering during Li insertion in these materials and remarkable fast ion diffusion across phosphorene. Moreover, we show that the encapsulation of cathode particles with 2D materials can be an innovative approach to suppress the oxygen release in the high voltage cathodes. 2D graphene is impermeable to oxygen molecules and would be an ideal layer to coat over the surface of LiCoO2 nanoparticles. The cells containing graphene encapsulation could run at higher voltage for longer cycles

PACRIM Symposium 29: Ceramics and Ceramic Matrix Composites for Next Generation Nuclear Energy

Ceramics and Ceramic Matrix Composites for Next Generation Nuclear Energy

Room: Regency B
Session Chair: Ming Tang, Clemson University

1:30 PM

(PACRIM-048-2021) Iron Phosphate Glass Waste Forms To Immobilize Salt Waste Stream via Composition-Property-Structure Correlations

M. Tang*[1]

1. Clemson University, Department of Materials Science & Engineering, USA

Glass waste forms are currently used to stabilize legacy high-level waste (HLW) materials, and glass is the baseline technology for treatment of HLW that would result from reprocessing commercial used nuclear fuel. Due to the low solubilities (< 1.5 mass% for silicate-based glass) of the chloride/fluoride ions and evolution of Cl_2/F_2 gas from the melt under operation temperature, it is not suitable to employ the borosilicate glasses as the host for the full-salt waste streams. Phosphate glass, especially iron phosphate glass, exhibited relatively good chemical durability; high solubility for many heavy metals (e.g., uranium, cesium, molybdenum), noble metals, and rare earths, at intermediate melting temperatures. In this study, we intend to develop and optimize highly durable and easily processable iron phosphate glass waste form to immobilize the salt streams from advanced nuclear fuel cycle by tailoring the composition of the glasses. For the compositional variation study, different Fe:P ratio vs waste loading in iron phosphate glass waste forms are fabricated; various glass modifiers are added into iron phosphate glass compositions to improve chemical durability and reduce melting temperature. The preliminary result shows promising for iron phosphate glass waste form being an improvement over other candidate waste forms.

1:50 PM

(PACRIM-049-2021) Science and technology of multifunctional ceramic coating for fusion reactor (Invited)

T. Chikada*[1]

1. Shizuoka University, College of Science, Academic Institute, Japan

Strict control of tritium migration is a critical issue for the realization of a fusion reactor. Tritium permeates fast through metal walls at elevated temperature, resulting in a crucial fuel loss and radiological hazard for the environment. Corrosion and erosion of structural materials by high-temperature tritium breeders such as solid lithium ceramic pebbles and liquid lithium alloys are also serious concerns. A promising solution to reduce tritium permeation and corrosion/erosion to an acceptable level is to coat a thin film as a multifunctional coating with tritium permeation barrier performance and corrosion protection property. Our efforts have been dedicated to investigating hydrogen permeation mechanism in ceramic coatings, resulting in the achievement of the world's largest permeation reduction at elevated temperatures. A multi-layer structure has been investigated to make the coatings have sufficient permeation barrier performance and compatibility with tritium breeders. Irradiation effects on tritium permeation in the coating have been also studied using heavy-ions and gamma-ray sources. The combined effects of hydrogen, corrosion, and irradiation for a further understanding of physical and chemical phenomena in fusion reactors are under study. In this presentation, recent advances and future challenges for the research and development of multifunctional ceramic coatings are overviewed.

2:20 PM

(PACRIM-050-2021) Development of advanced nuclear fuels for current and next generation reactors (Invited)

J. White*[1]

1. Los Alamos National Lab, Materials Science and Technology, USA

The DOE-NE Advanced Fuels Campaign has focused on the development of next generation nuclear fuels using a pragmatic approach to improve the safety and economics of nuclear reactors. Fuel development at LANL has utilized a multifaceted approach to fabricate and characterize fuel options for current and next generation reactors. This includes efforts to assess the thermophysical properties for accurate fuel performance code modeling, fabrication of test articles for small scale accelerated irradiation studies, evaluation of industrial scale up, and understanding degradation mechanism under accident scenarios. The focus of this talk will detail the motivation as well as research and development into fuels of interest within the AFC and industry collaborators including doped oxide fuels and high uranium density fuels as well as prospects of increasing the burn up limits for current generation light water reactors.

3:10 PM

(PACRIM-051-2021) Mechanical strength of particle agglomerates

J. Bayle*[2]; G. JOUAN[1]; P. Sandral-Lasbordes[1]; R. Ramachandramoorthy[3]

1. CEA, Nuclear fuel fabrication, France
2. CEA Marcoule, DES/ISEC/DMRC/SPTC/LSEM, France
3. Empa, Swiss Federal Laboratories for Materials Science and Technology, Switzerland

The study of the powder flow abilities in various processes consists in analysing the physical behaviour of these powders such as crushing, erosion, agglomeration, segregation. Digital tools are frequently used today to simulate the behaviour of powders in different processes. The numerical methods for making these simulations use mathematical models whose parameters have been identified from macroscopic characterizations. In many cases, a calibration is carried out on the powder models obtained in order to get as close as possible to the experimental results. Powders are generally made with an assembly of aggregates to form agglomerates. Macroscopic laws reflect the overall behaviour of the powder and the interactions between each agglomerate. However, these behavioural laws do not account for particle behaviour and very few in situ tests can be used to translate the microscopic behaviour of agglomerates. This is the main goal of this work which, through in situ measurements of the mechanical behaviour of agglomerates using a nano-indentation device, will make it possible micro-compressions until the agglomerates break down according to their particle size, shape and composition. These characterizations will be used to determine material parameters at the agglomerate scale, and simulations using the discrete element method will be particularly used to model particle behaviour at crushing.

3:30 PM

(PACRIM-052-2021) Flash Sintering of High Uranium Density Fuels: UO_2-UN Composites and the Effect of the Sintering Atmosphere

R. Ingraci*[1]; D. Byler[1]; K. McClellan[1]; E. Kardoulaki[1]

1. Los Alamos National Lab, Materials Science and Technology, USA

UO_2-UN composite, a high uranium density fuel alternative with improved thermal conductivity and expected oxidation resistance, was sintered in less than 30 min by controlled current-rate flash sintering using alternating electric fields. UO_2-UN pellets were heated under purified argon atmosphere up to 600 °C and then an electric field was applied triggering the flash sintering onset. Once the temperature of the pellet was higher than 1400 °C due to Joule heating, the sintering atmosphere was switched to N_2+6%H_2 increasing the sintering rate. By adding N_2 to the furnace atmosphere at high temperatures the disassociation of molecular N_2 from uranium could be minimized. However, the final pellet composition was affected by the presence of N_2(g) during the cooling down, between 1352 °C and 950 °C, which caused the conversion of UN to uranium sesquinitride (U_2N_3). Moreover, reactions between UO_2 and UN at high temperatures and reducing atmospheres generated uranium oxynitrides $U(ON)_{2-x}$ phases, which destabilized during the cooling down with the excess of N been accommodate by the formation of U_2N_3. Nevertheless, by tuning the UO_2-UN ratio, the flash sintering parameters, and the gas flow during sintering, it was possible to obtain dense pellets with good mechanical integrity, drastically reducing the sintering time and temperature.

3:50 PM

(PACRIM-053-2021) Development of Accident Tolerant Oxide Fuel Grain Growth Kinetic Models

T. L. Ulrich*[1]; D. M. Frazer[2]; J. White[1]

1. Los Alamos National Lab, Materials Science and Technology (MST-8), USA
2. Idaho National Laboratory, Advanced Characterization and PIE, USA

With the increasing demands for clean energy, the continuous improvement of the efficiency and safety of the operating nuclear reactor fleet, remains a top priority for the nuclear energy community. For the last six decades it has been demonstrated that increasing the grain size of UO_2 has improved fission gas release behavior and is expected to increase the viscoplastic behavior of the fuel at operating temperatures. However, large grain size UO_2, has been cause by the addition of dopants, which can make decoupling of dopant effect vs large grain size effect difficult. This work aimed at developing a method that enhances the grain size of undoped UO_2 which could then be used as a reference for fission gas behavior in enhanced UO_2 accident tolerant fuel. By changing the sintering atmosphere, a 3 fold increase in grain size was achieved compared to standard UO_2. These results will be discussed in the context of the impact of crystalline defects on the sintering behavior of UO_2 relative to the respective commercial dopants. This work was funded by the Department of Energy's Accident Tolerant Fuel Campaign.

4:10 PM

(PACRIM-054-2021) Flash sintering of uranium dioxide

E. Kardoulaki*[1]

1. Los Alamos National Lab, MST-8, USA

Field assisted sintering techniques (FAST) have been studied to identify their suitability as novel fabrication routes for nuclear fuels. FAST include spark plasma sintering (SPS) and flash sintering (FS), both variations of sintering methods that require the application of an electric field and current across a sample to sinter it at reduced times and temperatures, compared to conventional sintering methods. In this work, UO_2, which was selected as a reference material, has been sintered to high densities within minutes using FS. Densification of UO_2 has been previously achieved via traditional FS and has resulted in high densities, however, the resulting microstructures were significantly cracked, featured porosity and were not consistent for pellets flashed under the same conditions. These drawbacks are induced by the abrupt increase in power during traditional FS, as UO_2 features poor thermal shock resistance and, therefore, significant cracking can occur during high power surges. To enable high density pellets with structural integrity and consistent microstructures, FS using a current control method has been implemented. This process can minimize thermal shock and produce high density, structurally sound pellets in under half an hour at 600 °C. In this work, a study of the parameters that produce the highest density pellets and the best microstructures is carried out with AC current controlled FS.

PACRIM Symposium 31: Advanced Functional Materials, Devices, and Systems for Environmental Conservation, Pollution Control, and Critical Materials

Phosphors and Optical Ceramics for LEDs

Room: Plaza A

2:00 PM

(PACRIM-055-2021) Analysis of energy-structure relationship for Eu^{2+} in garnet-type oxides based on first-principles calculations (Invited)

K. Ogasawara*[1]

1. Kwansei Gakuin University, Department of Chemistry, Japan

Since the $4f^7$-$4f^65d^1$ transition energy of Eu^{2+} in crystals strongly depends on the local structure around Eu^{2+}, it is important to understand the relationship between the energy levels and the local structure for theoretical design of novel optical materials based on Eu^{2+}. In order to control the local structure, garnet-type oxides are useful as host crystals since there are three types of cation sites and the local structure around the impurity Eu^{2+} ion can be modified widely by cation substitution. Considering that the fractional coordinates of three types of cations are fixed, the crystal structure of garnet-type oxides can be specified by four structural parameters, which are the lattice constant and the fractional coordinates of oxygen (x, y, z). In this work, in order to visualize the energy-structure relationship of Eu^{2+} occupying the dodecahedral site in garnet-type oxides, two-dimensional energy maps were constructed by performing first-principles relativistic configuration-interaction (CI) calculations for a series of EuO_8 clusters constructed by changing two selected structural parameters. Then the relationship between the $4f^7$-$4f^65d^1$ transition energy and the local structure depending on the four structural parameters was clarified by the combination of these energy maps.

2:30 PM

(PACRIM-056-2021) Surface-modification of titania nanoparticles and their application to polymer-based optical hybrid materials (Invited)

Y. Sugahara*[1]

1. Waseda University, Department of Applied Chemistry and Kagami Memorial Research Institute for Science and Technology, Japan

Polymer-based hybrid materials with oxide nanoparticles have been developed as optical materials. To achieve transparency required for optical materials, it is important to suppress aggregation of oxide nanoparticles in polymer matrices. Their surface modification with organic compounds is therefore necessary for their homogeneous dispersion. Since the refractive indices of common polymers are in the range of 1.3-1.7, the preparation of hybrid materials with high reflective indices, which have various applications, has been developed using oxide nanoparticles, including titania. We have developed the surface modification of titania nanoparticles and their use as nanofillers for preparing hybrid materials with high reflective indices. For surface modification, we utilized phosphorous coupling agents, such as organophosphonic acids and esters of phosphoric acids. Surface-modified titania nanoparticles was achieved via several reactions, including surface modification of titania prepared by the non-hydrolytic sol-gel route with octylphosphonic acid, surface modification using an interfacial reaction in a biphasic system consisting of an aqueous dispersion of titania and toluene dissolving oleyl phosphate, and atom transfer radical polymerization of methyl methacrylate from surface-modified titania nanoparticles.

Critical Materials / Recovery and Recycling of Rare Metals

Room: Plaza A

3:20 PM

(PACRIM-057-2021) Processing and evaluation of hydrophobic rare-earth molybdenum complex oxides with antibacterial and antiviral activities (Invited)

A. Nakajima*[1]

1. Tokyo Institute of Technology, Materials Science and Engineering, Japan

Today, coronavirus (SARS-CoV-2) has spread among humans all over the world, and the need and demand for antiviral materials are increasing. Recently, we prepared $La_2Mo_2O_9$ by combining La in rare-earth elements with molybdenum oxide using a polymerizable complex method. We demonstrated that this material exhibits hydrophobicity, antibacterial activity against both gram-negative (E. coli) and gram-positive (S. aureus) bacteria, and antiviral activity against non-envelope type (bacteriophage Qβ) and envelope type (bacteriophage Φ6) viruses, simultaneously. In the initial study, activity against Φ6 was found to be inferior to that against E. coli, S. aureus, and Qβ. However, very recently, the antiviral activity of this material against Φ6 was successfully improved by substituting Ce with La. Detailed procssing condition of the materials and evaluation results including expected mechanisms of the properties will be presented.

3:50 PM

(PACRIM-058-2021) Development of new chemical recycling method of inorganic wastes by using property of glass (Invited)

T. Nanba*[1]; Y. Benino[1]; S. Sakida[2]

1. Okayama University, Material and Energy Science, Japan
2. Okayama University, Environment Management Center, Japan

In the author's research group, chemical recycling methods of various inorganic slags such as municipal waste slag and steel slag have been developed, in which slags vitrified at higher temperatures are soaked in acid, obtaining SiO_2-rich solid. In the recovered solid, TiO_2 and P_2O_5 present in the slags are condensed, and it is hence expected that this method can be also used as recycling of TiO_2 and P_2O_5. Then, we have tried various methods to recover phosphorus from the slags. In this paper, research outcomes for recovery of valuable elements and removal of hazardous elements will be introduced.

4:20 PM

(PACRIM-059-2021) Construction of resource-recycling manufacturing process using advanced coating process (Invited)

T. Tsuchiya*[1]; T. Nakajima[1]; N. KIJIMA[1]; Y. Fujishiro[1]

1. National Institute of Advanced Industrial Science and Technology (AIST), Japan

In order to significantly reduce greenhouse gases by 2050, a drastic change from a society based on mass production, mass consumption and disposal is essential. In recent years, recycling has been promoted in various fields to reduce resources and environmental load, but recycling alone does not lead to innovative reduction of resources and environmental load. Innovative that dramatically improves energy consumption (energy saving) and resource utilization rate (resource saving) throughout the life cycle of materials and components from the perspective of maintaining the global environment and continuing development toward a prosperous future society. It is necessary to develop a design / process that realizes the 3Rs. In this paper, we will introduce following new processes such as, 1. Reduce from conventional components(Resistor, Phosphor) by improving the durability and function of parts (reduction of manufacturing energy: advanced coating technology), 2. Design / remanufacturing process for transparent conductive materials: Tin doped Indium oxide(ITO) film and Smart window: VO2 films (smart reuse: (recycling resources / improving LCA environmental efficiency)) that enables reuse and upgrade instead of recycling raw materials.

4:50 PM

(PACRIM-060-2021) Preparation of crystalline oxide thin films on plastic substrates by sol-gel transfer technique: What allows the transfer (Invited)

H. Kozuka*[1]; M. Kubota[1]; Y. Tanida[1]; K. Niinuma[1]; T. Yamada[1]

1. Kansai University, Dept Chem Mater Eng, Japan

We have proposed an original technique that realizes 50 - 700 nm thick, crystalline oxide thin films on plastic substrates. The technique consists of (i) the deposition of a polymer layer on a silicon substrate, (ii) the deposition of a precursor gel film on it, (iii) the firing to convert the gel film to a crystalline oxide film, and (iv) the transfer of the film to a plastic substrate. The transfer is achieved by heating the film on a hot plate at around 200°C, and pressing a plastic substrate on it. The significant point of the technique is that the firing step ensures the crystallization and densification of films. For instance, a polycarbonate substrate coated with a rutile film fired even at 1000°C can be prepared. The polymer underlayer is indispensable for the transfer of the fired oxide film while it is decomposed and lost during firing. Then why does the "lost" polymer underlayer allow the oxide film to be transfer? We have studied the effects of (i) the polymer layer thickness, (ii) the microstructure and porosity of the oxide film, and (iii) the roughness of the silicon substrate on the oxide film transferrability. Finally we found that the practical contact area between the fired oxide film and the mother silicon substrate is the key for the transfer.

5:20 PM

(PACRIM-061-2021) New multi-functional properties by metal-ceramics nano-composite films (Invited)

H. Masumoto*[1]

1. Tohoku University, Frontier Research Institute for Interdisciplinary Sciences (FRIS), Japan

Development of the multi-functional film materials magnetic metal nano-composite structures, consisting of the ferromagnetic granules with the size of several nanometers dispersed in an insulating ceramics matrix, have played crucial roles for the understanding of spin-dependent transport phenomena, such as spin-torque transfer phenomena or higher-order tunneling with Coulomb blockade, and application in magnetic sensors, and nondestructive examination, etc. In our group, a new magneto-dielectric effect was discovered in granular nano-composites, and we named it the Tunneling Magneto-Dielectric (TMD) effect [1]. The TMD nano-composites consist of magnetic granules with the size of several nanometers sparsely dispersed in an insulator matrix, which enables the quantum tunneling possible as the tunnel barrier is thin enough. With applied AC field and magnetic field (H), the oscillation rate depending on the height of the tunnel barrier and relative magnetization orientation of the granules may coincide with the frequency of AC field, thus producing a peak dielectric change $\Delta\varepsilon'/\varepsilon'0$ at a specific peak frequency. In this paper, I describe the fabrication method of magnetic metal-ceramic nanocomposite thin films, and their relationship with the structure, properties and TMD characteristics. [1] N. Kobayashi, H. Masumoto et al. Nat. Commun. 5, 4417 (2014).

PACRIM Symposium 35: Advanced Additive Manufacturing Technologies for Bio-applications; Materials, Processes, and Systems

Advanced Additive Manufacturing Technologies for Bio-applications; Materials, Processes, and Systems I

Room: Regency C

Session Chairs: Martin Schwentenwein, Lithoz GmbH; Woo Soo Kim, Simon Fraser University

1:30 PM

(PACRIM-062-2021) Dense ceramics by additive manufacturing : typical defects and their effects on mechanical properties (Invited)

L. Gremillard*[1]; M. Maillard[1]; S. Fournier[1]; H. Reveron[1]; V. Garnier[1]; G. Baeza[1]; J. Chevalier[1]

1. INSA, Materials, Engineering and Science, France

More and more ceramic pieces are fabricated using additive manufacturing (AM) techniques, that are inexorably coming to maturity. Conventional processes (pressing, casting...) have been studied and optimized for centuries, to the point where a mere glimpse at the fracture surface of a failed ceramic piece may point to specific step of the process. Not so (yet) for additive manufacturing. Indeed, the possibility brought by additive manufacturing processes to deposit mater where it matters is somewhat hindered by the appearance of new kind of defects directly related to the process. These defects obviously impact the strength of the printed pieces. During this talk, three different AM processes will be examined: direct ink writing, stereolithography and fused filament fabrication. The main characteristic defects will be pointed out, as well as their effects on mechanical strength and solutions to avoid them.

2:00 PM

(PACRIM-063-2021) CerAMfacturing of Silicon nitride for medical applications (Invited)

U. Scheithauer*[1]; E. Schwarzer-Fischer[1]; E. Zschippang[2]; W. Kunz[2]

1. Fraunhofer IKTS, Shaping, Germany
2. Fraunhofer IKTS, Materials, Germany

Silicon nitride (Si_3N_4) is a high-performance ceramic material. The entirety of its outstanding properties such as strength, toughness, hardness, even at very high temperatures makes it superior compared to most technical ceramics. New applications such as in the medical sector are coming into focus for Si_3N_4. In addition to its mechanical properties, Si_3N_4 also has a unique surface chemistry. Studies show that Si_3N_4 surface chemistry can promote the healing of soft and bone tissue, inhibit the proliferation of bacteria and kill viruses. These benefits permit it to be used in a wide array of different disciplines inside and outside of the human body including orthopedics, dentistry, virology, agronomy, and environmental remediation. We qualified Si_3N_4 for different CerAM (Ceramic Additive Manufacturing) technologies at Fraunhofer IKTS. Our presentation focuses on our work in the field of CerAM VPP/LCM (a lithographical method), as this process currently achieves the highest resolution (minimum wall thickness $< 200\ \mu m$). In addition to the development of suitable suspensions, the determination of process parameters for the shaping and thermal processing is in the foreground. The presentation summarizes the current developments and illustrates them using various demonstrators. Initial investigations showed very good biocompatibility as well as cell spreading on various surfaces.

2:30 PM

(PACRIM-064-2021) 3D Printing of Octacalcium Phosphate for Bone Substitutes (Invited)

S. Kim*[1]

1. Hudens_Bio, R&D, Republic of Korea

Many studies have been done on the development of artificial bone substitutes over the past 30 years, but clinically satisfactory products have not yet been developed. Among various calcium phosphates, many researchers have studied the potential use of octacalcium phosphate(OCP) as a bone graft material because it is known as a precursor of biological apatite. A newly developed OCP synthesis method that can be mass-produced unlike the existing method will be introduced. Hydroxyapatite (HA) and biphasic calcium phosphate (BCP) which are currently widely used, were selected and implanted into the rabbit tibia defects. The degree of resorption and the ability to new bone formation were compared after 4 and 12 weeks. Based on the results of animal experiments, the resorption properties of these three bone substitutes and the mechanism of new bone formation will be presented. Since the stability of the OCP phase is very sensitive to heat, a new 3D printing method for the production of bone graft materials at room temperature has also been proposed. The low-temperature 3D printing method will be very useful for the fabrication of patient-specific bone scaffolds, the incorporation of cells, drugs, and proteins, and formation of porous structures.

Advanced Additive Manufacturing Technologies for Bio-applications; Materials, Processes, and Systems II

Room: Regency C

3:20 PM

(PACRIM-065-2021) Towards the Development through Additive Manufacturing of Osteoregenerative Hybrid Ceramic/Polymer Scaffolds Suitable for Load-Bearing Applications (Invited)

P. Miranda*[1]; C. Paredes[1]; F. J. Martínez-Vázquez[1]; A. Pajares[1]

1. University of Extremadura, Spain

Developing scaffolds from materials with bonelike properties that are capable to interact with the living tissue and actively induce bone regeneration is the holy grail in the field of bone tissue engineering. It would allow overcoming all the main limitations of currently available procedures for repairing large bone lesions, namely: the limited amount of material and need of secondary surgical sites in the case of autografts; risk of immunogenic response and disease transmission from donor in the case of allografts; and fixation problems of current bioinert prostheses. In this work, the latests results are presented of a research effort oriented towards the development of such scaffolds through the combination of bioceramic additive manufacturing —direct ink writing and digital light processing— with biopolymer impregnation. The resulting hybrid scaffolds, through a careful control of the microarchitecture, exhibit optimized mechanical performance in terms strength, and especially of toughness, while preserving their osteoregenerative capacity. The results presented provide guidelines for further optimization of the mechanical and biological performance of hybrid scaffolds for bone tissue engineering applications, paving the way for attaining the long sought-after scaffold materials for load-bearing bone tissue engineering.

3:50 PM

(PACRIM-066-2021) 3D printing of complex bioceramic structures in a reversible hydrogel bath

Y. Choi*[1]; A. Sung[1]; N. Raja[1]; H. Park[1]; H. Yun[1]

1. Korea Institute of Materials Science, Republic of Korea

In conventional bioceramic 3D printing, a 3D ceramic structure is fabricated by extruding ceramic paste or solidifying ceramic slurry. However, as human bones are anatomically complex and have overhang structures that can be unfavorable for 3D printing, support structures must also be printed. This increases printing time and wasted materials and creates a risk of collapse or breakage of the structure during the process of removing the supports following 3D printing. In this study, a hydrogel-based bath was developed for the self-standing printing of 3D bioceramic structures without supports. The bath consists of a temperature-sensitive hydrogel, within which the cementation reaction is induced as soon as the ceramic ink is printed and which successfully facilitated the fabrication of 3D bioceramic structures. These structures were confirmed as having similar characteristics, including crystallinity and biological activity, to those printed using the conventional method. In addition, hybrid structures were fabricated by simultaneously printing bioceramic ink and cells through the bath system, and high cell viability due to the humid environment of the bath was confirmed. This study demonstrates that the bath system enables fabrication of anatomically complex human bones and could improve regeneration capacity through simultaneous printing with cells.

4:10 PM

(PACRIM-067-2021) Additive Manufacturing of Bioceramic Scaffolds by Combination of FDM and Slip Casting

R. Gadow*[1]; S. Esslinger[1]

1. Institute for Manufacturing Technologies of Ceramic Components and Composites, University of Stuttgart, Germany

The use of bioceramics like calcium phosphates or bioactive glasses for the regeneration of critical bone defects, as they can occur for example after serious injuries or diseases, is intensively researched worldwide. The advantages of the additive manufacturing technology make it possible to process these ceramics into customized patient-specific implant devices, so called scaffolds. In this study the possibilities to produce bioceramic scaffolds by using the fused deposition modelling (FDM) technique in combination with ceramic slip casting is introduced and described. In the first step the polymer models, which represent the negative geometry of the ceramic component, are printed by a commercial FDM printer. In the next step a calcium phosphate powder slurry is casted into the polymer molds according state of the art ceramic processing technologies. After debinding and sintering the mechanical and structural properties of the scaffolds are characterized and the biocompatibility is tested via simulated body fluid. First tests indicate that the mechanical properties as well as the biocompatibility of these scaffolds are similar to scaffolds that are produced by further additive manufacturing technologies, for example binder jetting or foam casting.

4:30 PM

(PACRIM-068-2021) 3D printed architectured solid and architectured sensors (Invited)

W. Kim*[1]

1. Simon Fraser University, School of Mechatronic Systems Engineering, Canada

Advanced 3d printing technologies can provide opportunities of single building process of multiple materials and multi-directional integration of the complex 3D structures seamlessly. In this talk, I'll discuss recent research activities on the design and fabrication of architectured solids and sensors at the SFU Additive Manufacturing Lab, Canada. As a representative example, the origami-based architectured solids and architectured sensors will be highlighted. 3D origami structures can be categorized as one of the architectured solids because it has the repeated cellular solids in the monolithic structure. Therefore, we named 3D architectured solid based sensors as an architectured sensor. Applications including energy storage, mechanically reinforced 3D structures, and dynamic sensing systems will be discussed with selected 3D printed architectured solids and architectured sensors.

5:00 PM

(PACRIM-069-2021) Multi-functional bone mimetic structure of calcium phosphate-bioglass ceramic by multi-material additive manufacturing

I. A. Sutejo*[2]; J. Kim[1]; Y. Choi[1]; H. Park[1]; H. Yun[2]

1. Korea Institute of Materials Science, Advanced Biomaterials Research, Republic of Korea
2. Korea University of Science and Technology / Korea Institute of Materials Science, Republic of Korea

Ever-changing nature was the prove of it adapting to the environment providing the efficient solution to the surrounding. Human bone architecture in specific was showing the structured skeletal build to support body and external mechanical load from other activities while promoting metabolic activities internally. Therefore, combine dense-porous model mimicking the cancellous bone that provide path for cell and nutrient circulation envelop by robust dense cortical bone was artificially built through additive manufacturing. Engineering process of these structure was further improved with the new multi-material printing system. This enabling different material to be utilized concurrently in one structure and merging its contrast properties together. Calcium phosphate based and bioglass ceramic were chosen as the materials for its chemical resemblance and supportive function. The properties of each material particularly compressive strength, in vitro bioactivity and degradability were investigate as the standard to select the material for either dense and porous part. Lastly, the synthetically created multi-ceramic structure was inspected again to confirm the multi-function feasibility.

5:20 PM

(PACRIM-070-2021) Development of near-field light curing three-dimensional printing technology with solvent-free zirconia slurry to fabricate dental restoration device

C. D. Jiang*[1]

1. National Taipei University of Technology, Mechanical Engineering, Taiwan

This study aims to develop a near-field light curing three-dimensional printing technology based on the top-down method to solve the problem caused by poor fluidity of the slurry. In the slurry preparation, HDDA resin is used to blend 3Y-TZP zirconia powder to prepare a photocurable zirconia slurry with a powder content of 70 wt%. The experimental results prove that the proposed immersion-light engine emits layer pattern to solidify slurry and it can generate a vacuum state during the lifting up process resulting in high viscosity slurry refilling back automatically. The first holding temperature of the two-stage sintering treatment is 380 °C according to the result of thermogravimetric analysis and the second holding temperature is determined at 1530 °C. The volume shrinkage analysis of the sintered unit cube shows that the shrinkage ratios of length, width and height are 29.9%, 29.7% and 30.6%, respectively. The results of hardness test, Archimedes analysis and three-point bending test show that the sintered part has the average hardness of 1224 (HV), the density of 91.8%, and the bending strength of 641.04 MPa, respectively. A high-fitness zirconia premolar and a three-unit zirconia dental bridge are also successfully printed.

5:40 PM

(PACRIM-071-2021) The influence of particle size distribution on the behaviour of ceramic-reinforced photo-curable resins for stereolithography

W. Yared*[1]; R. Gadow[2]

1. Institut for Manufacturing Technologies of Ceramic Components and Composites, Additive Manufacturing, Germany
2. Institut for Manufacturing Technologies of Ceramic Components and Composites, Germany

The intensity of scattered light in a ceramic-reinforced resin is a function of the dimensions, the mass and the concentration of the ceramic particles, along with the optical properties of these particles in relation to that of the polymeric matrix. This contribution is a study of the influence collective of the aforementioned factors on the scattering of light in photo-curable resins. The particle size distribution was measured using a laser diffraction particle size analyser based on the Mie theory, and the scattering of light was quantified with a UV/VIS spectrometer equipped with an integrating sphere accessory. It was possible to separate scattered light from absorbed and transmitted light, and small changes in the particle size and concentration induced noticeable changes in the quantum of scattered light, with smaller particles leading to large scattering of low-wavelength light. The developed resins are intended for use in stereolithography additive manufacturing, which adds challenging mechanical and temporal requirements on the curing behaviour of the resin. Thus, a photo-rheometer was employed to establish a correlation between polymerization rate and conversion levels versus particle size and concentration.

PACRIM Symposium 38: Nanotechnology in Medicine

Nanotechnology in Medicine I

Room: Stanley
Session Chair: David Medina-Cruz, SynCell Biotechnology

1:30 PM

(PACRIM-072-2021) Nanomaterials for the COVID-19 Pandemic: What Went Wrong and Will We Be Better Prepared Next Time? (Invited)

T. Webster*[1]

1. Northeastern University, USA

The COVID-19 pandemc has closed schools, universities, places of work, and an entire way of life. Novel solutions to prevent, diagnose, and treat COVID-19 have emerged from using nanomaterials. This talk will review varoius nanomaterials that are being used to stop SARS-CoV-2 adsorption to every day surfaces and deactive the virus if attached. It will also cover how nanomaterials functionalized to attached to SARS-CoV-2 are being used to improve at home diagnosis. Lastly, it will cover how various nanomaterials are being used to develop a vaccine and other therapies for treating COVID-19. In this manner, this talk will show how nanomaterials may be the platform technology needed for the current and future viral outbreaks.

2:00 PM

(PACRIM-073-2021) Nanotechnology and Advanced 4D Bioprinting for Complex Tissue Regeneration (Invited)

L. Zhang*[1]

1. The George Washington University, Mechanical and Aerospace Engineering, USA

As an emerging biomanufacturing technique, 4D bioprinting is highly innovative to fabricate self-assembly structures which can dynamically change their shapes over the time. Current 4D bioprinting based biomanufacturing technologies are hindered by the lack of advanced smart "inks". The main goal of our research is to create novel biomimetic nano or smart bioinks to fabricate the next generation of 4D multi-responsive tissue constructs ((such as cardiovascular tissue, neural tissue and osteochondral tissue). For this purpose, we have 4D bioprinted a multi-responsive smart structure triggered by different stimulations (i.e., solvent immersion and temperature modulation) using a natural shape memory polymer. Our results show that these bioprinted scaffolds have not only excellent cytocompatibility properties but also well controlled structure transformation and shape recovery. Moreover, recently we created novel light-triggered 4D bioprinted tissue constructs via incorporating nanomaterials into our stimuli-responsive smart materials. A proof-of-concept 4D bioprinted brain model was fabricated and the resultant construct exhibited a dynamically and remotely controllable transformation triggered by light in a spatiotemporal manner, which promoted excellent stem cell growth and differentiation, thus promising for complex tissue/organ regeneration.

2:30 PM

(PACRIM-074-2021) Designing Nano-Drugs by Pulsed Laser Ablation in Liquids (Invited)

G. Guisbiers*[1]

1. University of Arkansas at Little Rock, Physics & Astronomy, USA

Currently, antibiotic resistance and cancer are the two of the most important public health problems killing more than ~1.5 million people annually, showing that antibiotics and current chemotherapeutics are not as effective as they were in the past. Nanotechnology is presented here as a potential solution. However, current protocols for the traditional physicochemical synthesis of nanomaterials are not free of environmental and social drawbacks, often involving the use of toxic catalysts. This talk will present a novel process called « Pulsed Laser Ablation in Liquids » (PLAL) that synthesize nanostructures free of any surface contamination. The effect of the pulse duration, wavelength, repetition rate, solvent and target on the synthesized nanoparticles will be determined. And finally, the biomedical properties of the nanoparticles synthesized by PLAL will be discussed.

Nanotechnology in Medicine II

Room: Stanley
Session Chair: Thomas Webster, Northeastern University

3:20 PM

(PACRIM-075-2021) DNA-Inspired Nanomaterials for Improved Biocompatibility and Electrical Conductivity (Invited)

Y. Chen*[1]

1. University of Connecticut, USA

Families of electrically conductive coatings have aimed to facilitate the integration of medical devices such as conductive polymers, carbon nanotubes and bioactive hydrogels. Many of these materials possess biological, functional or electrochemical disadvantages that impede on device efficacy. Developing a new family of coating material based on DNA-nanotechnology that preserves device functionality while enhancing cell integration has become a top priority for biological interface study. Herein, we developed Janus Base Nanotubes (JBNTs); a family of electrically conductive biomimetic nanotubes assembled from DNA base pair units capable of densely coating metal or ceramic devices. Our results demonstrate that JBNTs present excellent electrical conductivity because of their long-distance electron dislocation via π-π rosette stacking. JBNTs also enhance interface integration including cell differentiation and functions. We also showed JBNTs extremely biocompatible chemistry in comparison to popular microelectrode coatings. We anticipate these results will serve as a benchmark for the continued development and study of JBNT to enhance interface dynamics.

3:50 PM

(PACRIM-076-2021) Biological nanocoatings for deployment in healthcare settings (Invited)

D. Medina-Cruz*[1]; T. Webster[2]

1. Novaurum Biosciences, USA
2. Northeastern University, Chemical Engineering, USA

Growing sectors, like healthcare, require advancements in their materials, and these often come from the nanoscale through features that are accomplished by physicochemical processes with tradeoffs associated with lack of cost-effectiveness and release toxic by-products into an exhausting environment. To prevent these drawbacks, green nanotechnology was born with the idea of using living organisms and natural biomolecules to produce valuable nanoscale products in a cost-effective and eco-friendly manner. One of the novel green nanotechnology-based approaches uses mammalian cells to produce nanostructures following detoxification mechanisms inherited from their prokaryotic origins throughout evolution. To exploit this approach, human cells have been used for the biologically derived synthesis of noble metallic nanoparticles that grow around the cells and create a nano-biocomposite that keeps the shape of the cells. Such novel material is made of an interconnected network of petrified cells and extracellular matrixes that can grow virtually anywhere. The technology, named AuPEC (Augmented Protection and Enhancement Coating), has been studied as a way to enhance the eprformance of medical devices, providing 1) improved cell attachment and proliferation; 2) prevention of biofilm formation and microbial contamination; and 3) avoiding physicochemical corrosion of the coated materials.

4:20 PM

(PACRIM-077-2021) Functionalized Nanocarriers for tumor specific localization and delivery of therapeutic molecules (Invited)

S. Ilyas[1]; A. M. Renner[1]; S. Mathur*[1]

1. Institute of Inorganic Chemistry, Chemistry, Germany

With a size smaller than a living cell, nanocarriers became one of the most promising therapeutic tools in a variety of biomedical application. In particular, the new optical, physical and chemical properties of nanomaterials offer a new pathway of treating cancer without affecting healthy cells. Thereby, the hyperthermia properties of magnetic or optical active nanocarriers can be used to induce specifically natural cell death. Nevertheless, an immense challenge for using nanoparticles in cancer therapy is to ensure a highly specific uptake of nanoparticles for cancer cells. Therefore, we present monodispersed and highly selective nanocarriers for example iron oxide and gold nanorods were vectorized by attaching estrogen molecules at their surface by chemo-specific reactions. The as-vectorized particles were also loaded with therapeutic agent (such as drugs or chelator-radiotracers) to investigate drug and/or radiation based therapeutic efficacy. The nanocarriers showed enhanced cellular uptake for MCF7 breast cancer cell lines, which was quantified by FACS analysis. In addition, in tumor bearning mice, these functionalized nanocarriers showed promising accumulation at the tumor site as compared to the healthy organs. Our data shows, that as-presented probes hold potential to be used as targeting specific drug delivery nanocarriers.

4:50 PM

(PACRIM-078-2021) Tapping Natural Resources to Develop Nanomedicine (Invited)

S. Ghosh*[1]

1. School of Science, RK. University, Department of Microbiology, India

Nanoscale materials with exotic shape and size with tunable physico-chemical and opto-electronic properties can be fabricated in the laboratory using various physical and chemical methods. However, hazardous reaction conditions and toxic chemical reagents used in these methods render the synthesized nanoparticles unsuitable for biomedical applications. Hence, several environmentally benign biological routes have been designed to synthesize gold, silver, platinum, palladium, copper and bimetallic nanoparticles using virus, bacteria, fungi, algae and medicinal plants. As a part of our growing interest in the novel nanomedicine, we have developed methods to used various medicinal plants found in nature like Dioscorea bulbifera, Gnidia glauca, Plumbago zeylanica, Barleria prionitis, Gloriosa superba and many more for synthesizing nanospheres, nanohexagons, nanotriangles, and nanorods. Careful optimization of time, temperature, pH, concentration of metal salts may help in the synthesis of nanomaterials with well defined shape and size. Further, naturally occurring bioactive principles like curcumin, diosgenin, etc. may stabilize and enhance the anticancer, antioxidant, antidiabetic, antimicrobial, and antibiofilm activities of the biogenic nanoparticles. In view of the background, it can be concluded that there is huge scope for further research towards translation of the biogenic nanoparticles as potential nanomedicine in future.

5:20 PM

(PACRIM-079-2021) Exploiting Intrinsic Properties of Lipids in Nanomedicine: Current Trends and Future (Invited)

M. Rehman*[1]

1. Quaid-i-Azam University, Department of Pharmacy, Pakistan

More than 8000 new drug molecules have failed to reach clinical stage due to poor pharmacokinetics profile including low aqueous solubility and unfavorable biological condition. Lipid based drug delivery systems have long been used to enhance aqueous solubility of lipophilic drugs. Emergence of lipid based nanoparticles have further extended this field by offering tunable drug release, solubility enhancement and ability to cross biological membranes. Therefore, recent research is focused on exploration of intrinsic properties of lipids as well as synthesis of new lipid derivatives for drug delivery applications. Exploration of natural lipids is preferred over chemical synthesis due to safety, non-immunogenicity, low cost and adaptation of environment friendly techniques in pharmaceutical industry. In this regard, we have shown that isoelectric point of phospholipid nanoparticles makes them an interesting candidate for pH sensitive drug delivery after oral administration. Similarly, lipid mixtures can be prepared to target drug delivery to body tissues exhibiting different body temperature. Temperature dependent behavior of lipid mixtures has been studied in food industry for around two centuries in preparation margarine and other butter derivative, its application in drug delivery may also offer unique benefits over other lipid nanoparticles.

PACRIM Symposium 39: Biomimetics and Bioinspired Processing of Advanced Materials

Biomimetics and Bioinspired Processing

Room: English Bay

1:30 PM

(PACRIM-080-2021) Controlling the stability of amorphous calcium carbonate (Invited)

Z. Zou*[1]; Z. Fu[1]

1. Wuhan University of Technology, China

Amorphous calcium carbonate (ACC) is widely known as a metastable precursor for the formation of crystalline biominerals with unique morphologies and remarkable mechanical properties. ACC also has great potential in biomedical applications because of its excellent biocompatibility and good moldability. However, ACC is usually very unstable under physiological environment because of its high solubility. Therefore, it is important to understand the stabilization mechanisms of ACC. Previously, we have shown an opposite particle size effect on the stability and crystallization of ACC in water and under heating in air. Here, I will present our recent work unravelling the effect of intrinsic properties and additives on the stability of ACC. Especially, we show that while poly(Aspartic acid) (pAsp) is effectively inhibiting calcite nucleation at the surface of ACC particle, inorganic ions like PO43- and OH- ions act to modify the ion mobility therefore delaying crystal propagation. Moreover, we show that only 1.5% pAsp (amino acid monomer to Ca molar ratio) could dramatically increase the crystallization temperature from 141 to 350 °C. Thus, new insights on controlling the stability and crystallization processes of amorphous materials are provided and it may have great implications for biomedical applications.

2:00 PM

(PACRIM-081-2021) Water Controls the Crystallization and Densification of Crystal-Like Amorphous Calcium Carbonate

Q. Wang*[1]; Z. Zou[1]; Z. Fu[1]

1. Wuhan University of Technology, China

Amorphous calcium carbonate (ACC) is widely known as a metastable precursor in forming crystalline calcium carbonate biominerals. Both stabilized biogenic and synthetic ACCs are hydrated and its stabilization and crystallization are deeply involved with water. In this study, a novel crystal-like ACC is synthesized by inducing hexahydrate calcium carbonate (Ikaite) to dehydrate in organic solvent, such as ethanol. The as-obtained ACC retained the morphology of ikaite with clear crystalline phase and smooth surface, besides, a significant number of interconnected nano-pores are generated inside the particle during dehydration. Furthermore, by controlling water content in crystallization process, such ACC can be easily transformed to pure vaterite or calcite either by pressure or under heating, which is usually difficult to control in solution. The peculiar structure and high specific surface area make the ACC promising in various applications such as drug release and sewage treatment etc. Moreover, the densification caused by ACC crystallization under pressure provides a new insight in synthesizing biological ceramics.

2:20 PM

(PACRIM-082-2021) Self-assembly of silk fibroin direct hierarchical calcium phosphate formation

Y. Huang*[1]; Z. Zou[1]; H. Xie[2]; Z. Fu[1]

1. Wuhan University of Technology, Material Science and Engineering, China
2. Wuhan University of Technology, Chemical Engineering and Life Science, China

Natural biominerals are usually composite materials produced through mineralization of inorganic crystals within an organic matrix.[1] Silk fibroin is known to be capable of directing the nucleation and growth of hydroxyapatite crystals. [2] Here, we used silk film as the substrate to induce the formation of calcium phosphate. We show that the self-assembly of silk fibroin in solution induces the formation of monetite crystals with hierarchical structure, which are assembled by well aligned single crystal of monetite. The crystal contains a relatively high amount of silk fibroin and transmission electron microscopy images demonstrate the presence of silk fibers within the crystal, which are not stable under electron beam. Therefore, the self-assembly of silk fibroin during the crystallization process is critical for the formation of such hierarchical structure.

2:40 PM

(PACRIM-083-2021) Bioinspired preparation of virus-based finely organized nanostructures with specific functions

F. Wan*[1]; Z. Fu[1]

1. Wuhan University of Technology, State Key Laboratory of Advanced Technology For materials synthesis and Processing, China

Natural materials have brought resourceful inspirations to materials design and synthesis. Biosilica widely exists in nature and has abundant reserves. Numerous species contain silicified cell walls with unique patterns, such as diatoms and sponges. During the process of biosilicification in vivo, biological organisms can generate diverse complex and delicate silica nanostructures under mild conditions. To explore the biosilica formation process and fabricate silica with an exquisite microstructure for specific functions, a bacterial phage (M13) is used as a biotemplate to synthesize silica with diverse morphologies: cylinders, hexagonal prisms, assemblies of smaller cylinders and nanowires. A facile ethanol bath method is conducted to coat the nanowires with nitrogen-containing carbon. Carbon-coated SiO_2 nanowires with mesochannels (C@msSiO_2NWs) are first used as anode materials for LIBs. Attributed to the uniform carbon coating and parallel mesochannel structure, the electronic conductivity and capacity to accommodate volume variations were significantly improved. In the electrochemical performance test, the composites calcined at 750 °C (C@msSiO_2NWs-750) show an impressive capacity of 653 mA h g^{-1} at a current density of 500 mA g^{-1} and stability.

3:20 PM

(PACRIM-084-2021) Collagen-Directed Mineralization of Inorganic Materials with Periodically Patterned Nanostructure

W. Fang*[1]; H. Ping[1]; Z. Fu[1]

1. Wuhan University of Technology, China

Collagen fibrils have hierarchical and periodic structures, which provide space for intrafibrillar growth of oriented hydroxyapatite nanocrystals in bone and contribute to the good mechanical properties of bone. Due to the quarter-staggered array of collagen molecules, the collagen fibrils present hierarchical structure with periodic gap and overlap patterns. However, bioinspired synthesis of non-native inorganic materials inside collagen fibrils has not been reported and detailed forming processes of crystals inside collagen fibrils remain poorly understood. The negatively charged amorphous precursor phase infiltrates collagen fibrils through the gap zones creating an intricate periodic mineralization pattern. Later, the nanocrystals initially filling the gap zones only expand gradually into the remaining space within the collagen fibrils. Atomic probe tomography reconstruction also demonstrates the band patterns of nanocrystals. The dispersed distribution of C and H atoms means that the tropocollagen molecules may promote the organized growth of nanocrystals. The mineralized collagen fibrils present high bending strength and biocompatibility. Understanding the mineral growth processes in collagen may facilitate the development of tissue engineering and tissue repair.

3:40 PM

(PACRIM-085-2021) Mussel directed synthesis of SnO_2/graphene composite for Lithium-Ion Batteries

W. Chi*[1]; Z. Fu[1]

1. Wuhan University of Technology, State Key Laboratory of Advanced Technology for Materials Synthesis and Processing, China

Living organism is a complex "biosynthesis factory" for the fabrication of materials under environmentally benign conditions, which is worth learning in extending synthesis techniques. Inspired by the formation process of pearls, living mussels have been used to direct the formation of nitrogen-doped anatase TiO_2. Here, instead of only synthesizing single-phase materials, we extend the technique for the fabrication of SnO_2/graphene oxide composite. The tin chloride hydroxide hydrate precursor transforms into SnO_2 nanocrystals in situ on the surface of graphene oxide with good dispersion and high mass loading to form a homogeneous microstructure under the direction of mussels at ambient temperature. The SnO_2/graphene oxide composite exhibits good lithium storage performance as the anode electrode of lithium-ion batteries. This study demonstrates the possibility and practicability of synthesizing anthropogenically useful composite materials in living organism, which is advantageous to extend synthesis methods and reduce energy consumption.

4:00 PM

(PACRIM-086-2021) Bioinspired extraordinary mechanical properties

B. Wijerathne*[1]; Z. Sun[1]

1. Queensland University of Technology, School of Chemistry and Physics, Australia

Many man-made fundamental structures and related mechanics are well established owing to inspirations from nature. Due to the equilibrium and higher sustainment abilities of biomaterials, biomimicry and bioinspiration make a pedestal for more advancements in the world. Thus, this study will describe the structural and mechanical properties responsible for the hard-pointed leaves, which are commonly known as narrow-leaf century plant and exhibit outstanding mechanical properties, particularly at the tip region. We have performed a detailed study on the structure-mechanical properties relationships by investigating the microstructures and their effects on their mechanical performance. We hope that this study will give us inspiration in designing high-performance materials and composites.

4:20 PM

(PACRIM-087-2021) Bioinspired metal oxide nanomaterials for sustainable technologies

Z. Sun*[1]

1. Queensland University of Technology, School of Chemistry, Physics and Mechanical Engineering, Australia

Learning from nature takes ideas from natural species to develop novel functional materials. Based on this concept, bio-inorganic materials (biomineralization), bio-inspired structured materials, bio-nanomaterials, hybrid organic/inorganic implant materials (bonelike composites), and smart biomaterials have been developed. In our research, by mimicking the well-ordered multiscale structures of natural interfaces or surfaces, many inorganic nanomaterials with bio-inspired structures and functions have been designed. For example, we designed fly-eye inspired superhydrophobic anti-fogging nanomaterials that have a low adherence force to water droplets and thus resist fogging-induced ice build-up even when exposed in extreme environments. The development of the fish-scale bio-inspired inorganic nanostructures, which have similar multiscale structures and multiple-functions to that of the natural targets, greatly extends the applications of bio-inspired materials to fields from micromechanical devices to heavy-duty machines as protective coatings against mechanical damage and chemical corrosion, as optical elements in optical devices and photovoltaic energies, and as low-drag or low-friction surfaces in gaseous, liquid, and solid media, etc. Therefore, we believe that the design of bio-inspired nanostructures could provide a new approach for develop novel multifunctional nanomaterials in the future.

PACRIM Symposium 40: 6th International Richard M. Fulrath Symposium, "Frontiers of Ceramics for a Sustainable Society"

Ceramics for Sustainable Energy and Environmental Systems I

Room: Plaza B

1:30 PM

(PACRIM-088-2021) Novel H_2-triggered chemical valve function of Co-doped amorphous SiO_2/γ-Al_2O_3 composite membranes (Invited)

Y. Iwamoto*[1]

1. Nagoya Institute of Technology, Department of Life Science and Applied Chemistry, Japan

Co-doped amorphous SiO_2-modified γ-Al_2O_3 composites were synthesized by using liquid precursors. The Fenton reaction during the H_2O_2-catalyzed sol-gel synthesis utilized in this study lead to the partial formation of Co(III) in addition to Co(II) within the composites. HRTEM and HAADF-STEM analyses for the composite powder sample with a composition of Al:Si:Co = 85:10:5 showed the amorphous state of the Co-doped SiO_2 which modified γ-Al_2O_3 nanocrystalline surfaces. In-situ XAFS spectroscopic analysis suggested reversible redox reactions of Co species in the composite powder sample during heat-treatment under H_2 at 500 °C followed by subsequent cooling to RT under Ar. Further analyses by in-situ IR spectroscopy combined with cyclic TPR/TPD measurements, and XPS analysis revealed that the alternating Co(III)/(II) redox reactions were associated with OH formation (hydrogenation)-deformation (dehydrogenation) of amorphous aluminosilicate matrix formed in-situ at the SiO_2/γ-Al_2O_3 hetero interface, and the redox reactions were governed by the H_2 partial pressure at 250 to 500 °C. As a result, a supported mesoporous γ-Al_2O_3/Co-doped amorphous SiO_2/mesoporous γ-Al_2O_3 three-layered composite membrane exhibited H_2-triggered chemical valve property: mesopores under H_2 flow (Open) and micropores under He flow (Closure) at 300 to 500 °C (S. Tada et al., materials, 13, No. 23, 5345 (2020)).

2:00 PM

(PACRIM-089-2021) Nano-Volumetrically Resolved Grain Boundary Effects in Thin Film Solar Cells (Invited)

J. Song[1]; Y. Zhou[2]; L. A. Ortiz[1]; B. Huey*[1]

1. University of Connecticut, Materials Science and Engineering, USA
2. Hong Kong Baptist University, Physics, Hong Kong

Ceramics and glasses are crucial to photovoltaics in obvious ways for structural support, packaging, passivation layers, and transparent conducting films. Especially for thin film solar cells, the importance of more traditional ceramic microstructural features such as grain boundaries, twin boundaries, and other defects are also increasingly recognized for their roles in even enhancing photovoltaic properties. Leveraging our recent development of Tomographic AFM, uniquely enabling fully 3-dimensional nano-resolved local properties such as photoconductivity, our investigations literally provide a new perspective on grain boundary effects in these materials systems whose continued improvement is crucial to achieving a sustainable society. Photoconductivity variations between, and within, individual grains and grain boundaries are presented and analyzed for both CdTe thin films and the rapidly evolving family of halide perovskites. Voltage and illumination-intensity dependencies can furthermore reveal local changes in mobility and efficiency. Inspired by classic grain boundary investigations of more conventional functional ceramics, these locally resolved studies suggest pathways for engineering future photovoltaics with improved efficiencies for light generation or solar energy production.

Ceramics for Sustainable Energy and Environmental Systems II

Room: Plaza B

3:20 PM

(PACRIM-090-2021) Engineering macro-porous ceramics with modulated pore configurations (Invited)

M. Fukushima*[1]

1. National Institute of Advanced Industrial Science and Technology (AIST), Japan

This presentation reviews our advanced methodology for engineering macroporous ceramic components, that can create wide range of porosity up to 98 vol% and various pore configurations such as unidirectional, bridging, bamboo-like, spherical and gradient porosities, by employing carefully selected gelation-freezing, blowing and sintering conditions. This presentation intends to give an overview of their distinctive characteristics such as fluid permeability, mechanical strength, machinability, electrochemical performance, piezoelectric property and thermal conductivity of the produced ceramic components with varied morphology and compositions. In addition, the relationship between the microstructure and mechanical/thermal properties has been, in which a homogenization method has been conducted with microscopic models created from three dimensional images, collected by X-ray computed tomography (CT) and stress or temperature distributions in macroscopic samples by finite element method. The simple, ecofriendly, and versatile approaches proposed tailor porous architectures with engineered porosity and yields macro-cellular component with distinctive characteristics suitable for a variety of industrial applications.

3:50 PM

(PACRIM-091-2021) Multifunctional Ceramic Materials for Rechargeable Li-ion Batteries (Invited)

V. Pol*[1]

1. Purdue University, Chemical Engineering, USA

ViPER (Vilas Pol's Energy Research) laboratory at Purdue University focuses its research activities on the development of high capacity ceramic based electrode materials, ionic conductors or separators for longer cycle life safer lithium ion and next generation solid state batteries. The talk will emphasize and discuss on the ViPER group's recent efforts on the structural, morphological, compositional and electrochemical properties of various fascinating ceramics and their effects on the electro-chemistries with transformative technological aspects.

4:20 PM

(PACRIM-092-2021) Analysis of hydrogen degradation in Ni-MLCCs (Invited)

Y. Iwazaki*[1]

1. TAIYO YUDEN CO., LTD., Japan

The recent widespread of electronic components in various fields have pushed BaTiO3-based Ni-MLCCs to become more important passive components. Although the reliability of MLCCs has been discussed for a long time, it is still an important issue and more reliability improvement is required. Degradation due to moisture is a typical reliability issue in BT-based Ni-MLCCs and further control is required. The typical degradation mechanism due to water is thought to be caused by ionic/electron conductions due to the separation of hydrogen in H2O molecules into protons (H+) and electrons (e-). However, it is extremely difficult to detect hydrogen because it is a light element and has no inner-shell orbitals, and the hydrogen degradation mechanism is not fully understood yet. We have clarified the possible atomic configuration of hydrogen through first-principles calculations. It was also confirmed that some of the electrons released from hydrogen did not spread uniformly and tended to be localized as small polarons, and the electron localization as a small polaron may be effectively used for identifying the degraded area in Ni-MLCC. In the presentation, we will discuss the results of these first-principles calculations and the possibilities of hydrogen detection.

Advanced Ceramic Technologies in AI, IoT, and Big Data

Room: Plaza B

5:20 PM

(PACRIM-093-2021) Ultra-Thin Piezoelectric MEMS (Invited)

T. Kobayashi*[1]

1. National Institute of Advanced Industrial Science and Technology (AIST), Japan

Piezoelectric MEMS devices are promising for the application to structural health monitoring and human health care for their characteristics of zero power sensing. For such application, the piezoelectric MEMS devices can be integrated onto flexible substrate so that they are attached on curved surface like human skins and concrete structures. Recently, we have developed novel fabrication technology to integrate ultra-thin piezoelectric MEMS devices onto flexible substrate. The ultra-thin piezoelectric devices have been fabricated so that they are connected to Si substrate by hinges. They are transferred onto flexible substrate and wired by screen printed silver paste. Based on the fabrication technology, we have developed the ultra-thin PZT/Si sensor array sheet for bridge health monitoring and the ultra-thin piezoresistive Si strain sensor integrated on textile for muscle monitoring with mechanomyogram. Moreover, we have developed ultra-thin PZT/Si actuator array for haptics. The device can present various tactile sensations to the finger by changing not only the vibration strength of the actuator but also the pattern of the actuator that vibrates. We expect that new application fields of piezoelectric MEMS will be further opened up by integrating the sensor and actuator functions of ultra-thin piezoelectric MEMS.

PACRIM Symposium 42: Young Investigator Forum - Next-Generation Materials for Multifunctional Applications and Sustainable Development, and Concurrent Societal Challenges in the New Millennium

Energy: Advances in Fundamental Science of Emerging Energy Materials I

Room: Prince of Wales

Session Chair: Daniele Benetti, Institut National de la Recherche Scientifique

1:30 PM

(PACRIM-094-2021) Enhanced Performance and Stability of 3D/2D Tin Perovskite Solar Cells Fabricated with a Sequential Solution Deposition (Invited)

E. Jokar*[1]

1. National Chiao Tung University, Applied Chemistry, Taiwan

Tin-based perovskite solar cells (PSC) have attracted considerable attention due to their rapid progress as an alternative to solve the toxic problem of their lead-based analogs. Here to grow the 2D/quasi-2D layer, we introduced hexafluoro-2-propanol (HFP) instead of isopropanol (as a common solvent in lead perovskite) to grow low dimensional perovskite on top od 3D perovskite in the second step. In the second step to deposit six cations, such as Butylammonium, hexylammonium, allylammonium, cyclohexylammonium, phenyethylammonium and anilinium (AN), on top of the 3D perovskite layer. This step introduces a general sequential method to produce a stable 3D/2D bilayer for tin-based PSC application, for which the AN device attained remarkable PCE 10.6% with significantly enhanced open-circuit voltage (V_{OC} = 0.66 V). Here, the chemistry of bulky organic cation was studied systematically.

2:00 PM

(PACRIM-095-2021) Reining in the Compositional Flexibility of Halide Perovskites for Durable Photovoltaics (Invited)

D. P. Fenning*[1]

1. University of California, San Diego, Dept. of Nanoengineering, USA

Halide perovskites have attracted widespread attention for solar cells because of their reported defect tolerance, in stark contrast to traditional semiconductors. This talk will focus on insights from our recent X-ray and electron microscopy investigations of the relationship between the heterogeneities and defects that do appear in thin film hybrid perovskites and optoelectronic performance and stability. Using model experiment materials from single-crystal thin films to state-of-the-art complex alloys, we reveal a wide-ranging local chemistry and structure in perovskite films and link these nanoscopic variations to charge collection. I will focus on insights drawn from our recent study of non-stoichiometry and second-phase formation in alloy perovskite solar cells to provide feedback to homogenize the film chemistry, improve performance, and increase stability. By elucidating and ultimately reining in the microscopic landscape of halide perovskites, we aim to systematically accelerate the development of these materials for ubiquitous photovoltaics.

2:30 PM

(PACRIM-096-2021) The Use of Third Component in Organic Solar Cells (Invited)

N. Doumon*[1]; F. Rosei[1]; L. Yang[2]

1. Institut National de la Recherche Scientifique, Énergie Matériaux Télécommunications Research Centre, Canada
2. Jilin Normal University, Key Laboratory of Functional Materials Physics and Chemistry of the Ministry of Education, China

Organic photovoltaics (OPVs) are a viable technology to meet the increasing world energy demand. For commercial viability, issues related to their low efficiency, high instability, and short lifetime must be resolved. The dawn of non-fullerene acceptors (NFA) enabled OPVs to break the 10% efficiency barrier. Thus, in the past few years, hundreds of NFA were synthesized for applications in OPVs, facilitating a current record efficiency of above 18%. This impressive development in NFAs also begot a significant shift in interest in ternary organic solar cells (TOSCs). The development of many novel organic molecules, be it donors (D) or acceptors (A), has created room for three-component organic solar cells for high efficiency instead of the traditional two-component bulk heterojunction ones. This has led to a 17.4% efficiency for TOSCs in the space of 2-3 years. However, there has been little work on the stability of this new type of materials and devices. This work mostly studies the photostability of two workhorse acceptor-based solar cells, namely ITIC and [70]PCBM blended with either PBDB-T or PTB7-Th polymer. Irrespective of the polymer or the initial efficiency, the [70]PCBM-based devices are more photostable than the ITIC ones. Finally, we look at D:A:A ternary blend made of the PBDB-T polymer and the two acceptors as an early indication of a path to making organic solar cell devices more stable.

Energy: Advances in Fundamental Science of Emerging Energy Materials II

Room: Prince of Wales

Session Chair: Daniele Benetti, Institut National de la Recherche Scientifique

3:20 PM

(PACRIM-097-2021) Tuning the activity of perovskite oxides for oxygen electrocatalysis (Invited)

K. A. Stoerzinger*[1]

1. Oregon State University, School of Chemical, Biological, and Environmental Engineering, USA

The intermittent nature of renewable energy sources requires a clean, scalable means of converting and storing energy. One Earth abundant storage option is water electrolysis, storing energy in the bonds of O_2 and H_2, later extracting electricity by the electrochemical reaction of gasses in a fuel cell. Perovskite oxides (AMO_3) can be comparably active to precious metal-based catalysts for oxygen electrocatalysis in alkaline solutions. We will present studies of model oxide electrodes grown by pulsed laser deposition (PLD) and molecular beam epitaxy (MBE) on single crystal substrates that display a known crystallographic orientation, surface area, path for charge transport, and strain. Electrochemical measurements on these heterostructures can establish the intrinsic activity of oxide catalysts in a way that cannot be realized with polydisperse nanoparticle systems, and we use these findings to rationally design the perovskite composition and structure to maximize activity. This fundamental insight will build understanding necessary for the design of active, earth-abundant electrocatalysts for efficient conversion of renewable energy into chemical fuels.

3:50 PM

(PACRIM-098-2021) When Molecules Met Graphene: Generation of Novel Hybrid van der Waals Heterostructures (Invited)

E. Orgiu*[1]

1. Institut National de la Recherche Scientifique (INRS), EMT Centre, Canada

The rise of graphene and related 2D materials makes it possible to form heterostructures held together by weak interplanar van der Waals (vdW) interactions. Whilst nature offers a finite number of 2D materials, an almost unlimited variety of molecules can be designed and synthesized with predictable functionalities. The possibilities offered by systems in which continuous molecular layers are interfaced with inorganic 2D materials to form hybrid organic/inorganic van der Waals heterostructures are emphasized. Moreover, specific molecular groups can be employed to modify intrinsic properties and confer new capabilities to 2D materials. During my talk, I will give a brief overview of how molecular self-assembly at the surface of 2D materials can be mastered to achieve precise control over position and density of (molecular) functional groups, paving the way for a new class of hybrid functional. In particular, within such vdW heterostructures, currently assembled by mechanical superposition of different layers, periodic potentials naturally occur at the interface between the 2D materials. These potentials significantly modify the electronic structure of the individual 2D components within the stack and their alignment. Furthermore, I will show how dynamic light-switchable supramolecular lattices can be created on graphene thus imparting novel functionalities to the pristine material.

4:20 PM

(PACRIM-099-2021) In situ transient absorption spectroscopy of molecular aggregation (Invited)

C. Y. Wong*[1]; Z. S. Walbrun[1]

1. University of Oregon, USA

The electronic structure and exciton dynamics of the molecules and polymers that form the active layer in organic electronic devices can change dramatically during solution deposition and thermal annealing. As solvent vaporizes, molecules aggregate and become electronically coupled, sometimes dramatically changing the exciton dynamics and thus the suitability of the material for electronic devices. Similar changes occur during thermal annealing, as the relative orientation of molecules change. The exciton dynamics of molecules in solution and in films of aggregates can be measured using transient absorption spectroscopy. However, the progression of exciton dynamics during film formation and thermal annealing is unknown, as measurements typically cannot be performed quickly enough to collect accurate dynamics of the evolving molecular aggregates. The exciton dynamics of evolving material systems can be measured by increasing the speed of data collection. Single-shot transient absorption spectroscopy using tilted pulses can measure transient spectra in 8 ms, with an adequate signal-to-noise ratio achieved in just a few seconds. The exciton dynamics of intermediate aggregation states are revealed during the formation of an organic film and during thermal annealing. This type of measurement provides insight into the complex process of organic film formation and how the excitonic properties of a film emerge.

4:50 PM

(PACRIM-100-2021) Luminescent materials: From synthesis to applications, with a focus on rare-earth doped glass-ceramics for spectral conversion and lighting (Invited)

F. Enrichi*[1]

1. CNR-ISP, Institute of Polar Sciences, National Research Council, Italy

In this presentation I will give an overview of my 20 years of experience in the synthesis and application of luminescent materials, focused on fluorophore doped or rare-earth (RE) doped glasses, ceramics and glass-ceramics mainly prepared by the sol-gel technique. The sol-gel technique allows the preparation of different kind of functional materials with controlled structural and optical properties, from fine powders to thin films and coatings, resulting in high purity and precise composition of the final product. Luminescent materials are widely used in many applications, such as phosphors for lighting and displays, luminescent nanoparticles as medical biomarkers, waveguides for lasers and optical amplifiers, up- and down-conversion layers for energy harvesting in solar cells. Among them, I will discuss one subject in more details, which is the optimization of down-converting glass and glass-ceramic films co-activated by Tb^{3+}/Yb^{3+} ions to enhance the efficiency of silicon solar cells, by converting poorly absorbed regions of the solar spectrum into the frequencies of maximum absorption of the cell. Moreover, the role of Ag nanoaggregates to boost the luminescence of RE-ions will also be highlighted as a field of research still in progress, with significant impact not only for photovoltaics but also for lighting.

5:20 PM

(PACRIM-101-2021) Construction of broad solar spectrum-responsive nanostructures for environmental and energy applications (Invited)

Q. Zhang*[1]; M. Chaker[1]; D. Ma[1]

1. INRS, Canada

Photocatalysis is considered to be one of the most promising technologies for tackling energy and environmental issues by directly harvesting and utilizing solar energy. However, most of the photocatalysts can only capture photons in the ultraviolet (UV) and blue ranges, with a large portion of solar energy remaining unutilized. Herein, I will present some of our recent development in using plasmonic nanoparticles (NPs), upconversion materials and constructing heterojunctions to realize broadband solar harvesting for environmental and energy applications. One example is our synthesized nanocomposite based on plasmonic Au NPs, lanthanide-doped $NaYF_4$, and $g\text{-}C_3N_4$. It shows enhanced UV-, visible- and near-infrared (NIR)-light photocatalytic activity in the degradation of organic pollutants in wastewater. The used Au and rare-earth metals are still expensive. To utilize solar energy more cost-effectively, a 0D/2D heterojunction based on NIR-responsive quantum dots loaded $g\text{-}C_3N_4$ nanosheets was constructed, which showed a record NIR photocatalytic activity in wastewater treatment. The use of heavy-metal based QDs constitutes a big concern in the environmental application. To address this concern, a completely metal-free 2D/2D heterojunction of black phosphorus/$g\text{-}C_3N_4$ was synthesized. The 2D/2D heterojunction showed high activity and long-term stability in solar H_2 evolution.

Tuesday, December 14, 2021

PACRIM Symposium 4: Novel, Green, and Strategic Processing and Manufacturing Technologies

Novel, Green, and Strategic Processing II

Room: Georgia A

Session Chair: Tatsuki Ohji, National Institute of Advanced Industrial Science and Technology (AIST)

8:30 AM

(PACRIM-102-2021) Why is Locally Activated Solution Processing Preferable for Continuous Material Production than General Hydrothermal Processing using Autoclaves? (Invited)

M. Yoshimura*[1]

1. National Cheng Kung University, Mater. Sci. & Eng., Taiwan

Recently various nano particles have been synthesized via hydrothermal and/or solvothermal reactions. Hydrothermal reactions need high temperature(T) and Pressure(P). T would be effective for kinetics via KT. However P would affect as LogP like Concentration(C) as Log C, where C can be changed in 5-8 order of magnitudes but P is difficult to change even 2-3 orders. Particularly in "Autoclaves" pressure increases would be more difficult and cost than temperature. Thus Autoclaves, a batch systems, seem to be not always best choice of hydrothermal reactor. Pipe-line or capillary tube may be better for continuous production. Moreover, most of Hydrothermal/Solvothermal reactions can occur at the interface between liquids and solids, therefore, they can be activated by Chemically, Electrochemistry, Sono-, Mechano-, Microwave-, Laser-, Photo-, Plasma-, etc., where those local activation could be applicable to 1D,2D and 3D patterning directly in Solution. These locally activated solution processes will be developed in quite widely and successfully. Complex formation in solution species and on solid surfaces should be important for them. In those routes:Soft, Solution Processing should be developed more widely to eliminate uses of resourses and energies then wastes and exhaust heats(= Entropies) for sustainable society.

9:00 AM

(PACRIM-103-2021) On the fabrication of silica-chalcogenide three-dimensional micro-glass composites

E. Casamenti*[1]; G. Torun[1]; L. Borasi[2]; M. Lautenbacher[1]; A. Mortensen[2]; Y. Bellouard[1]

1. EPFL, IEM, Switzerland
2. EPFL, IMX, Switzerland

Chalcogenide glasses combine a wide optical transmission window up to 12 μm and beyond in the infrared spectrum with a high refractive index and nonlinear optical properties making them interesting materials for infrared broadband optics. Here, we explore a novel process to form glass/glass composites made of arbitrarily shaped chalcogenide glass volumes encased within a fused silica substrate. Specifically, we use femtosecond laser-assisted chemical etching to carve out three-dimensional molds that are later pressure-infiltrated with a molten chalcogenide glass. This process enables the fabrication of 3D micro-optical components forming a hybrid composite with optically transparent mid-IR elements embedded in a mechanically strong and chemically stable fused silica support frame. Depending on the composition of the chalcogenide glass to be infiltrated, this process can require special attention both in terms of safety and avoiding crystallization during the solidification; yet we demonstrate that in the case of As_2S_3 it offers a particularly interesting pathway for forming mid-IR waveguides with high refractive index contrast as well as rigid 3D-micro mid-IR components.

9:20 AM

(PACRIM-104-2021) Densification, microstructure tailoring and properties of $Ta_{1-x}Hf_xC$ based ceramics

J. Yin*[1]; X. Liu[1]; Z. Huang[1]

1. Shanghai Institute of Ceramics of the Chinese Academy of Sciences, China

Owing to the same crystal structure with NaCl, continuous $Ta_{1-x}Hf_xC$ (0<x<1) solid solution could be obtained with TaC and HfC across the whole compositional ranges. They are promising to serve as thermal protection system (TPS) materials for various applications above 2000°C. However, densification of $Ta_{1-x}Hf_xC$ ceramics is difficult because of strong covalent bonding and low self-diffusion coefficient. A relative density of ternary $Ta_{1-x}Hf_xC$ ceramics sintered using Cr_3C_2 additive reached >97.5% by spark plasma sintering (SPS). The densification mechanism of Cr_3C_2 was to create transient liquid phase sintering when increasing temperature in $Ta_{1-x}Hf_xC$ matrix. Phase transition and decarburization, from orthogonal phase (Cr_3C_2, Cr_7C_3) to rock-salt structured ($Cr_{23}C_6$), would happen. A promising route to brittle-ductile transition was proposed and verified experimentally while retaining high hardness of the rock-salt structure $Ta_{1-x}Hf_xC_y$ by density functional theory (DFT) in conjunction with special quasi-random (SQS) structures. The Pugh's ratio (k) and Poisson's ratio (v) values of $Ta_{0.8}Hf_{0.2}C_{0.8}$ were 0.58 and 0.26 implying a brittle to ductile transition point. A stronger peak appeared at the non-zero region (Fermi level) with carbon-deficiency that strengthened the metallic bonding characteristic by virtue of density of electronic states (DOS) and charge density analysis.

10:00 AM

(PACRIM-105-2021) Densification mechanism and microstructure of alumina by ultra-high pressure and low temperature sintering

H. Xu*[1]; Z. Fu[1]; W. Ji[1]

1. Wuhan University of Technology, State Key Laboratory of Advanced Technology for Materials Synthesis and Processing, China

Fully dense ceramics with limited grain growth can be sintered effectively at relative low temperatures under high-pressure assisted sintering. However, there is a paucity of in-depth research on the densification mechanism and microstructure evolution. The sintering temperature and the starting threshold temperature of grain growth (T_{sg}) were determined by the applied pressure and grain size, as they were both observed to increase with grain size and to decrease with applied pressure. Densification with limited grain coarsening occurred under an applied pressure of 200 MPa at 1,050°C for the 220 nm Al_2O_3 powder and 1,400°C for the 3 μm Al_2O_3 powder. The grain boundary energy, residual stress, and dislocation density of the ceramics sintered under high pressure were higher than those of the samples sintered by non-pressure sintering. Plastic yield occurring at the contact area of the adjacent particles was proved to be the dominant mechanism for sintering under high pressure and we established a mathematical model based on the plasticity mechanics. Furthermore, the alumina sintered under ultra-high pressure could decrease the grain size and we obtained fully dense nano-sized alumina with excellent mechanical properties by plastic deformation of both grain boundaries and inner grains and studied microstructure evolution.

10:20 AM

(PACRIM-106-2021) Alignment of tubal pores in textured ceramics by magnetic-field assisted colloidal processing (Invited)

T. S. Suzuki*[1]; S. Azuma[2]; T. Uchikoshi[2]; K. Yoshida[1]

1. National Institute for Materials Science, Ceramics Processing Group, Japan
2. National Institute for Materials Science (NIMS), Japan

Porous materials are used in a wide variety of applications, such as absorber, separator, medium for cellular culture. Pore structure is important for their properties including mechanical properties. In this study, alignment of the tubal pore was attempted to control in ceramics with crystallographic orientation by the strong magnetic field-assisted colloid processing. When a strong magnetic field is applied to the particles in a stable suspension, the particles were rotated to an angle minimizing the system energy by a magnetic torque. Slurries including B4C powder, Nylon 66 as a pore former, and PEI as a dispersant was prepared via planetary mixing and ultrasonication. The slurry was consolidated by slip casting in a rotating magnetic field of 12 T. We achieved simultaneously to control the crystalline orientation of B_4C and align the pore-forming agent by the rotating condition of a magnetic field. After burning off N66, alignment of the residual tubal pores was observed. The c-axis of B_4C and the tubal pore were aligned perpendicular to the rotating plane. After densifying by SPS, the textured B_4C with the oriented tubal pore can be obtained from this green compact. This ceramic is constructed with oriented tubal pore-phase, and dense- c-axis alignment B_4C phase.

10:50 AM

(PACRIM-107-2021) Crystal oriented $Ba_2NaNb_5O_{15}$ ceramics fabricated by colloidal processing in magnetic field and subsequent sintering (Invited)

S. Tanaka*[1]

1. Nagaoka University of Technology, Materials Science and Technology, Japan

The crystal orientation technique is effective for producing polycrystalline ceramics of a substance having a large crystal anisotropy. $Ba_2NaNb_5O_{15}$ (BNN) is a crystal with a tungsten bronze structure and exhibits electro-optical effects and nonlinear optical characteristics, so it is expected to be applied to optical devices. Single crystals are usually used, but large crystal synthesis is not easy. When this is made of polycrystalline ceramics, cracks occur due to the anisotropy of the thermal expansion coefficient. The objective of this study is to orient the BNN crystal particles by the colloidal molding method in a high magnetic field and to increase the density by sintering. First, BNN particles were synthesized to prepare a green body by a rotating magnetic field, and the c-axis uniaxial orientation was confirmed. However, in the normal sintering of oriented BNN, pores developed anisotropically due to the anisotropic grain growth and the evaporation of alkali metal oxides, and it was difficult to obtain a high-density sintered body. Therefore, the excessive addition of alkali metals and the fiinement of BNN particles were investigated. Furthermore, in order to increase the density, HIP sintering conditions were examined. As a result, the relative density after HIP sintering increased to 99.7%.

11:20 AM

(PACRIM-005-2021) Advanced functionally graded yttria stabilized zirconia–lanthanum zirconate thermal barrier coatings

M. Yaghtin*[1]; A. Yaghtin[1]; Z. Tang[2]; T. Troczynski[1]

1. University of British Columbia, Materials Engineering, Canada
2. Northwest Mettech Corporation, Canada

Lanthanum zirconate (LZ) has shown superior thermophysical properties among the candidates to augment/ replace yttria stabilized zirconia (YSZ) in modern plasma-sprayed thermal barrier coatings (TBC). LZ has excellent phase stability up to 2680 °C and a much lower thermal conductivity (~1 W/mK) than YSZ (~2 W/mK). However, LZ has a lower (~9.4 × 10^{-6} K^{-1}) coefficient of thermal expansion (CTE), compared to that of YSZ, bond coat and nickel alloy substrates. This research is focused on process development for advanced functionally graded YSZ-LZ thermal barrier coatings FG-TBC, by the combination of water-based Suspension Plasma Spraying (SPS) and Solution Precursor Plasma Spraying (SPPS) processes. SPS/SPPS technology requires careful control of the process and feedstock parameters. Axial III Mettech plasma torch has been utilized throughout this work to deposit FG-TBC with fine porous and vertically-textured microstructures. The composition of coating layer varied over the thickness from 100% YSZ at the bond coat interface to 100% LZ at the outer surface of FG-TBC. Rheology and surface tension of the YSZ suspensions, LZ solutions and their mixtures were studied and optimized for the SPS/SPPS process. Microstructural and elemental analyses revealed fine columnar microstructures and gradual change of chemical composition through the thickness for the coatings.

PACRIM Symposium 6: Advanced Powder Processing and Manufacturing Technologies

Controlled Composites or Pore Structure

Room: Oxford

8:30 AM

(PACRIM-108-2021) Evolution of coarse defects in zirconia-alumina composite ceramics (Invited)

S. Tanaka*[1]

1. Nagaoka University of Technology, Materials Science and Technology, Japan

We have studied on visualization of development of ceramic microstructures based on the direct observations. When the powder compact shaped from granules and sintered, an internal structure based on the deformed granules is formed in the ceramics. Coarse defects govern the mechanical properties of ceramics, so this is a structure that requires special attention. It was shown that the evolution of coarse defects in ceramics produced using alumina granules occurred with non-uniformly sintering and proceeded with the progress of sintering. Similarly, in zirconia ceramics, coarse defects generated and grew due to the packed structure of the deformed granules. On the other hand, when the zirconia granules containing the alumina particles were used, the sintering of the zirconia particles was inhibited by the alumina particles. It is known that the grain growth of zirconia crystals is pinned by alumina grains, but observations have revealed that it is also effective in suppressing the growth of coarse defects. Further, when alumina granules containing a small amount of zirconia particles were press-molded and sintered, the zirconia particles inhibited the sintering of the alumina particles. The grain growth of alumina crystals-grains were pinned by the zirconia grains, and the effect of zirconia grains on suppressing the growth of coarse defects was also observed.

8:50 AM

(PACRIM-109-2021) Development of synthetic bone-based nanofilter for particle collection and concentration from water

C. Hassam*[1]; V. Proust[4]; D. Berthebaud[2]; T. Uchikoshi[3]

1. National Institute for Materials Science (NIMS), LINK IRL 3629, Japan
2. CNRS, LINK IRL 3629, Japan
3. National Institute for Materials Science (NIMS), Japan
4. CEA, France

Hydroxyapatite (HAp) is hydrothermally synthesised into different structures and deposited onto silicon carbide (SiC) via electrophoretic deposition (EPD). This generates all-inorganic filter membranes, with controllable properties, aimed towards water filtration applications. Wastewater monitoring has emerged as a critical technology for disease outbreak monitoring, with detection possible prior to the emergence of other symptoms. However, wastewater-based virus detection is impeded by its low concentration in a complex matrix. This necessitates the sampling and concentration of large quantities of raw material, and is a critical step that must be improved for its future application. HAp is a naturally occurring mineral with a hexagonal crystal structure exhibiting differently charged planes. By careful synthetic control this enables production of charge-controlled HAp nanoparticles. Using EPD, these particles can be deposited in films with controllable thickness, controlled orientation, and concentration dependent porosity. By combining HAp with SiC, an already important filter material, we are developing low-cost, all-inorganic, pH and temperature resistant filter membranes suited for application in wastewater sampling. This research investigates HAp synthesis conditions, and the impact of conditions used for their dispersion, deposition, processing and subsequent application.

Particle and Powder Design and Synthesis

Room: Oxford

10:00 AM

(PACRIM-110-2021) Problems Relevant to Microstructural Evolution during Fabrication and Applications about Metastable Si-B-C-N Ceramics and Their Matrix Composites (Invited)

D. Jia*[1]; D. Li[1]; Z. Yang[1]; Y. Zhou[1]

1. Harbin Institute of Technology, China

Using inorganic powders (such as Si, C, B, BN, etc.) instead of organic precursor as raw materials, the inorganic processing route based on mechanical alloying, one of the non-equilibrium processing technique, coupled with sequential sintering, although apparently very 'hard' compared to the 'soft' polymer precursor method, is actually a simple and effective way to prepare metastable SiBCN ceramics and CMCs with the uniform microstructures and superior properties. It has been used to obtain dense metastable Si-B-C-N monoliths and structural parts stable at high-temperatures providing new experimental data and therefrom a more detailed understanding of the intrinsic properties of metastable Si-B-C-N materials. In this report, the latest development of the metastable SiBCN ceramics and CMCs obtained by the inorganic processing route is presented, with the intent of summarizing key findings, uncovering major trends and providing guidance for future efforts. Focus is on the basic principles for material design and preparation, solid-state amorphization mechanisms of mechanical alloyed amorphous SiBCN powders, microstructural features and evolutionary process of as-sintered metastable monoliths, synergistic toughening/strengthening and resistance to thermal shock and ablation for CMCs, crystallization behavior of amorphous SiBCN monoliths.

10:20 AM

(PACRIM-111-2021) Fabrication of core-shell chopped C_f @phenolic resin composite powder for laser additive manufacturing of C_f/SiC composites (Invited)

J. Yin*[1]; X. Chen[1]; X. Liu[1]; Z. Huang[1]

1. Shanghai Institute of Ceramics of the Chinese Academy of Sciences, China

Laser additive manufacturing is promising technique on the preparation of complexed-shaped SiC composites. High-quality powders are critical for high-precision laser printing. In this work, core-shell C_f @phenolic resin (PR) composites with excellent flowability for selective laser sintering of C_f/SiC composites were fabricated by surface modification using 3-aminopropyltriethoxy silane coupling agent (KH550) in combination with planetary ball milling. PR coated uniformly on the fiber surface to form a core-shell structure. The effect of PR on the morphology, elemental composition, interfacial interactions and flowability of the core-shell composite powder were investigated in detail. Results indicated that the composite powder exhibited not only excellent flowability but also good laser absorptivity within the near-infrared band.

10:40 AM

(PACRIM-112-2021) High-entropy oxynitride $TiZrHfNbTaO_6N_3$ synthesized by high-pressure torsion for photocatalytic hydrogen production

P. Edalati*[1]; X. Shen[2]; M. Watanabe[3]; T. Ishihara[2]; M. Arita[4]; M. Fuji[1]; K. Edalati[3]

1. Nagoya Institute of Technology, Department of Life Science and Applied Chemistry, Japan
2. Kyushu University, Department of Applied Chemistry, Faculty of Engineering, Japan
3. Kyushu University, WPI, International Institute for Carbon-Neutral Energy Research (WPI-I2CNER), Japan
4. Kyushu University, Department of Materials Science and Engineering, Faculty of Engineering, Japan

Photocatalytic water splitting is an ideal solution for clean production of hydrogen fuel under sunlight. Metal oxynitrides are promising photocatalysts which are used for hydrogen production due to their low bandgap, but they suffer from poor stability. High-entropy ceramics are a new family of materials with high stability due to their low Gibbs free energy. By considering the features of oxynitrides and high-entropy ceramics, we developed a high-entropy oxynitride with low bandgap and high stability for photocatalytic hydrogen production. The high-entropy oxynitride with a composition of $TiZrHfNbTaO_6N_3$ and a structure of FCC+monoclinic is synthesized by high-pressure torsion followed by oxidation in air and nitriding in ammonia. The material has a better light absorbance compared with related binary and high-entropy oxides and can generate photocurrent and photocatalytic hydrogen with a high chemical stability.

11:00 AM

(PACRIM-113-2021) High-pressure TiO_2-II phase synthesized by high pressure torsion for photocatalytic CO_2 conversion

S. Akrami*[1]; M. Watanabe[2]; T. Ling[3]; T. Ishihara[3]; M. Arita[4]; M. Fuji[1]; K. Edalati[2]

1. Nagoya Institute of Technology, Department of Life Science and Applied Chemistry, Japan
2. Kyushu University, WPI, International Institute for Carbon-Neutral Energy Research (WPI-I2CNER), Japan
3. Kyushu University, Department of Applied Chemistry, Faculty of Engineering, Japan
4. Kyushu University, Department of Materials Science and Engineering, Faculty of Engineering, Japan

Conversion of CO_2 to fuels or CO is a solution to overcome the global warming crisis. TiO_2 with the anatase and rutile phases is a conventional material used for photocatalytic CO_2 conversion. In this study, the high-pressure TiO_2-II phase is synthesized by high-pressure torsion (HPT) and used for photocatalytic CO_2 conversion for the first time. The TiO_2-II phase shows better photocurrent density and higher photocatalytic CO_2 conversion activity compared with anatase. Subsequent annealing after the HPT process, resulted in enhanced photocurrent and photocatalytic activity in comparison to both initial and HPT-processed samples. This enhancement is due to the annihilation of oxygen vacancies formed in bulk by the HPT method. The samples containing the TiO_2-II phase had low electron-hole recombination rate, low bandgap, and better band positions, which were the reasons for their higher photocatalytic activity for CO_2 conversion.

11:20 AM

(PACRIM-114-2021) Drying Agent for Lowering Hydrogen Reduction Temperature of Iron Oxide Powder

T. Tsuchida*[1]; J. Fukushima[1]; M. Tobise[1]; Y. Hayashi[1]; H. Takizawa[1]

1. Tohoku University, Japan

Lowering the reduction temperature of oxide can be applied for synthesis of metal nanomaterials, produced by reduction of oxide nanomaterials while retaining the shape, and search for new compounds such as oxide layer compounds produced by reduction of perovskite materials while preventing the thermal decomposition. Hydrogen gas is well known as one of the reducing agents for oxide. However, it is difficult to lower the reduction temperature in the reaction system in which the total pressure is limited to 1 atm such as tube furnace. To solve the problem, we focused on the control of the partial pressure of water vapor. In general, hydrogen reduction temperature of oxide highly depends on the ratio of the partial pressure of hydrogen to that of water vapor as shown in Ellingham-Richardson diagram. Accordingly, the reduction of oxide can be interrupted by the water vapor generated in the reaction. Therefore, we consider that the reduction temperature of oxide can be lowered by introducing the drying agent into the reaction system, which can be efficiently absorb the water vapor generated in the reaction. In this study, we attempted to lower the hydrogen reduction temperature of the iron oxide powder in the tube furnace by controlling the water vapor pressure using a drying agent.

11:40 AM

(PACRIM-115-2021) Gas-Phase Functionalization of Multi-Kilogram Quantities of ca. 150 nm Silica Particles by a Fluidized Bed Reactor

B. Moeini*[1]; M. D. Argyle[1]; M. R. Linford[1]; M. Al-Bagoury[2]

1. Brigham Young University, USA
2. Elkem ASA, Norway

In this talk, we discussed the design and operation of a large (ca. 0.2 m × 2.0 m) muti-kilogram fluidized bed reactor for the gas-phase silanization of ca. 150 nm silica particles that run at atmospheric pressure. This process requires control of the reagent gas and reactor temperature, the introduction of the silane, the gas flow rate, and the time of the reaction. Special features of this device include a preheating system for the reagent just before it enters the reactor, control of the flow of the reagent through the reactor, and a steady stream of inert gas through the reactor. The flows in this system are adjusted to optimize the degree of surface modification. Silanes used had the form $RSi(OEt)_3$ or $RSi(OMe)_3$, where 'Et' and 'Me' denote an ethyl or a methyl group, respectively, and R refers to a chemical moiety that gives some special chemical activity to the silane. In one case, R is $-CH_2CH_2CH_2NH_2$. X-ray photoelectron spectroscopy (XPS), thermogravimetric analysis (TGA), vibrational spectroscopy (Raman), and Brunauer–Emmett–Teller (BET) analysis were used to characterize the modified particles

PACRIM Symposium 10: Sol-Gel Processing and Related Liquid-Phase Synthesis of Ceramics

Liquid-Phase Synthesis/Sol-Gel Process

Room: Regency E

8:30 AM

(PACRIM-116-2021) Continuous Flow Synthesis of Zeolites: Recent Progresses and Future Perspectives (Invited)

T. Wakihara*[1]

1. The University of Tokyo, Japan

Zeolites are typically synthesized via the hydrothermal treatment operated in batch, which suffers from several drawbacks. In recent years, the continuous flow process has experienced a fast development as it offers many merits over the batch. Switching the synthesis of zeolites from the batch operation to the continuous flow one, however, is a daunting challenge because the synthesis involves a high solid content and takes considerably long time to yield a zeolite with desired structure and properties. Recently, our group has established an ultrafast route for zeolite synthesis. The synthesis periods for a series of industrially important zeolites were shortened to tens of or several minutes, which are in stark contrast to several days that are required. The continuous flow synthesis of zeolites was further established based on the ultrafast synthesis. In this contribution, we present our recent progresses in this regard, with a focus on discussing how the continuous flow synthesis of zeolites could be developed and its potential consequences.

9:00 AM

(PACRIM-117-2021) Acid-catalyzed tetraalkoxysilane-water binary solutions: Liquid-state NMR characterization and silica thin film processing (Invited)

K. Kajihara*[1]; K. Hiruta[1]; K. Kanamura[1]

1. Tokyo Metropolitan University, Department of Applied Chemistry for Environment, Graduate School of Urban Environmental Sciences, Japan

Silica films were dip-coated from solutions prepared by cosolvent-free (solventless) acid-catalyzed hydrolytic polycondensation of tetramethoxysilane (TMOS)-water and tetraethoxysilane (TEOS)-water binary systems. These solutions were examined by liquid-state ^{1}H and ^{29}Si NMR spectroscopy and quantitatively analyzed based on a general reaction formula of tetraalkoxysilane deduced in this study. Crack-free ~500 nm-thick sintered silica films were obtained from the TEOS-based solutions aged at 80 °C, and similar films were also prepared from solutions stored for 30 d at room temperature. Water molecules were nearly absent both in the TMOS and TEOS systems, whereas only the latter system contained many unreacted SiOH groups, whose hydrophilicity is essential in producing uniform thin films on glass substrates. Interestingly, these SiOH groups were inert and their slow polycondensation enables the extension of pot life. Films prepared from the TMOS system were non-uniform and cracked because the lack of residual SiOH groups impaired wettability to glass substrates.

9:30 AM

(PACRIM-118-2021) Engineering optically transparent mesoporous silica for thermal insulation (Invited)

D. Butts*[1]; P. McNeil[1]; B. Dunn[1]

1. University of California, Los Angeles, Materials Science and Engineering, USA

Heat loss through buildings' windows accounted for ~30% of total commercial and residential building energy consumption in 2013, constituting 3.5% of primary U.S. energy consumption annually. Window material innovations, however, have largely been limited to thin films for conventional windows that require high optical transparency or to opaque aerogels synthesized using expensive supercritical drying techniques for applications such as skylights. We have found that ambiently dried silica aerogel monoliths can simultaneously achieve high optical transparency and low thermal conductivity, proving to be an ideal candidate for energy saving windows. Synthesis conditions were controlled such that silica monoliths of ~ 70% porosity can achieve a narrow pore size distribution, < 15 nm, of interconnected mesopores which leads to transmittance > 97% and low thermal conductivity (0.04 W/mK). A mechanism of phonon-boundary scattering based on pore size and fractal dimension is proposed to account for the low thermal conductivity.

10:20 AM

(PACRIM-119-2021) Sol-Gel Synthesized Metal Oxide Layers for Printable Thin-Film Transistor Applications (Invited)

S. Jeong*[1]

1. Kyung Hee University, Republic of Korea

Thin film transistor (TFT) is one of essential building blocks for various optoelectronic device applications. Among functional materials for TFTs, the metal oxide semiconductor has been considered a promising channel material, owing to its high-performance TFT characteristics. In particular, a solution-processed deposition method can provide a variety of appealing advantages, such as simplicity, low cost, and high throughput, allowing for high performance, low-cost electronics. The various sol-gel chemical routes (including a precursor-derived chemistry, a self-combustion reaction, a chemical additive-based methodology and a photo-chemical reaction) are introduced in enabling the high performance, soluble metal oxide transistors. The wet patterning techniques, both of jet printing and soft printing, which can be employed with soluble metal oxide semiconductors, are also presented, clarifying the prerequisites in making patterns by a soft patterning method.

10:50 AM

(PACRIM-120-2021) Structural Analysis of Proton Conducting Zirconia-Phosphosilicate Hybrid Membranes

L. Joseph[1]; L. C. Klein*[1]

1. Rutgers University, MS&E, USA

Proton (H^+) transport in membranes and solid electrolytes leads to their potential use in fuel cells and chemical sensors. Sol-gel oxide composites are a class of materials that exhibit proton conductivity at room temperature and above. The requirements for oxide proton conductors are that they are chemically and thermally stable in the temperature range 200-400°C. At the same time, the oxides cannot permit gas crossover. To prevent crossover, phenyl triethoxysilane (PhTES) has been substituted for part of the silica in the zirconia-phosphosilicate gels. The amount of substitution and the amount of zirconia have been varied. Effects of the substitutions on proton conduction in dry and humid samples have been measured using impedance spectroscopy and I-V curves. The molecular structure of the gels has been investigated with FTIR, to identify hydroxyls and phenyl groups, and the presence of bridging oxygen bonds. The presence of hydroxyls and water in the structure assists the hopping mechanism for the protons.

11:10 AM

(PACRIM-121-2021) Pore Structure Conditioning and Characterization in Sol-Gel Derived Hybrid Organic-Inorganic Solid Electrolytes

V. Keshishian[1]; G. Wang[1]; J. Kieffer*[1]

1. University Of Michigan, USA

To simultaneously achieve high ionic conductivity and mechanical stiffness in solid-state electrolytes (SSE), we pursue a composite materials design. We develop hybrid organic-inorganic electrolytes, in which a silica backbone, formed by sol-gel synthesis, provides a mechanically rigid backbone. The fluid in this nano-porous structure is subsequently replaced with polymer solutions. The polymer grafts onto the backbone via reactive groups, and is thus anchored into structure to establish the ion conducting phase. This unique approach allows us to decouple the influence of mechanical properties on ionic transport properties of materials and achieve both high elastic stiffness and ionic conductivity. We discovered that the network structure of a gel-cast material can be further conditioned by influencing the structural evolution during drying. By changing sample aspect ratio, various degrees of anisotropy and spatial gradients are achieved in the network topology, as revealed by nano-mechanical characterization of the materials using Brillouin light scattering. Since there is a strong correlation between the adiabatic elastic modulus and the activation energy of ion hopping, this anisotropy also affects ionic conductivity. We elaborate on strategies to harness this structural conditioning to create better-performing SSE. (Acknowledgement: NSF-DMR 1610742)

PACRIM Symposium 15: Engineering Ceramics and Ceramic Matrix Composites: Design, Development, and Applications

Advanced Processing of Ceramic Matrix Composites

Room: Regency D
Session Chair: Amjad Almansour, NASA Glenn Research Center

8:30 AM

(PACRIM-122-2021) SiC_F/SiC ceramic matrix composites fabricated by Fused Filament Fabrication (FFF) (Invited)

H. Klemm*[1]; J. Abel[1]; A. Michaelis[1]; M. Singh[2]

1. FhG IKTS Dresden, Germany
2. Ohio Aerospace Institute, USA

Ceramic matrix composites with SiC short fibers have been fabricated by Additive Manufacturing (AM) using a thermoplastic approach. By means of Fused Filament Fabrication (FFF) or Fused Deposition Modelling (FDM) a ceramic part is shaped layer by layer using a thermoplastic filament which is extruded through a heated nozzle. According to the degrees of freedom for movement of the printing head (and printing bed) green bodies with complex geometries can be fabricated. In this study a ceramic feedstock with SiC fibers (Hi Nicalon, UBE SA3), SiC powder with sintering additives and a thermoplastic binder system was used. Tailored filaments with approximately 1.8 mm in diameter were developed, fabricated by extrusion and printed by FFF to achieve CMC with short and long SiC-fibers. After binder burnout ceramic matrix composites were fabricated by polymer infiltration and pyrolysis (PIP) and liquid silicon infiltration (LSI). Various composites with unidirectional fiber orientation and different geometries were obtained. Finally, the materials fabricated were characterized regarding their microstructure and mechanical properties.

9:00 AM

(PACRIM-123-2021) Versatile processing of different ceramic fibers based on silazanes (Invited)

G. Motz*[1]

1. University of Bayreuth, Ceramic Materials Engineering, Germany

Non-oxide ceramic fibers are very attractive and essential for the fabrication of lightweight thermostructural materials like metal matrix (MMCs) and ceramic matrix composites (CMCs), due to their superior mechanical properties, such as high tensile strength > 2 GPa and stiffness, combined with excellent thermal stability and oxidation resistance to temperatures above 1000 °C. The disadvantages are the high price, the limited commercial availability and the restricted number of fiber types which are mostly based on carbosilanes leading to ceramic SiC fibers. As an alternative we have been developing ceramic SiCN fibers for about 15 years based on inexpensive silazanes that are available in large quantities using relatively simple processing strategies. In this contribution I give an overview of the processing, characterization, and potential of very different types of ceramic fibers based on polysilazanes. Beside thin ceramic SiCN fibers we developed thick fibers with diameter up to more than 100 μm, very interesting for MMC applications. The combination of the silazane with acrylonitrile led to a C/SiCN fiber type with outstanding oxidation stability for use in C/C composites. Furthermore the chemical modification of the silazane with iron complexes resulted in a superparamagnetic ceramic FeSiCN fiber suitable for structural and functional applications like shielding and defect detection in CMCs.

9:30 AM

(PACRIM-124-2021) Development of SiCf/SiC Composite for gas turbine components (Invited)

J. Park*[1]; D. Kim[1]; W. Kim[1]

1. Korea Atomic Energy Research Institute, Nuclear Materials Development Division, Republic of Korea

SiC_f/SiC composite components can provide many advantages for high-performance engines such as reduced cooling air requirements, simplified component design, reduced support structure weight, improved fuel efficiency, reduced pollutant emissions, extended service life and higher thrust. Since 2012, the national task of developing SiC_f/SiC composites for gas turbine components has been carried out in Korea, focusing on the development of manufacturing processes such as CVI, LSI, PIP and EPD+HP (Electrodeposition and Hot press). In this presentation, the manufacturing process of SiCf/SiC composite tube through the CVI process and their C-ring characteristics of the tube, and oxidation and ablation behavior of the composite plate produced by the EPD+HP process are described. The tube preform was prepared by a filament winding method, PyC interlayer was coated by CVD, and a SiC matrix was filled using methyltrichlorosilane as the starting material. The matrix filling of composite plate was performed using commercial β-SiC powder with 12 wt% Al_2O_3-Y_2O_3 sintering additive by AC-electrophoretic deposition (EPD). Twenty layers of SiC fiber with β-SiC embedded in the matrix using EPD were stacked to carry out the hot pressing, which was conducted under an argon atmosphere for 2 h at 1750°C and a pressure of 20 MPa.

Mechanical Properties of Ceramic Matrix Composites

Room: Regency D
Session Chairs: Javier Mena Garcia, Pennsylvania State University; Günter Motz, University of Bayreuth

10:20 AM

(PACRIM-125-2021) Effect of Steam Exposure on the Tensile Behavior of 2700°F EBC-CMC System (Invited)

A. S. Almansour*[1]; J. D. Kiser[1]; D. Gorican[4]; K. K. Lee[2]; J. Setlock[3]

1. NASA Glenn Research Center, Ceramic & Polymer Composites Branch, USA
2. NASA Glenn Research Center, Environmental Effects and Coatings Branch, USA
3. University of Toledo at NASA Glenn Research Center, Environmental Effects and Coatings Branch, USA
4. HX5, LLC at NASA Glenn Research Center, Ceramic & Polymer Composites Branch, USA

In water vapor-rich hot jet engine environments, SiCf/SiC ceramic matrix composite (CMC) components react with oxygen and water vapor, forming a thermally grown oxide (TGO) silica (SiO_2) layer. This silica layer reacts with water vapor to form volatile silicon hydroxide ($Si(OH)_4$) which degrades the composite by causing surface recession. Therefore, environmental barrier coatings (EBCs) are needed in order to provide elevated temperature environmental stability via protection of the CMC against surface recession. In this work, several layers of oxide-based EBC were applied on Hi-Nicalon™ Type S SiC fiber-reinforced chemical vapor infiltrated (CVI) SiC minicomposites via slurry spraying. Coated minicomposites were exposed to a steam environment for 50 and 100 hours at temperatures ranging from 2200 to 2600°F without mechanical loading. Minicomposites were then tested in monotonic tension at room temperature with the use of acoustic emission to assess surface and subsurface damage mechanisms in the EBC-CMC system. Micromechanical modeling was used to partition stresses on EBC layers, SiC/SiC minicomposites, and CVI-SiC matrix at the onset of damage in EB-coated minicomposites. Matrix cracking strength of unexposed coated samples was compared to that of steam-exposed coated samples. It was found that the degradation of the CVI-SiC and the formation of TGO reduced the matrix cracking strength.

10:50 AM

(PACRIM-126-2021) Full-field 3D characterization of crack initiation and propagation in CMCs over a range of temperatures (Invited)

D. Liu*[1]

1. University of Bristol, United Kingdom

Due to its application in extreme environments, the deformation and fracture of CMCs have to be studied under temperatures relevant to its service condition. Synchrotron X-ray computed micro-tomography (XCT) combined with in situ mechanical loading and heating has been proven to be a powerful tool for this purpose. During such tests, a series of XCT scans will be acquired during the loading process to capture the evolution of the 3D microstructure which permits subsequent full-field digital volume correlation analysis of the strain distribution in the scanned volume. The failure mechanisms, both at ambient and elevated temperatures (e.g., 1000°C and 1200°C) can be derived accordingly. A range of materials, including unidirectional and 2D woven Ox-Ox and SiC CMCs have been studied. The initiation and propagation of cracks with respect to the local microstructures have been captured and they were found to be dependent on the testing temperature. Post-test characterizations of the residual stresses and local thermal/mechanical properties were conducted to provide a fuller understanding of the high temperature behaviors of these CMCs.

11:20 AM

(PACRIM-127-2021) Development of mechanical machining processes for fabric reinfoced C/C-SiC through material removal mechanisms

A. Rösiger*[1]; R. Goller[1]

1. University of Applied Sciences, Mechanical Engineering, Germany

Surface finishing processes are required to generate the final shape of CMC (Ceramic Matrix Composite) components. Diamond grinding is still the most effective machining method for manufacturing C/C-SiC components like ceramic brake discs. Turbine vanes, shrouds and combustion parts are new applications which have 3d shapes and designs. Therefore the milling process becomes an interesting option for more economic and reliable production. C/C-SiC show heterogeneous machining behavior with high dependancy on the fibre orientation. Typical brittle material removal mechanisms of the hard SiC matrix in combination with the weak fibre/matrix interface leads to micro chipping on the surface. In this study grinding and milling tools are compared and used to machine a 2d fabric reinforced C/C-SiC with different speed and efficiency. During machining the applied loads were measured by a piezoelectric 3d sensor. The influence of the different machining conditions is evaluated regarding the surface quality (cracks, roughness). The roughness of the machined surfaces was measured by optical focus variation method and evaluated according DIN EN ISO 25178. While the feed rate of grinding process is limited, especially milling with high material removal rate shows high surface roughness. To overcome these limitations, different strategies for the machining of C/C-SiC are introduced.

PACRIM Symposium 20: Multiferroic Materials, Devices, and Applications

Multiferroic Materials, Devices, and Applications III

Room: Cypress

Session Chair: Liuyan Zhao, University of Michigan

8:30 AM

(PACRIM-128-2021) Domain-wall-induced electromagnons in multiferroics (Invited)

L. Bellaiche*[1]

1. University of Arkansas, Physics, USA

The high demand for the development of multifunctional devices as well as the interesting mechanisms behind the coexistence of ferroic order parameters within the same phase in multiferroic materials has launched a new wave of research in recent years. The unique properties of some multiferroics such as magnetoelectric (ME) coupling, which allows the cross control of electrical dipole moments by applying magnetic fields (or conversely magnetic moments by electric field), open opportunities for designing novel sensors and memory devices with significantly lower power consumption and higher efficiency. The existence of domains as well as excitations such as electromagnons award multiferroic systems with richer multifunctionalities that pave the way towards enhancing ME responses as well as designing novel devices such as configurable magnonic circuits. Here we report dynamical couplings between magnons and motions of ferroelectric domain walls that lead to the emergence of new hybrid quasi-particles, namely localized electromagnons. These quasi-particles also induce THz resonances in magnetoelectric responses, which can be exploited in reaching strikingly large ME conversion and designing more reliable and ultrafast ME devices using, e.g., local probes.

9:00 AM

(PACRIM-129-2021) Understanding and optimizing ferroelectric switching in $BiFeO_3$ (Invited)

J. Iniguez*[1]

1. Luxembourg Institute of Science and Technology and University of Luxembourg, Luxembourg

In order to produce low-power devices based on multiferroic $BiFeO_3$, one critical challenge is attaining low-voltage ferroelectric switching while maintaining control over $BiFeO_3$'s magnetic moment. Doping $BiFeO_3$ with rare-earth lanthanides appears as a promising strategy to accomplish this, and my group is pursuing this possibility by theoretical simulation. More precisely, we combine phenomenological Landau-Devonshire (LD) theory and density functional theory (DFT) to investigate the effect of La doping on the structural and dynamic properties of $La_xBi_{1-x}FeO_3$ (LBFO). First, we compute the parameters of the LD potential via an expansion relative to the reference cubic phase as a function of all the relevant degrees of freedom: polarization, antiphase rotations of oxygen octahedra, and strain. For several concentrations of La, we determine the LD coefficients by requiring that we accurately reproduce the DFT energies and the distortion amplitudes for a set of relevant LBFO polymorphs. Then, we use this LD potential in Landau-Khalatnikov simulations of ferroelectric switching, to predict how La doping affects the trajectories and electric field thresholds for polarization reversal. In this talk I will describe our approach and most recent findings. In collaboration with N. Fedorova (LIST), D.E. Nikonov and I.A. Young (intel). Funded by the Semi. Res. Corp. and Intel, contract 2018-IN-2865.

Multiferroic Materials, Devices, and Applications IV

Room: Cypress

Session Chair: Liuyan Zhao, University of Michigan

10:00 AM

(PACRIM-130-2021) Ferroelectric control of spin-charge interconversion: A new approach for magnetoelectric coupling (Invited)

M. Bibes[1]; S. Varotto*[1]

1. Unité Mixte de Physique CNRS/Thales, France

After 50 years of exponential increase in computing efficiency, new schemes must be devised to contain the ever-increasing power consumption of ICT systems. This requires the introduction of non-traditional materials and new state variables. As recently highlighted, the remanence associated with collective switching in ferroic systems is appealing to reduce power consumption. A promising approach is spintronics, which relies on ferromagnets to provide non-volatility and to generate and detect spin currents. However, magnetization reversal by spin transfer torques is a power consuming process. This is driving research on multiferroics to achieve a low-power electric-field control of magnetization, but practical materials are scarce and magnetoelectric switching remains difficult to control. In this talk, we will propose an alternative strategy to achieve low-power spin detection, in a non-magnetic system. We harness the electric-field-induced ferroelectric-like state of $SrTiO_3$ to manipulate the spin-orbit properties of a two-dimensional electron gas, and efficiently convert spin currents into positive or negative charge currents, depending on the polarisation direction. This non-volatile effect opens the way to the electric-field control of spin currents and to ultralow-power spintronics, in which non-volatility would be provided by ferroelectricity rather than by ferromagnetism.

10:30 AM

(PACRIM-131-2021) Making EuO multiferroic by epitaxial strain engineering (Invited)

S. Kamba*[1]; V. Goian[1]; R. Held[2]; E. Bousquet[3]; A. Melville[2]; P. Ghosez[3]; N. A. Spaldin[4]; D. Schlom[2]

1. Institute of Physics, Czech Academy of Sciences, Department of Dielectrics, Czechia
2. Cornell University, Department of Materials Science and Engineering, USA
3. Physique Théorique des Matériaux, Q-MAT, CESAM, Université de Liège, Belgium
4. Materials Theory, ETH Zurich, Switzerland

Infrared spectroscopy is effective tool for the study of polar soft phonons driving the displacive ferroelectric phase transitions because it can reveal the phase transitions without any structural investigations or without any electrodes required for dielectric measurements. We will demonstrate application of IR spectroscopy for already published studies of strain-induced ferroelectric phase transition in $SrTiO_3$ and $EuTiO_3$ and newly for investigation of strained EuO ferromagnet. Our first-principles DFT calculations using a hybrid functional predicted strain-induced ferroelectric phase transition in ferromagnetic EuO with a tensile strain superior to 5%. For that reason, we investigated different EuO films grown using molecular beam epitaxy on various substrates inducing strain from 0 to 6.4 %. To achieve strain higher than 5 %, we had to prepare and characterize $(EuO)_2/(BaO)_2$ superlattices on (001) LSAT substrates. In this case the EuO and BaO are nominally strained by +6.4 % and -1.2 %, respectively. We found that the EuO optical phonon is drastically shifted down in comparison with the phonon in unstrained EuO and exhibits softening towards 100 K and a hardening below this temperature. This is a typical phonon behavior near ferroelectric phase transition. The system remains ferromagnetic, i.e. the superlattice becomes multiferroic at low temperatures below 30 K.

11:00 AM

(PACRIM-132-2021) Real-time observation of dynamic modulations over a ferro-rotational order (Invited)

L. Zhao*[1]

1. University of Michigan, USA

Ferro-rotational order, the antisymmetric components of the second-rank electric quadrupolar order, has been experimentally probed and characterized using nonlinear optics recently. However, the realization of dynamic manipulation of the ferro-rotational order awaits to be explored because the coupling between multipolar orders and electromagnetic fields is nonlinear and nontrivial. In this talk, using the commensurate charge density wave (CCDW) in 1T-TaS_2 as the archetype, we demonstrate the dynamic control over the ferro-rotational order. We first confirm the ferro-rotational nature of CCDW in 1T-TaS_2 by performing temperature-dependent rotation anisotropy-second harmonic generation (RA-SHG). We then show in real time the dynamic modulation of this ferro-rotational CCDW order, using time-resolved-RA-SHG (tr-RA-SHG). We find that this ultrafast modulation manifests itself as the breathing and the rotation of RA-SHG patterns at three different frequencies in the neighborhood of the previously reported CCDW amplitude mode frequency, with the mode of the highest (lowest) frequency primarily in the breathing (rotation) channel and the middle one in both channels. We further reveal a sudden shift of these three frequencies and a dramatic increase in the breathing and rotation magnitudes across a critical pump fluence of ~ 0.5 mJ/cm^2, by performing fluence dependent tr-RA-SHG.

11:30 AM

(PACRIM-133-2021) Revisit Aurivillius Ferroelectrics: From Structures to Functionalities (Invited)

J. Zhang*[1]

1. Beijing Normal University, China

Exploration of new materials is crucial for modern information and energy technologies. Recently, complex oxides with multiple functionalities (high Tc superconductivity, emergent magnetism, ferroelectricity, ultrahigh piezoelectricity etc.) show great potentials in the applications of information processing and energy interconversion. Aurivillius oxides with layered structure such as Bi2WO6, Bi4Ti3O12 etc. exhibit fatigue-free ferroelectricity and intriguing band structure, making them promising for new-generation data storage and photocatalysis. In this presentation, taking Bi2WO6 as an example, I shall share some of our insights about its layered structure, the clamping-free ferroelasticity, emergent surface states, and so on. Further design of high-efficient and integratable magnetoelectric and photoelectric devices will be discussed.

PACRIM Symposium 21: Crystalline Materials for Electrical, Optical, and Medical Applications

Piezo/Ferro-electric Materials

Room: Regency C
Session Chair: Tetsuo Tsuchiya, National Institute of Advanced Industrial Science and Technology (AIST)

9:00 AM

(PACRIM-134-2021) Piezoelectric properties of printable ceramic-polymer unimorphs on metal substrates

T. Siponkoski*[1]; H. M. Jantunen[1]; J. Juuti[1]

1. University of Oulu, Microelectronics Research Unit, Finland

The piezoelectric properties of printable Poly(vinylidenefluorine-trifluoroethylene)-lead zirconate titanate polymer-ceramic composite on steel and brass foil substrates are investigated. In addition, the dielectric properties from 1 kHz to 1 MHz were measured. Samples are stencil printed directly on metal foils and cut into cantilevers. The highest temperature in fabrication was 120 °C. The results show increasing effective transverse piezoelectric coefficient (d_{31eff}) when stiffer substrate metal (steel) was used as a substrate. The highest d_{31eff} achieved with these unimorph cantilevers was about 22 pm/V, which was 38% higher compared to samples having lower Youngs modulus metal substrate and substantially higher than reported for similar piezoelectric composites on polymer substrates. Due to the high effective d_{31} and low relative permittivity of the composite, the developed cantilevers could be used in sensor or energy harvesting applications.

Semiconductors

Room: Regency C

10:00 AM

(PACRIM-135-2021) Advanced flexible thin films prepared by ultraviolet laser-assisted chemical solution processing for IoT society (Invited)

T. Tsuchiya*[1]; Y. Uzawa[1]; T. Nakajima[1]; J. Nomoto[1]; I. Yamaguchi[1]; Y. Kitanaka[1]

1. National Institute of Advanced Industrial Science and Technology (AIST), Japan

To construct low carbon and IoT society, it is necessary to develop a high performances new green device and sensor in post 5G information communication system is necessary. To fabricate the new devices, multi-material process would be inevitable. For these purposes, we have developed the photo-induced chemical solution process(PICSP) for the preparation of the patterned metal oxide thin film on organic, glass and single crystalline substrates. In this presentation, we demonstrate a preparation of flexible ceramics for the optical and electrical applications such as sensor, post 5G components, SiC power device and OLED. By using PICSP, flexible RuO_2 films on plastics was successfully obtained. The resistivity of the flexible RuO_2 film was $4.8\times10^{-4}\Omega$cm. In addition, high conductive film on polyimide substrate was successfully obtained for a post 5G information communication system. Also, for the development of the flexible display, it is necessary to develop a flexible transparent conductive film. By hybridizing the solution with nanoparticles, a transparent conductive film with 46 Ω / sq. and transmittance of 82% (550 nm) was produced on the PET substrate. Based on these technologies, we also succeeded in lowering the resistance and increasing the work function, and high bright flexible OLED (10000 cd/m^2).

10:30 AM

(PACRIM-136-2021) Fabrication and Characterization of Nanomaterial-based Sensor Devices by Metal Organic Decomposition with Printed Solution Method (Invited)

T. Sugahara*[1]

1. Osaka University, Japan

Sensing technologies such as gas sensor, and etc. are very important of the IoT devices in CPS(cyber physical systems). Sensing with tenuous volatile organic compounds (VOC) gas which contain in human exhalation is important issue for next generation health-care. In recent years, gas sensors are much attention research topics to synthesis oxide semiconductor nano-materials through fabrication conventional devices product process. In this study, we are demonstrating fabrication of some oxide (include MoO_3, and TiO_x) nanostructure gas sensor device for VOC gas with a very simple printed coating process, and the VOC (such as 1PrOH, EtOH, MeOH, IPA and ACE) gas sensing properties are evaluated by the MoO_x and TiO_x sensor devices

11:00 AM

(PACRIM-137-2021) Carrier modulation in monolayer MoS_2 through ferroelectric domain patterns (Invited)

J. Fernandez-Tejedor[1]; D. Gallego[1]; P. Molina[1]; P. Ares[1]; J. Gomez-Herrero[1]; L. E. Bausa[1]; M. Ramirez*[1]

1. Universidad Autonoma de Madrid, Spain

The atomic thickness nature of one layer (1L) transition metal dichalcogenides (TMDs) offers a unique opportunity to modulate the optoelectronic properties of 2D materials by means of the surrounding environment. In this context, ferroelectric crystals as substrates for TMDs provide a means for electrostatic doping due to their remnant polarization. In this work, we have studied the photoluminescence (PL) of 1L MoS_2 deposited on the polar surface of an antiparallel ferroelectric domain structure by means of confocal microfluorescence. The possibility of lateral modulation of the relative PL intensity generated by excitons and trions recombination is demonstrated. The result is explained by the different electrostatic doping imposed by the alternate polarity of the ferroelectric domains, which modify the Fermi level of the 1L MoS_2. Accordingly, periodical arrays of lateral p-n homojunctions located on the ferroelectric domain boundaries can be easily obtained. The results open new routes for the design of simple and cost-effective electronic and optoelectronic devices based on 1L MoS_2 a direct bandgap semiconductor.

11:30 AM

(PACRIM-138-2021) Solid-phase crystallization of Sn-doped In2O3 films deposited on glass substrates by reactive plasma deposition (Invited)

T. Yamamoto*[1]; H. Makino[1]

1. Kochi University of Technology, Materials Design Center, Research Institute, Japan

We have been investigating the effects of solid-phase crystallization on structural, electrical, and optical properties of transparent conductive oxide films such as In2O3-based films for use in electrodes of photovoltaic cells and LCD panels. Solid-phase crystallization of In2O3-based films has been studied by many authors because it can be a useful method to obtain high carrier transport. When annealed, amorphous films undergo crystallization through nucleation and grain growth. In this study, amorphous conductive 50-nm-thick Sn-doped In2O3 (ITO) films were deposited on glass substrates with no intentional heating by reactive plasma deposition with direct current arc discharge. When annealed at a heating rate of 5°C/min in air, the formation of crystallites were observed at a temperature of 190°C by X-ray diffraction(XRD) measurements. The XRD profiles and electron back scattering diffraction (EBSD) images show that crystallization is favored in the <111> directions of the crystallites. The EBSD study yielded that an increase in temperature up to 205°C promotes grain growth while maintaining very small misorientation within each crystallite. In such films, a disorder parameter, the energy width of the band tail (DE), remained almost constant.

PACRIM Symposium 22: Microwave Dielectric Materials and Their Applications

Dielectric Materials and Metamaterials for Microwave Applications

Room: Plaza C

8:30 AM

(PACRIM-139-2021) Direct Laser Lithography of Microwave Structures as Additive Manufacturing (Invited)

S. Kirihara*[1]

1. Osaka University, Joining and Welding Research Institute, Japan

Ultraviolet laser lithography was newly developed as a direct forming process of fine ceramic components with micro geometric patterns. As an additive manufacturing technique, cross sectional layers were created through dewaxing and sintering by UV laser drawing on spread resin paste including ceramic nanoparticles, and solid models were sterically printed by layer laminations and interlayer joining. Alumina particles of 300 nm in average diameter were dispersed in to photo sensitive liquid resins at 50 % in volume fraction. The resin paste was spread on a glass substrate at 50 μm in layer thickness by a mechanically moved knife edge. An ultraviolet laser beam of 355 nm in wavelength was adjusted at 10 μm in diameter and scanned on the surface. Irradiation power was increased to 1.0 W for enough solidification depth. The half wavelength of the incident ultraviolet ray should be comparable with the nanoparticles gaps in the resin paste, and electromagnetic field can be resonated and concentrated through Anderson localization. In this investigation, through computer aided smart manufacturing, design and evaluation (Smart MADE), the alumina or titania structures with about 99 % in volume fraction were successfully processed to create thermoacoustic or electromagnetic devices for supersonic or terahertz waves modulations.

9:00 AM

(PACRIM-140-2021) Microwave dielectric properties of Spinel-structured ceramics (Invited)

A. Kan*[1]; S. Takahashi[2]; H. Ogawa[3]

1. Meijo University, Department of Vehicle and Mechanical Engineering, Japan
2. National Institute of Technology, Fukui College, Department of Mechanical Engineering, Japan
3. Nagoya Industrial Science Research Institute, Department of Research, Japan

The spinel-structured ceramics are known as a dielectric materials with low dielectric constant and high Qf values; these ceramics are considered to be one of the appropriate candidates for the application to high-frequency devices such as millimeter-wave radars and wireless communication systems. Spinel with nominal composition of $A^{2+}B^{3+}{}_2O_4$ can be generally explained by the general formula, i.e., $(A^{2+}{}_{1-\lambda}B^{3+}{}_{\lambda})^{IV}[B^{3+}{}_{2-\lambda}A^{2+}{}_{\lambda}]^{VI}O_4$, where A^{2+} and B^{3+} represent the die- and tri-valent cations in the tetrahedral and octahedral sites when the crystal structure has a cubic with Fd-3m (No. 227). The degree of inversion, λ, which is equivalent to the fraction of A^{2+} cation in the tetrahedral site, varies from 0 (normal spinel) to 1 (inverse spinel). In the normal spinel (λ = 0). However, the relationship between cation distribution and microwave dielectric properties of spinel has not been characterized to date; therefore, this study focused on the characterization of crystal structure and microwave dielectric properties of spinel. The Qf values of $MgAl_2O_4$ ceramic were significantly increased from 85100 to 201690 GHz when λ varied from 0.25 to 0.42. The preferential site occupation of Al in tetrahedral site plays an important role in enhancing the Qf. Further preferential site occupation in tetrahedral site was observed for $Mg_{0.4}Al_{2.4}O_4$ ceramic (λ = 0.60) and the Qf was 232300 GHz.

9:50 AM

(PACRIM-141-2021) Temperature-dependent τ_f of microwave dielectric composites (Invited)

L. Li*[1]; S. Yang[1]; X. Chen[1]

1. Zhejiang U., Zijingang Campus, Department of Materials Science and Engineering, China

The temperature coefficient of resonant frequency (τ_f) is a key parameter for microwave dielectric ceramics, and it is usually treated as a constant over a wide temperature range. However, the present theoretical prediction shows that the resonant frequency (f_0) changes non-linearly with temperature, and τ_f exhibits significant temperature dependence for the microwave dielectric composites with small τ_f, even when f_0 changes linearly with temperature for the constituting phases. This is attributed to the temperature dependence of permittivity and τ_f for the constituting phases that has usually been ignored, and determined by the dielectric mixing rule that the permittivity of the composite obeys. The theoretical prediction is confirmed by the experimental results in $0.94MgTiO_3$-$0.06CaTiO_3$ composite, and can explain the different non-linear temperature dependence of f_0 for Al_2O_3-TiO_2 composites and layered ceramics with small τ_f. It is suggested to measure f_0 at more temperatures with a smaller interval during the t_f measurements for both microwave dielectric composites and single-phase ceramics, so that the possible non-linear variation of f_0 with temperature and temperature dependence of t_f can be revealed.

10:20 AM

(PACRIM-142-2021) $BiVO_4$ based microwave dielectric ceramics (Invited)

D. Zhou*[1]

1. Xi'an Jiaotong University, School of Electronic Science and Engineering, China

Monoclinic $BiVO_4$ possesses a high microwave permittivity 68 and Qf 8,000 GHz, which make it a strong candidate for microwave device applications. Monoclinic $BiVO_4$ undergoes a second order reversible ferroelastic phase transition to tetragonal scheelite at 255 °C, which can be induced by external pressure to below room temperature. Furthermore, it was found that single or complex substitutions (10 mol. %) for V^{5+} in $BiVO_4$ using trivalent or hexavalent ions with larger ionic radius, such as Fe^{3+}, Mo^{6+} and W^{6+}, increased the atomic packing factor of monoclinic scheelite structure, which resulted in an increase in the so-called "internal or chemical pressure", meanwhile lowered the phase transition temperature from 255 °C to below room temperature. Microwave dielectric properties of scheelite-structured $BiVO_4$ are enhanced by using Li, Mo and Fe, Mo substitutions according to the general formulae $(Li_{0.5x}Bi_{1-0.5x})(Mo_xV_{1-x})O_4$ and $Bi(Fe_{x/3}Mo_{2x/3}V_{1-x})O_4$ with Qf > 8,000 and >13,000GHz, respectively, for compositions near the ferroelastic phase boundary. Besides, TiO_2, in which Ti^{4+} resides in octahedral coordination, was found to be immiscible with pure $BiVO_4$, forming a thermodynamically stable $0.45BiVO_4$–$0.55TiO_2$ composite with permittivity 86, Qf = 9,500 GHz and near-zero TCF. In the present talk, recent progress in modifications of $BiVO_4$ based solid solution and composite ceramics will be introduced in detail.

PACRIM Symposium 24: Solid Oxide Fuel Cells and Hydrogen Technologies

Electrode Materials and Degradation/Oxygen Ion and Mixed Conductors

Room: Georgia B

Session Chair: Konrad Swierczek, AGH University of Science and Technology

8:40 AM

(PACRIM-143-2021) A highly efficient and stable bifunctional oxygen electrode for SOFC and SOEC at reduced temperatures (Invited)

K. Lee*[1]

1. Korea Advanced Institute of Science and Engineering (KAIST), Mechanical Engineering, Republic of Korea

Solid oxide cells (SOCs) are promising energy production and storage devices due to their high efficiency and low pollutant emissions. A key barrier to achieve high performance of SOCs is high electrode polarization resistance which attributes to the sluggish kinetics of oxygen reduction reactions (ORRs) and oxygen evolution reactions (OERs) at reduced temperatures. One of the potential oxygen electrodes is a perovskite $La_{0.6}Sr_{0.4}Co_{0.2}Fe_{0.8}O_{3-\delta}$ (LSCF6428) with mixed ionic and electronic conduction (MIEC) because of its considerable conductivities and good chemical compatibility with conventional electrolyte materials such as ceria-based oxides. However, upon long-term operation under fuel cell conditions, the LSCF6428 electrode is usually subjected to severe degradation of ORR kinetics owing to chemical instability. It has been well known that the tailoring the chemical composition of the perovskite material strongly modulates their electrochemical activities and oxygen ion transport properties. In this study, we synthesized the $La_{0.2}Sr_{0.8}Co_{0.8}Fe_{0.2}O_{3-\delta}$ (LSCF2882) and systematically investigated its characteristics on intrinsic properties. In addition, we evaluated oxygen transport kinetics and electrochemical performance of LSCF2882 as an oxygen electrode for reversible SOC applications.

9:10 AM

(PACRIM-144-2021) The impact of catalyst infiltration on the electrochemical performance of the cobalt free $SrTi_{1-x}Fe_xO_{3-\delta}$ (STF) oxygen electrode for Solid Oxide Cells

A. Mrozinski*[1]; S. Molin[1]; J. Karczewski[2]; B. Kamecki[2]; P. Jasinski[1]

1. Gdansk University of Technology, Laboratory of Functional Materials, Faculty of Electronics, Telecommunications and Informatics, Poland
2. Gdansk University of Technology, Faculty of Applied Physics and Mathematics, Poland

High oxygen reduction reaction (ORR) rate is one of most required features of effective intermediate temperature Solid Oxide Cells that limits the performance at decreased temperatures. One the approaches for fabrication of high performance electrodes is the infiltration of porous electrodes by active electrocatalysts. In our work we have decided to study Co-free mixed ionic and electronic conductor: $SrTi_{1-x}Fe_xO_{3-\delta}$ (STF). The surface modification of the STF backbone was carried out by employing different materials as the infiltrate: $(La_{0.6}Sr_{0.4})_{0.98}CoO_{3-\delta}$ (LSC), PrO_x or $(La_{0.75}Sr_{0.25})_{0.95}MnO_{3-\delta}$. The preliminary results at 700 °C and 600 °C show that the Area Specific Resistance (ASR) of the STF70 can be improved at least by a factor of 2 by only a small addition of LSC due to changing electrochemical processes occurring during ORR. Detailed electrochemical analyses were performed in order to describe the rate limiting steps if the electrochemical processes on the non-infiltrated and infiltrated electrodes.

9:30 AM

(PACRIM-145-2021) Theoretical Analysis of Oxygen Ion Transport in Doped Ceria: Effect of Vacancy Trapping

M. Kilic[1]; J. Lee[1]; K. Lee*[1]

1. Korea Institute of Science and Technology, Republic of Korea

Trivalent dopants in CeO_2 lattice generate the oxygen vacancies in the lattice, that is vital to oxygen diffusion. However, the dopant also has two detrimental effects on the oxygen vacancy migration: trapping the oxygen vacancies to the dopant and blocking the vacancy migration by increasing the energy barrier for the vacancy transport. We studied the diffusion of oxygen in Gd doped CeO_2 with La or Y co-dopants by using molecular dynamics (MD) simulation and compared them with the vacancy trapping and blocking behavior by the dopants. The trapping was characterized by the coordination number (CN) of cations statistically analyzed in large scale. The blocking caused by the dopants were investigated by the first principle nudged elastic band analysis. We found that the oxygen diffusion measured by the mean square displacement of oxygen in doped CeO_2 is enhanced when the trapping of oxygen vacancies is reduced by the co-doping. In contrast, the change in the energy barrier against oxygen migration in the co-doped lattice is not consistent with the diffusion behavior. This result evidently shows that the oxygen ion transport is governed by the trapping of oxygen vacancies.

PACRIM Symposium 27: Advanced Materials and Technologies for Electrochemical Energy Storage Systems

Electrode/Electrolyte Interphase

Room: Balmoral

8:40 AM

(PACRIM-146-2021) Designing electrolyte and interphase for high-voltage and safe batteries (Invited)

Y. Yamada*[1]; A. Yamada[1]

1. The University of Tokyo, Department of Chemical System Engineering, Japan

Safety is of prime importance for large-scale rechargeable batteries. A conventional electrolyte for lithium-ion batteries is composed of $LiPF_6$ and organic carbonates, which can create a stable solid electrolyte interphase (SEI) on graphite anodes. However, the organic carbonates are highly flammable and cause severe fire accidents. Here we introduce electrolyte design strategies to high-voltage and safe batteries. First, increasing the salt concentration is introduced to tune SEI chemistries in both nonaqueous and aqueous electrolytes, thus widening their potential windows. A remarkable finding is that the electrochemical alloying reaction of Al with Li is achieved in concentrated aqueous electrolytes. This high-concentration strategy also enables us to utilize a nonflammable organic phosphate solvent (e.g., trimethyl phosphate, TMP) in combination with graphite anodes, which otherwise causes severe side reactions during the first charge. Second, a multifunctional solvent molecule, fluorinated cyclic phosphate (TFEP), is designed and synthesized. This solvent has a fused molecular structure of ethylene carbonate and organic phosphate, thus showing both excellent SEI-forming ability and intrinsic nonflammability. Importantly, this solvent does not need high salt concentrations for stable SEI formation, thus overcoming the issues of concentrated electrolytes.

Solid Electrolytes

Room: Balmoral
Session Chair: Mickael Dollé, Université de Montreal

10:00 AM

(PACRIM-147-2021) A material search of lithium ionic conductors based on crystal structures and compositions (Invited)

K. Suzuki*[1]; R. Kanno[2]

1. Tokyo Institute of Technology, Department of Chemical Science and Engineering, Japan
2. Tokyo Institute of Technology, All-Solid-State Battery Unit, Japan

Lithium ionic conductors were searched for developing all-solid-state lithium batteries. Two directions for the material search were examined in this study. Composition based material search using the recommender system provided the two novel materials with new composition and crystal structures in Li-Ge-P-O and Li-Zn-Ge-O systems. However, the ionic conductivity of those materials was lower than 10^{-5} S cm^{-1}. On the other hand, the structure-based material search using γ-Li_3PO_4 framework suggested the possibility of a high-ionic conducting phase with Li-Ge-P-V-O and Li-Ge-V-Si-O systems. Especially, $Li_{3.55}(Ge_{0.45}Si_{0.10}V_{0.45})O_4$ with the highest conductivity (5.8×10^{-5} S cm^{-1}) functioned as a solid electrolyte of the all-solid-state battery. Both direction of the material search discovered novel materials, which could contirbute to expanding the material search field or enhancing the battery performance of the all-solid-state lithium batteries.

10:30 AM

(PACRIM-148-2021) Enabling solid-state batteries at ambient temperature by hybrid composite electrolytes based on UV cross-linked polymer matrixes (Invited)

M. Falco[1]; S. Saffirio[1]; F. Smeacetto[1]; C. Gerbaldi*[1]

1. Politecnico di Torino, Department of Applied Science and Technology - DISAT, Italy

Nowadays, rapid switch to safe, solid-state battery designs is fundamental, and most promising approaches include in situ preparation of polymer electrolytes using simple processes. Ionic mobility is directly linked to polymer chain mobility. It can be controlled through crystallinity reduction, by addition of plasticisers, ceramic fillers, and proper crosslinking. Here, an overview is offered of recent developments on innovative polymer electrolytes for Li-/Na-based batteries by means of UV-induced photopolymerization, which is easily scalable, rapid, highly efficient and green, being solvent-free. Crosslinking allows the incorporation of high amount of RTIL or tetraglyme and metal salt (FSI^-, $TFSI^-$ anion), leading to polymer electrolytes with remarkable homogeneity and robustness. To increase the cycling ability at (sub-)ambient temperatures, recent efforts are focused on the formulation of hybrid composite polymer electrolytes (CPEs), where the ceramic superionic conducting material is homogeneously embedded in the polymeric matrix. CPEs are stiff while preserving flexibility, easily processed, and can be conceived to attain improved ionic transport and interfacial contact with the electrodes, showing enhanced cycling in real cell configuration. The Si-DRIVE project received funding from the EU's H2020 R&I programme under GA 814464.

11:00 AM

(PACRIM-149-2021) Assessing Electrochemical Stability Windows of $Li_{1+x}Al_xM_{2-x}(PO_4)_3$ (M=Ge,Ti) NASICON-type Solid Electrolytes for All Solid-State Lithium Batteries

Y. Benabed*[1]; M. Rioux[1]; S. Rousselot[1]; G. Hautier[2]; M. Dollé[1]

1. Université de Montréal, Chemistry, Canada
2. Université catholique de Louvain, Chemistry, Belgium

Lithium-ion batteries have been developed to power a growing number of portable applications and to meet the needs of an increasingly mobile society. A new generation of lithium batteries has been recently developed to meet high expectations in terms of safety, stability and capacity: All-Solid-State Lithium Batteries (ASSLB), where the conventional liquid electrolyte is replaced by a safer and more stable ceramic, polymer or glass solid electrolyte (SE). ASSLB are partly developed with the prospect of using high potential materials as positive electrode and lithium metal as negative electrode. This is only possible through SE stated large electrochemical stability windows (ESW). Nevertheless, values for these ESW are very divergent in the literature published through the last decades. Recently, several studies have come to specifically decry the frequent overestimation of SE electrochemical stabilities[1,2]. Establishing a robust procedure to determine SE's accurate ESW has become of the utmost importance. In this work, we focus on assessing the ESW of the widely used NASICON-type solid electrolytes Li1.3Al0.3Ti1.7(PO4)3 (LATP) and Li1.5Al0.5Ge1.5(PO4)3 (LAGP). We adress the importance of selecting the right experimental setup and explore the link between experimental and interpreted thermodynamic results.

11:20 AM

(PACRIM-150-2021) Effects of process parameters on lithium-ion conductivity of garnet-type solid-state electrolyte films deposited by suspension plasma spray

I. Koresh*[1]; Z. Tang[2]; T. Troczynski[1]

1. University of British Columbia, Materials Engineering, Canada
2. Northwest Mettech Corp., Canada

In the last decade, the research of all-solid-state Li-ion battery (ASSLIB) has increased significantly due to the foreseen advantageous properties of this technology. An ASSLIB which contains a Li metal anode could increase the energy density by 40% relative to the currently used liquid-state Li-ion battery. In this study, we investigate a novel approach to prepare garnet-type solid-state electrolyte (SSE) films by the suspension plasma spray process. This process offers several advantages that make it a promising low-cost, rapid and robust SSE production method. A cubic garnet-type $Li_{6.75}La_3Zr_{1.75}Nb_{0.25}O_{12}$ (LLZNbO) ethanol-based suspension was plasma sprayed on stainless steel substrates at different torch power levels and spray distances. Dense films (~5.2 g/cc) of ~20 μm thickness were deposited in 7 minutes of spraying. A maximum Li-ion conductivity of 7.3×10^{-6} S cm^{-1} at 22 °C was obtained for the samples sprayed at 30.9 kW and spray distance of 25 mm. These samples contained ~20 wt% of LLZNbO, ~13 wt% of Li-deficient phase $La_2Zr_2O_7$, and the balance of ~67 wt% of the amorphous phase. We relate the relatively high Li-conductivity to the combination of the favoured phase composition and relatively high Li number density of 2.2×10^{22} cm^{-3}. Higher torch power and/or longer spray distance lead to Li-deficient and higher content of $La_2Zr_2O_7$.

11:40 AM

(PACRIM-151-2021) Powder Aerosol Deposition Method: A pathway for the large-scale production of solid oxide electrolyte films for lithium metal batteries?

T. Nazarenus*[1]; J. Exner[1]; Y. Sun[1]; J. Kita[1]; R. Moos[1]

1. University of Bayreuth, Depertment for Functional Materials, Germany

All Solid-State Batteries (ASSBs) with a lithium metal electrode are of high interest for the next generation of lithium ion batteries due to increased capacity and safety. The ceramic garnet solid electrolyte $Li_7La_3Zr_2O_{12}$ is a promising alternative to currently used liquid electrolytes due to its non-flammable character and wide electrochemical stability window in combination with a high ionic conductivity. Despite the positive aspects of this type of solid electrolyte, an economic mass production of cells is a major challenge. The production of 5-20 μm thick films in combination with high processing temperatures and high dwell times lead to cost-intensive components as well as high process times. Furthermore, co-sintering of the solid electrolyte and cathode active materials >600°C forms isolating interface layers and reduces the performance of the ASSB. In contrast to all other ceramic process technologies, the Powder Aerosol Deposition (PAD) method allows the fabrication of adhering dense ceramic films in the thickness range of several μm at room temperature. In this work, the industrial scalable and moisture free processing of oxide solid electrolyte films for ASSBs via PAD is shown. Furthermore, the electrochemical properties of the films are investigated by temperature dependent impedance spectroscopy as well as cycling experiments.

PACRIM Symposium 31: Advanced Functional Materials, Devices, and Systems for Environmental Conservation, Pollution Control, and Critical Materials

VOCs / Ion-conducting Ceramics I

Room: Plaza A

10:00 AM

(PACRIM-152-2021) Noble-metal-free Catalysts for Toluene Combustion Based on Apatite-type Lanthanum Silicate

K. Matsuo*[1]; N. Nunotani[1]; N. Imanaka[1]

1. Osaka University, Japan

Toluene is one of volatile organic compounds (VOCs), which cause air pollution. One effective method for the toluene removal is the catalytic combustion. Although platinum-based catalysts have high oxidation ability, platinum is expensive and limited resource. Therefore, the development of noble-metal-free catalysts is desired. Among the noble-metal-free catalysts, $LaCoO_3$ is effective for the toluene combustion; however it requires the elevated temperature of 410°C to oxidize toluene completely. In order to enhance the catalytic activity, it is important to introduce promoters which can supply oxygen species toward active sites in $LaCoO_3$. In our previous study, we demonstrated that $La_{10}Si_5CoO_{27-\delta}$ with apatite-type structure possessed high oxygen release and storage abilities, due to its one-dimensional tunnel structure suitable for oxide ion migration and the redox property owing to the valence change of $Co^{2+/3+}$. Hence, we developed $LaCoO_3/La_{10}Si_5CoO_{27-\delta}/\gamma-Al_2O_3$, composed of $LaCoO_3$ and $La_{10}Si_5CoO_{27-\delta}$ as activator and promoter, respectively, and $\gamma-Al_2O_3$ as a support having high surface area, and their activities were investigated. Among the catalysts prepared, the highest activity was obtained for 10wt%$LaCoO_3$/20wt%$La_{10}Si_5CoO_{27-\delta}/\gamma-Al_2O_3$, and the complete toluene combustion was realized at the temperature as low as 300°C.

10:20 AM

(PACRIM-153-2021) Catalytic Liquid-phase Oxidation of p-Cresol by Using $Pt/CeO_2-ZrO_2-SnO_2/SBA-16$

N. Nunotani*[1]; A. Supandi[1]; N. Imanaka[1]

1. Osaka University, Japan

p-Cresol is one of the phenolic compounds, frequently encountered in the wastewater stream. Since p-cresol has highly toxic and permeability effects, it is necessary to decompose p-cresol in wastewater as much as possible. Among various treatment methods, the catalytic liquid-phase oxidation is effective owing to its simple process, while the previous catalyst ($Ag/Al_2O_3-ZrO_2$) requires severe conditions of 160 °C and 15 atm for the complete p-cresol removal. To obtain the high efficiency under the moderate condition, we focused on the $CeO_2-ZrO_2-SnO_2$ solid solution as the promoter to supply active oxygen species toward the active sites. In addition, Pt with high oxidation ability was selected as the activator, and Pt and $CeO_2-ZrO_2-SnO_2$ were dispersed on the mesoporous silica SBA-16 (Santa Barbara Amorphous No. 16) support with large surface area. Thus, $Pt/CeO_2-ZrO_2-SnO_2/SBA-16$ was prepared, and its catalytic activities were investigated. The catalytic liquid-phase oxidation was carried out by mixing 1000ppm p-cresol aqueous solution (10 mL) and the catalyst (0.1 g) at 80 °C under the atmospheric air for 4 h. As a result, the complete p-cresol removal was achieved by using $Pt/CeO_2-ZrO_2-SnO_2/SBA-16$, which was higher than the Pt/SBA-16 case (removal percentage: 75%). This high activity is attributed to the acceleration of the oxidation ability of Pt by supplying active oxygen species from the promoter.

10:40 AM

(PACRIM-154-2021) Novel Catalytic Combustion-type Carbon Monoxide Gas Sensor Having High Selectivity Operating at Moderate Temperature (Invited)

N. Imanaka*[1]

1. Osaka University, Applied Chesistry, Japan

One of the significant demerits for conventional catalytic combustion-type carbon monoxide (CO) gas sensors (such sensors with Pt loaded Al_2O_3 or Pd loaded Al_2O_3 catalysts) is that they need relatively high operating temperatures over 400 degree C in order to realize complete CO gas oxidation. Because other gases such as methane, toluene and volatile organic compounds (VOCs) also burn out at such elevated temperatures, the sensors lacks in selectivity. Recently, we have succeeded in developing novel type of catalytic combustion-type gas sensor showing appreciably low-temperature operating characterstics by applying 10 wt% Pt loaded $Ce_{0.68}Zr_{0.17}Sn_{0.15}O_{2.0}$ composite as the oxidation catalyst and they sucessfully demonstrated the operation at 70 degree C, which is more than 300 degree C below compared to the conventional ones. Furthermore, by combining the superior thermoelectric material of aluminum nitride ceramics powder as an intermediate layer between the Pt heating element and the Pt loaded $Ce_{0.68}Zr_{0.17}Sn_{0.15}O_{2.0}$ composite, a novel catalytic combustion-type carbon monoxide gas sensor was sucessfully realized, posessing the superior sensing performance that drastically accelerated the senser response.

11:10 AM

(PACRIM-155-2021) Temporary reinforcement of ion conductive ceramics (Invited)

A. Kishimoto*[1]

1. Okayama University, Graduate School of Natural Science and Technology, Japan

Solid electrolyte used in the high temperature electrochemical device also has a role of supporting the structure as a partition wall. If the mechanical strength is high, the thickness required to ensure reliability can be reduced, but the strength peculiar to the solid electrical material exists. We propose a method to temporarily strengthen it so that it will not be damaged even if it receives sudden stress during the assembly of electrochemical devices.

PACRIM Symposium 34: Glass and Ceramics for Nuclear Waste Treatment and Sequestration

Waste Form Matrices-Synthesis and Characterization I

Room: Regency F
Session Chair: Hans-Conrad zur Loye, University of South Carolina

8:30 AM

(PACRIM-156-2021) Directed Synthesis of New Actinide Containing Oxides and Fluorides as Potential Waste Forms (Invited)

H. zur Loye*[1]

1. University of South Carolina, Chemistry and Biochemistry, USA

The development of the next generation waste form materials – new families of structures that can accommodate multiple radioactive elements – can be achieved via flux and hydrothermal crystal growth approaches. The use of high temperature solutions has proven to be an effective method for exploring the synthesis of new compositions and for systematically varying the elemental content of complex materials, including oxides and fluorides. Using this high temperature halide flux crystal growth approach, a number of complex U(IV) and U(VI) uranates and salt inclusion materials (SIM), such as $K_2USi_6O_{15}$, $K_8(K_5F)U_6Si_8O_{40}$, $[NaK_6F][(UO_2)_3(Si_2O_7)_2]$, and $[CsCs_6Cl][(UO_2)_3(Ge_2O_7)_2]$, have been obtained in single crystal form and structurally characterized. Recently we synthesized the novel Pu(IV) silicate, $Cs_2PuSi_6O_{15}$ using a flux crystal growth method and its stability was predicted and confirmed from a combination of crystal chemical reasoning and DFT calculations. Low temperature syntheses can also be used to obtain complex U(IV) and Ce(IV) containing fluorides. Given the similar ionic radii of Ce^{4+} and Pu^{4+}, we used the cerium fluoride phases as a guide to predict which of the plutonium fluorides could be formed, and our initial work on this system has successfully resulted in the novel compositions $Na_3FePu_6F_{30}$, $Na_3GaPu_6F_{30}$ and $Na_3AlPu_6F_{30}$.

9:00 AM

(PACRIM-157-2021) Defect chemistry and radiation behaviour of Gd-doped UO_2

R. Mohun[1]; D. Bailey*[1]; S. Sun[1]; C. Gausse[1]; M. R. Cole[1]; H. Smith[1]; J. McCloy[2]; M. Weber[2]; M. C. Stennett[1]; N. C. Hyatt[1]; C. L. Corkhill[1]

1. The University of Sheffield, Department of Materials Science and Engineering, United Kingdom
2. Washington State University, USA

In this study, we investigated the irradiation effects on trivalent rare earth element (REE)-doped UO_2 in an attempt to predict whether the structural stability of spent nuclear fuels is altered by the radiation damage from alpha decay during the long-term storage in geologic repositories. The objective was first to quantify the defect chemistry when Gd dopant, at concentrations from 8 to 15 mol.%, is introduced in UO_2, and then to study how the intrinsic defect configuration is changed when these solid solutions are exposed to 1 MeV Kr-ion beam. For this purpose, several Gd-doped UO_2 pellets were prepared by the wet chemistry co-precipitation route, and SEM images revealed a significant modification of the initial UO_2 microstructure induced by the dopant incorporation. The lattice defect chemistry was then studied using XRD and Raman spectroscopy, and allowed to quantify the presence of both U^{5+} and oxygen vacancies formed to compensate for the charge. Post-irradiation Positron annihilation spectroscopy measurements were performed to compare the irradiation response of UO_2 and the Gd-doped pellets and also to investigate the effect of microstructure on the irradiation defects. In this talk, the mechanisms occurring near the defect-boundary interfaces will be discussed to highlight the precise role of dopant and grain boundaries modifying the defects behaviour in fluorite structures relevant to nuclear fuels.

Waste Form Matrices-Synthesis and Characterization II

Room: Regency F
Session Chair: Hans-Conrad zur Loye, University of South Carolina

10:00 AM

(PACRIM-158-2021) A Computational Database Approach to Discovering Hierarchical Nuclear Waste Forms (Invited)

M. Christian*[1]; V. Klepov[1]; K. Pace[1]; G. Morrison[1]; T. M. Besmann[1]; H. zur Loye[1]

1. University of South Carolina, USA

While high level nuclear waste is currently being effectively converted to glass for disposal, issues remain such as improving the loading of waste forms or handling the particularly difficult elements to sequester. Computational methods, such as density-functional theory (DFT), can be used to increase efficiency in discovery of novel waste forms that can help develop improve those systems. Database approaches to materials science, such as the Materials Project and the Open Quantum Materials Database, have aided in the discovery of new materials with targeted properties. This presentation will discuss two studies that focused on discovery of new transuranic compounds and how results successfully translated into potential new waste-forms. Trends regarding composition and structure will be discussed.

10:30 AM

(PACRIM-159-2021) Structural and rheological analysis of Nickel enriched peraluminous glasses

E. Hansen*[1]; D. Perret[1]; I. Giboire[1]; S. Mure[1]; C. Rapin[2]

1. CEA, DE2D, France
2. University de Lorraine, France

Part of the Research and Development program concerning high level nuclear waste conditioning aims to assess new glass compositions able to incorporate higher waste content with enhanced properties in terms of homogeneity and process ability. This study focuses on peraluminous glasses, defined by an excess of aluminium ions Al3+ in comparison with modifier elements such as Na+, Li+ ou Ca2+ ions. Peraluminous glasses have already proven to be able to incorporate up to 25 wt.% of rare earth, surrogate for actinides. The present study aims to investigate the effect of the incorporation of nickel as nickel oxide NiO on peraluminous glasses homogeneity and process ability. Quenched glasses show an incorporation limit greater than 12 wt. % of NiO, while slowly-cooled glasses show a limit of 5 wt. % of NiO. Nickel oxide content is also found to have a fluidification effect on the peraluminous glasses. When nickel oxide content is above its incorporation limit, resulting crystals are found to have a significant effect on the glass rheological properties.

10:50 AM

(PACRIM-160-2021) Impact of V_2O_5 on the structure and sulfur solubility in borosilicate glasses

R. Saini*[1]; R. Youngman[2]; A. Goel[1]

1. Rutgers University, Materials Sci. & Engg., USA
2. Science and Technology Division, Corning Incorporated, USA

Sulfur being an integral component of LAW poses challenges during vitrification due to its low solubility in borosilicate glass melts, thus, severely limiting the waste loading of the vitrified radioactive waste form. V_2O_5 has been shown to exhibit a significant impact on the sulfur solubility in borosilicate glasses. However, the underlying chemical and structural drivers governing this impact are not well understood. In this pursuit, a simplified LAW aluminoborosilicate glass composition, $30Na_2O$-$5Al_2O_3$-$15B_2O_3$-$50SiO_2$ (mol.%) has been chosen as the baseline glass and a series of (100-x)($30Na_2O$-$5Al_2O_3$-$15B_2O_3$-$50SiO_2$)-xV_2O_5 (V_2O_5 = 1-10 mol%) have been synthesized by melt- quench technique at 1200 °C in Pt-Rh crucible. NH_4VO_3 has been used as a precursor to incorporate V_2O_5 in the glass. The structure and sulfur solubility in these glasses has been studied by XRD, SEM, DSC-TGA, density measurements, ICP-OES, Raman, and ^{51}V, ^{27}Al, and ^{11}B MAS-NMR spectroscopies. The impact of V_2O_5 on the glass structure and sulfur solubility in borosilicate glasses will be presented.

11:10 AM

(PACRIM-161-2021) Thermochemical Modeling in Development and Assessment of Ceramic Nuclear Waste Forms (Invited)

T. M. Besmann*[1]; M. Christian[1]; J. Schorne-Pinto[1]; A. Mofrad[1]

1. University of South Carolina, Nuclear Engineering, USA

Thermodynamic values, namely Gibbs energies, have long been used to represent phase stability of nuclear waste forms, including characterizing leaching of species from waste glass and in determining the stability of simple ceramic waste compositions. With renewed interest in relatively complex ceramic phases (e.g., SYNROC) for sequestering certain problematic radionuclides, thermochemical representations are seen as key to the efficient identification and development of these systems. The presentation will review how thermochemical models of multi-element ceramic phases for waste systems are being effectively developed, including use of the results of first-principles calculation-generated values for completing the Gibbs energy models.

PACRIM Symposium 36: Advanced Multifunctional Bioceramics and Clinical Applications

Bioceramic Coatings / Bioglass

Room: English Bay

Session Chair: Qiong Wang, University of British Columbia

8:30 AM

(PACRIM-162-2021) Bioinspired hydroxyapatite coatings by Electrostatic Spray Deposition (Invited)

V. Muller[2]; L. Gremillard*[1]; S. Tadier[1]; C. Gaillard[1]; E. Djurado[2]

1. INSA, Materials, Engineering and Science, France
2. Université Grenoble Alpes, LEPMI, France

For an optimal efficiency, hip and knee total replacement implants have to be well integrated in the surrounding bone. To favour osseointegration, a calcium-phosphate coating is often required on the implants surface. Traditionally, these coatings are fabricated by plasma-spray, leading to well crystallized films in the most stable phases (mainly hydroxyapatite). Even though these plasma-sprayed coatings are efficiednt, their use is sometimes subject to controversy because of several drawbacks such as excessive thickness, possible delamination leading to local inflammations, and the overly stable nature of the constitutive materials that do not favour reactivity. We present here an alternative process, Electrostatic Spray Deposition, that enables the fabrication of thin, nanostructured hydroxyapatite coatings with tunable morphology (from dense to coral-like). The effects of precursors and processing parameters on the composition and morphology of the coatings will be described, as well as initial biological results.

9:00 AM

(PACRIM-163-2021) Hydroxyapatite Coatings Deposited by Axial Suspension Plasma Spraying (Invited)

N. Markocsan*[1]

1. University West, Dept. of Engineering Science, Sweden

Axial suspension plasma spray (ASPS) is a recent addition to the plasma spray technology to produce multifunctional advanced coatings. The present work explores the use of this process to coat thin, continuous and adherent Ca5 (PO4)3OH (hydroxyapatite, HAp) coatings. Current work presents the role of processing conditions on the evolution of coating's microstructure, phases, hardness and adhesion strength. Three different ASPS deposited HAp coatings on a Ti–6Al–4V substrate were investigated and compared with the atmospheric plasma spray (APS) deposited HAp coatings on a similar substrate. The effect of solute particle size and solid-loading in the suspension has been elicited on the evolution of microstructure and phases. Phase-characterization has shown a good retention of hydroxyapatite phase in the deposited coatings, whereas the adhesion strength of HAp-coating decreased with the addition of bioglass as feedstock material. This work concludes that ASPS can potentially deposit thin HAp coatings ($< 50\ \mu m$) with high adhesion-strength. This work also outlines the future directions moving forward to explore liquid feedstock i.e. suspension and solution precursor spraying for bio-coatings application using ASPS.

9:30 AM

(PACRIM-164-2021) Antimicrobial Cu-TiO2 and Ag-TiO2 coatings processed by plasma spray

Q. Wang*[1]; Z. Tang[1]; K. Yu[1]; T. Troczynski[1]; J. Kizhakkedathu[1]; R. Wang[1]

1. University of British Columbia, Canada

Transmission of diseases could occur indirectly by the contact of the surfaces contaminated by pathogens. Significant resources were allocated to the cleaning and disinfection of surfaces against the spreading of coronaviruses and other pathogens during Covid-19 pandemic. To reduce healthcare burden on surface disinfection and better fight against future pandemic, we developed two types of antimicrobial coatings (Cu-TiO2 and Ag-TiO2) on one of the most commonly used surface stainless steel (SS304) by using a plasma spray system equipped with a liquid/suspension feeder. The morphology, chemical compositions, crystal structure, adhesion and UV absorption of the coatings were characterized. The antibacterial property of these coatings against E-coli were evaluated. Results showed that up to 11.4 at% Ag and 30.8 at% Cu relative to Ti were homogeneously incorporated into the coatings. The effects of the incorporated Ag and Cu on antimicrobial property of the TiO2 coating will be presented together with detailed structural analyses. Antimicrobial Cu-TiO2 and Ag-TiO2 coatings could be potentially used on high touch surfaces to inactivate the pathogens so as to the reduce the transmission of related diseases.

10:10 AM

(PACRIM-165-2021) Hybrid Organic-Inorganic Biomaterials with Self-Healing Properties (Invited)

W. Fan*[1]; M. M. Smedskjaer[1]; D. Yu[1]; R. Youngman[2]

1. Aalborg University, Department of Chemistry and Bioscience, Denmark
2. Corning Incorporated, USA

In some biomaterial applications there is a need for the devices to withstand cyclic loading. A recent discovery of a self-healing hybrid system with interpenetrating-networks of organic and inorganic components is investigated further in this work, as there is a need to better understand their structure-mechanical property relations. Specifically, we report on a new series of silica-poly(tetrahydropyran)-poly(ε-caprolactone) (SiO_2-PTHP-PCL-diCOOH) materials, which are prepared through a three-step synthesis, including in situ cationic ring-opening polymerization, sol-gel reaction and polymer-silica condensation. We use THP as the main constituent of the organic phase, which can be polymerized under mild conditions. Additionally, we control the degree of silica-crosslinking structures and the organic-inorganic compositions. The thermal stabilities, densities as well as Young's modulus could also be regulated through such control. Of particular interest, the hybrid materials with certain organic polymer content above 73.29% show apparent self-healing ability, likely due to the reversible intermolecular forces and hydrogen bonding among the polymer chains. Finally, we discover that the PTHP-SiO_2 networks are stable in bio-like circumstances although PCL undergoes biodegradation. The present structural control approach could lead to the design of tailored functional hybrids.

10:40 AM

(PACRIM-166-2021) In-vitro reactivity and antibacterial studies of agro-waste derived biocompatible SiO_2-P_2O_5-MgO-CaO glasses

D. Kaur*[1]; M. Reddy[2]; O. pandey[1]

1. Thapar Institute of Engineering & Technology, School of Physics and Materials Science, India
2. Thapar Institute of Engineering & Technology, Department of Biotechnology, India

Rice husk and chicken eggshells are major agro-waste materials. The abundance of these materials creates serious environmental issues. Hence, there is a need to change these agro-waste materials into valuable products to reduce pollution. The main objective for the present study is the evaluation of the development of hydroxyapatite layer on the glass samples derived using agro-waste materials. These glass samples are prepared by the traditional melt-quench method. A new glass series comprising SiO_2-P_2O_5-MgO-CaO has been prepared where CaO is partially replaced by MoO_3. In this report, the source of SiO_2 and CaO is rice husk and chicken eggshells, respectively. Further, biocompatible studies were carried out with simulated body fluid. The development of hydroxyapatite layer on the glass samples was assessed by X-Ray Diffraction (XRD), Scanning Electron Microscopy (SEM), Energy Dispersive X-ray spectroscopy (EDX) and Fourier Transform Infrared (FTIR) spectroscopy techniques. Drug loading and antibacterial studies were carried out to evaluate the efficiency of as-prepared glass samples in biomedical applications.

11:00 AM

(PACRIM-167-2021) Surface modification of bioactive glasses to improve protein adsorption (Invited)

V. Gobbo*[1]; S. Spriano[2]; E. Verné[2]; v. Hytönen[1]; J. Massera[1]

1. Tampere University, Finland
2. Politecnico di Torino, Italy

Bioactive glasses (BGs) are a class of biomaterials characterized by high biocompatibility and bioactivity, i.e. able to induce hydroxyapatite precipitation when in contact with body fluids. BGs can be tailored to obtain a wide range of properties fitting with the application of interest. Furthermore, they can be functionalized with biomolecules in order to reduce the risk of undesired body reactions and bacteria adhesion. However, despite the tremendous body of work on BGs, the protein/BGs interaction is not yet well understood. Controlling proteins adsorption may enable to control cell fate. In this work fibronectin, avidin, bone sialoprotein and bone morphogenetic proteins have been adsorbed on four different BG compositions (S53P4, B25, SCNB and phosphate glass). Protein adsorption properties were improved through three different surface modification (72 hours-soaking in SBF, silanization and surface nanostructuration). BGs were characterized bare (pre-treated), after surface treatment and post protein adsorption by contact angle analysis, zeta potential titration curves, FTIR spectroscopy, AFM and SEM. In addition, the impact of surface treatment on the glass ions leaching was studied using ICP-OES spectrometry. The impact of surface chemistry and topography on protein adsorption was assessed by imaging and quantifying the amount and clustering of proteins at the materials' surface.

PACRIM Symposium 37: Material and Technology Needs for Medical Devices, Sensors, and Tissue Regeneration

Device, Sensor, and Tissue Regeneration I

Room: Stanley
Session Chair: Masanori Kikuchi, National Institute for Materials Science (NIMS)

8:30 AM

(PACRIM-168-2021) 3D Printing of Graded Tissue Engineering Scaffolds (Invited)

M. Wang*[1]

1. The University of Hong Kong, Department of Mechanical Engineering, Hong Kong

Many human body tissues, such as the osteochondral (OC) tissue, possess graded structures. The OC tissue is multilayered and consists of articular cartilage (AC), calcified cartilage and vascularized subchondral bone. For repairing OC defects that extend from AC to subchondral bone, complex, graded scaffolds need to be developed in OC tissue engineering. A variety of techniques have been studied for making graded tissue engineering scaffolds but most of them involves complex procedures and require considerable time in scaffold fabrication. 3D printing has distinctive advantages for constructing complex tissue engineering scaffolds including graded scaffolds, as it involves additive manufacturing rather than subtractive manufacturing and hence can produce complex structures. Automated 3D printing processes also minimize production time, eliminate human operational error/variation and ensure scaffold reproducibility. Therefore, with proper scaffold design, judicious selection/use of bioactive biomolecules (e.g., growth factors) and cells (including mesenchymal stem cells), and good formulation of inks / bioinks for printing, desirable graded scaffolds for regenerating complex, graded body tissues can be constructed. This talk will give an overview of employing 3D printing technologies for fabricating graded scaffolds. It will also present our research on 3D printing of graded scaffolds, either biopolymer- or bioceramic-based.

9:00 AM

(PACRIM-169-2021) Theranostics Based on Rare Earth Doped Nanoparticles (Invited)

F. Vetrone*[1]

1. Institut National de la Recherche Scientifique, Université du Québec, Centre Énergie, Matériaux et Télécommunications, Canada

In the last decade, the field of rare earth doped nanoparticles has progressed from the basic understanding of the photophysical properties governing their nanoscale luminescence to their use in a variety of applications, with considerable focus in biology and medicine. This interest stems primarily from the ability to stimulate these luminescent nanoparticles with near-infrared (NIR) light as well as their diverse emission wavelengths spanning the UV to the NIR regions. Therefore, with a single NIR excitation wavelength, it is possible to observe anti-Stokes emission, known as upconversion, or single photon (Stokes) NIR emission in the three biological windows (BW-I: 700-950 nm, BW-II: 1000-1350 nm, BW-III: 1550-1870 nm) where tissues are optically transparent. Here, we present methods for controlling the luminescence of these nanoparticles through core/shell nanostructures/nanoplatforms and demonstrate how their various emissions could be harnessed for applications in theranostics and nanomedicine.

Device, Sensor, and Tissue Regeneration II

Room: Stanley
Session Chairs: Min Wang, The University of Hong Kong; Fiorenzo Vetrone, Institut National de la Recherche Scientifique, Université du Québec

10:00 AM

(PACRIM-170-2021) Hydroxyapatite/Collagen Bone-Like Nanocomposite: Applications of Its Extreme Bioactivity (Invited)

M. Kikuchi*[1]

1. National Institute for Materials Science (NIMS), Bioceramics Group, Japan

Hydroxyapatite/collagen bone-like nanocomposite (HAp/Col), developed has a bone-like nanostructure and chemical composition and shows extreme biological activity similar to bone, i.e., it is completely incorporated into bone remodeling process to be substituted with bone in 3 months. Although a trigger of incorporation into bone remodeling process is still under investigation, the HAp/Col demonstrates up-regulation of osteogenic activities and osteoclastic acitivity in vitro. According to this high function to activate bone-related cell functions, the HAp/Col is already applied to bone void fillers and on going to be applied to various ways. The HAp/Col as a bone void filler, porous HAp/Col is practically used and demonstrates excellent clinical result in comparison to other bioresorbable ceramics in many cases. A coating of the HAp/Col on Ti showed acceleration of osseointegration by 3 times than bare Ti and biomimetically coated HAp surfaces. The coating has been improving by modified electrophoretic deposition utilizing Mg as an inorganic binder. Injectable self-setting paste of the HAp/Col with anti-decay property was fabricated by using silane coupling agent and showed the same bone tissue reaction as the bone void fillers. The HAp/Col also adsorbed antimicrobial agents, gentamicin sulfate and silver nanoparticles; thus it also applicable as antibacterial fillers, paste and coatings.

10:30 AM

(PACRIM-171-2021) Structure / property relationships in Biomaterials at the nanoscale (Invited)

F. Rosei*[1]

1. INRS, Canada

Nanostructuring materials allows to optimize their properties, by exploiting size effects. We created nanopatterns that act as surface cues, affecting cell behavior. Chemical oxidation creates nanoscale topographies, that improve biocompatibility. Our treatment provides a differential signal, selectively inhibiting fibroblast proliferation while promoting osteoblast growth in vitro. Related strategies for tissue regeneration and repair are also discussed. Further, we discuss the structure of silk and its remarkable properties. Improving antibacterial properties using laser/plasma strategies and growing graphene oxide coatings will be discussed. Finally, sensing and therapeutic approaches can be harnessed by exploiting the optical properties of nanocrystals, including Quantum Dots and upconverting nanoparticles.

11:00 AM

(PACRIM-172-2021) Biological safety of glass microspheres for in- & post-procedure x-ray-based visualization and targeting in malignant hyper-vascular hepatic neoplasia (Invited)

K. O'Connell*[1]; S. Mitchell[1]; J. Henley[1]; A. Headley[1]; D. Dobrowski[1]; D. Boyd[2]; S. Kehoe[1]

1. ABK Biomedical Inc., Canada
2. Dalhousie University, Applied Oral Sciences and School of Biomedical Engineering, Canada

Sterilized glass Eye90 microspheres ™ (ABK Biomedical) were characterized by X-ray diffraction, scanning electron microscopy, dynamic image analysis for sphericity and particle size. Toxicological Risk (TR) was established based on internationally recognized exhaustive extractable and leachable protocols (metals) with gravimetric and ICP-MS analysis; with the metals identified profiled to establish tolerable exposure and Margin of Safety (MOS) based on worst case clinical scenario. Cytotoxicity (L929 mouse fibroblasts) and hemocompatibility (direct and indirect assays in human whole blood) were established under Good Laboratory Practice (GLP) conditions per internationally recognized methodologies at an extract ratio of 0.2 g/mL, 37 and/or 50°C for 72h. Eye90 microspheres™ were contamination and corrosion free with >0.9 sphericity. Exhaustive extraction occurred at 144h and the TR assessment confirmed MOS >1 for each metal, identifying no toxicological risk. Cell viability of >80% and a hemolytic index of ≤2 indicate acceptable cytotoxicity and hemocompatibility. Achievement of these biological, chemical, and toxicological end points, provide support for the biological safety of Eye90 microspheres™ for in- & post-procedure x-ray-based visualization and targeting in the treatment of malignant hyper-vascular hepatic neoplasia.

PACRIM Symposium 40: 6th International Richard M. Fulrath Symposium, "Frontiers of Ceramics for a Sustainable Society"

Emerging Ceramic Materials and Technologies I

Room: Plaza B
Session Chair: John Ballato, Clemson University

8:40 AM

(PACRIM-173-2021) Stability and Meta-Stability of Zirconia Phases Would Result in Zirconia Mystery (Invited)

M. Yoshimura*[1]

1. National Cheng Kung University, Mater. Sci. & Eng., Taiwan

Zirconia(ZrO2) and its solid solutions have three modifications: Cubic(c), Tetragonal(t), and Monoclinic(m) ones. Zirconia can also form various metastable phases: c′(=t″),t′ and m′ with the same symmetry of c,t and m. According to our >40 years studies, we have established many phase diagrams including those stable and metastable phases for various ZrO_2-R_2O_3(R=rare earths), ZrO_2-CeO_2 as well as some HfO_2 systems. The metastable phases would be formed by some reasons, Diffusion-less phase transformations: c←→c′(=t″),t←→t′,m←→m′ and t″→t′ occur immediately during cooling and annealing. They are generally reversible due to diffusion-less nature even though they include metastable phases except for t″→t′. Diffusional phase transformations and/or phase changes: c→c+t,t→m+c require a long time heating and annealing because they need compositional changes due to slow cation diffusions, which is very slow ca. 10^5 times slower than anion diffusions. The slow cation diffusions would result in "frozen" cation sublattices, while accommodated anion sublattices due to fast anion diffusions. They would give often "ordered phases" in several ZrO_2 -R_2O_3 systems, typically R=Sc. Those ordered phases have "frozen" disordered cation sublattices and ordered anion sublattices. Thus they are metastable phases. "Frozen" phase boundaries with wide solubilities would occur at low temperature below 1000 C.

9:10 AM

(PACRIM-174-2021) Reliability of Lead Zirconate Titanate Piezoelectric Films (Invited)

S. Trolier-McKinstry*[1]

1. Pennsylvania State University, Materials Science and Engineering, USA

Use of lead zirconate titanate films for dielectric energy storage or for piezoelectric applications necessitates that they be able to sustain high DC electric fields. This paper will describe the coupled failure mechanisms that control these lifetimes for donor-doped, acceptor-doped, and films with graded doping. Curiously, the median failure time of PZT films can be increased on doping with either donors or Mn as an acceptor. It is found that lifetimes under DC fields are dominated by a combination of oxygen vacancy migration, coupled with changes in the interface barrier heights for conduction in thin films. This was assessed through a combination of highly accelerated lifetime testing, thermally stimulated depolarization currents, impedance spectroscopy, deep level transient spectroscopy, and electron energy loss spectroscopy. The relict defect chemistry of the film, including variations in this defect chemistry at interfaces strongly influences both the failure and the asymmetry of the failure as a function of the electrical polarity. As film thickness and/or the piezoelectric coefficients rise, the critical stress required for crack initiation decreases, and the failure modes can shift to electro-mechanically induced failure. Ball-on-three ball mechanical testing demonstrates that the critical flaw size in the failure of the stack becomes the thickness of the piezoelectric layer itself.

Emerging Ceramic Materials and Technologies II

Room: Plaza B

10:00 AM

(PACRIM-175-2021) Advances in optical fiber glasses for reduced heat generation (Invited)

J. Ballato*[1]; P. Dragic[2]

1. Clemson University, USA
2. University of Illinois at Urbana-Champaign, USA

Heat generation in active optical fibers represents an important sustainability opportunity. This is because the required thermal management has numerous practical consequences at a systems level, including the need for costly and energy inefficient active cooling as well as the thermal degradation of material properties, hence operational performance. This talk reviews the sources of heat generation in active optical fibers and fiber-based lasers and discusses advances in material approaches to reducing, if not entirely negating, heat generation in these systems. Examples will include recent experimental results on intrinsically low quantum defect optical fibers and laser cooling using anti-Stokes fluorescence.

10:30 AM

(PACRIM-176-2021) Reliable packaging technology for heat-resistant SiC power modules (Invited)

K. Suganuma*[1]; C. Chen[1]; Z. Zhang[1]; A. Suetake[1]

1. Osaka University, F3D, Japan

Metal bonded ceramic insulating substrates such as DBC and DBA are one of the essential components for high power electronics, which have been used for electric/hybrid vehicle, renewal energy generation/energy grids, data centers, communication infrastructures and etc. These substrates can provide reliable, heat-resistant property and excellent heat-dissipation capability for wide band gap power modules to be operated at high power and at high temperature. They will be suffered to harsh thermal cycling of a temperature swing beyond 300 degrees. In such thermal cycling, severe damage will be introduced to substrates. To avoid this, the authors group proposed an ideal substrate structure of aluminum bonded aluminum nitride substrate, which can survive in very severe thermal shock between -50 to 250 degree C. A demonstration by using SiC die bonded on a Al/AlN/Al substrate which is bonded on an complex shape Al cooling fin by utilizing Ag sinter joining paste showed an excellent stable cooling performance. The authors group also developed a new sensing technology for the package structure by using acoustic emission with the aid of AI data analysis.

11:00 AM

(PACRIM-177-2021) Mechanical response of $BaTiO_3$ single crystal in mesoscale by bending test using microcantilever beam specimens (Invited)

J. Tatami*[1]; H. Yamaguchi[1]; M. Iijima[1]

1. Yokohama National University, Japan

Because of low tensile strength of $BaTiO_3$, which is ferroelastic material, it was not possible to apply large stresses in bending and tensile tests. On the other hand, large bending stress can be applied in bending tests using micro-cantilever beam specimens that have been developed recently. In this study, the mechanical response of $BaTiO_3$ single crystal was evaluated using microcantilever beam specimens was performed. The top surface of $BaTiO_3$ single crystal was (001) and two types of specimens with <100> and <110> longitudinal directions were prepared. The stress-strain relationship is linear in the initial stage, but the microcantilever beam samples showed large nonlinear deformations before fracture. The Raman spectra of the samples subjected to high bending stress indicated that this nonlinear deformation resulted from domain switching and phase transformation of $BaTiO_3$. The bending strength was 6.2 GPa and 7.1 GPa for the samples with <100> and <110> longitudinal directions, respectively. SEM observation of the fracture surface showed that the initial cracks propagated perpendicular to the longitudinal direction. Deflection of the cracks was observed in the sample with <110> longitudinal direction, which may be attributed to the domain structure of $BaTiO_3$ under high bending stress.

PACRIM Symposium 42: Young Investigator Forum - Next-Generation Materials for Multifunctional Applications and Sustainable Development, and Concurrent Societal Challenges in the New Millennium

Computational Materials Prediction and Design Toward New Functional Materials

Room: Prince of Wales

Session Chair: Sahar Mahshid, Sunnybrook Research Institute

8:30 AM

(PACRIM-178-2021) Designing High Proton Conductive Phosphate Glasses with Topological Constraint Theory and Data-Driven Techniques

C. Wilkinson*[1]; M. Ono[2]; J. C. Mauro[1]

1. Pennsylvania State University, USA
2. Hokkaido University, Japan

With the growing need for renewable energy in recent years, hydrogen conducting materials could enable sources of green energy. One application of proton conducting materials is low-cost intermediate temperature fuel cells (ITFC), which must be able to conduct protons at a high conductivity (> 10^{-3} S/cm) at the temperatures being used (~200 Celsius) and must be intrinsically stable. Proton-conducting phosphate glasses are promising candidates for ITFC applications, as their performance can be improved by replacing the alkali content of the glass with protons in an electrochemical process called alkali-proton substitution (APS). However, the conductivity and stability are currently unsatisfactory. In this work, we have designed glasses that are stable, where APS can be implemented to give a high conductivity. This was done by using a combination of physical relationships (such as topological constraint theory (TCT) and the Stokes-Einstein relationship) with data-driven approaches. These combined methods were explored and resulted in four unique compositions. The methods and the results from some of the predicted compositions will be discussed.

8:50 AM

(PACRIM-179-2021) Python for Glass Genomics (PyGGi): Accelerating Glass Innovation with Artificial Intelligence and Machine learning

H. Singh Grover[1]; R. Ravinder[1]; M. Zaki[1]; S. Bishnoi[1]; M. Agarwal[1]; N. Krishnan*[1]

1. Indian Institute of Technology Delhi, Civil Engineering, India

Glasses are used in a variety of applications such as window glass, windshield of an automobile, fiber optics for communication, and bioactive implants in healthcare. Therefore, the discovery of a new composition of oxide glasses to the targeted property is of much importance. The traditional method used by industry to find new glass composition is based on a trial and error method which is neither economical nor scalable. Here, we present a software package, namely, Python for Glass Genomics(PyGGi), which is aimed at addressing some of the key challenges in glass science. PyGGi has, at present, three packages, namely, Bank, Seer, and Zen. PyGGi Bank is a large database of glass compositions and properties. PyGGi Seer is a machine learning-based predictive module that predicts up to 25 properties of glasses with different combinations of 232 compounds. PyGGi Zen is an optimization module that allows the user to obtain glass compositions with targeted properties subjected to various constraints in other properties and components. Overall, we believe that PyGGi will be a first step toward developing a community effort to accelerate the discovery of novel glasses for targeted applications.

9:10 AM

(PACRIM-180-2021) Accelerated Materials Discovery Based on Data Analysis and Machine Learning (Invited)

J. George*[1]

1. Université catholique de Louvain, Belgium

The stability of materials is a very important property for materials discovery. In recent years, we have explored several approaches for advancing the prediction of the stability of crystals (e.g., classical chemical heuristics and machine-learned interatomic potentials). For example, we have assessed a famous chemical heuristic – the Pauling rules – regarding their usefulness for structure prediction. They are only of limited predictive power, i.e. the rules 2 to 5 only work for 13 % of all tested oxides. Furthermore, we have shown that machine-learned interatomic potentials can be used to predict phonon properties of a range of silicon allotropes accurately and, therefore, they can also be used to assess the dynamic stability of materials. To do so, we have developed new schemes to build reference databases for machine learned interatomic potentials.

Health: Diagnostics and Therapy Towards Multifunctional Theranostics I

Room: Prince of Wales

Session Chair: Sahar Mahshid, Sunnybrook Research Institute

10:00 AM

(PACRIM-181-2021) Plasmonic copper sulfide nanoparticles afford negative contrast in optical coherence tomography (Invited)

R. Marin*[1]; J. Lifante[2]; L. Besteiro[3]; Z. Wang[3]; A. Govorov[3]; F. Rivero[4]; F. Alfonso[4]; F. Sanz-Rodriguéz[2]; D. Jaque[1]

1. Universidad Autónoma de Madrid, Física de Materiales, Spain
2. Instituto Ramón y Cajal de Investigación, Nanobiology Group, Spain
3. University of Electronic Science and Technology of China, Institute of Fundamental and Frontiers Sciences, China
4. Hospital la Pricesa, Unit of Cardiology, Spain

Optical coherence tomography (OCT) is diagnostic technique that affords optical biopsies relying on the light scattered by structures of interest within biological tissues. Ophthalmology, dermatology, and cardiology are among the fields where this technique is used as a diagnostic and surgical tool. Most OCT instruments work in the near infrared (NIR) wavelength range, where biological tissues are less opaque to light, thus allowing to reach deeper underneath the surface. Like for other imaging modalities, suitable contrast agents have been developed to help better visualizing structures of interest. All of them rely on strong scattering of the OCT probing light, hence yielding a bright signal in OCT scans (positive contrast). However, some tissues appear already bright in OCT images, and negative (dark) contrast agents would be better suited for differentiating them from the surroundings. Copper sulfide nanoparticles are the ideal candidates to address this need. In this talk, the use of plasmonic copper sulfide nanoparticles as first OCT negative contrast agents is presented. Strong absorption of the probing light and negligible light scattering are the key features of these particles, which enable this unprecedented OCT contrast modality. Among other scenarios, negative contrast in OCT opens the door to the detection of atheromatous plaques at a currently unmatched early stage.

10:30 AM

(PACRIM-182-2021) Cr^{3+} doped Nanoparticles as a Light-to-Heat Converting Agents

K. Elzbieciak-Piecka*[1]; L. Marciniak[1]

1. Institute of Low Temperature and Structure Research PAS, Poland

The importance of the interactions between light and nanoparticles, such as light to heat conversion is significant, especially when they could be applied in targeted therapies like photodynamic therapy or hyperthermia of tumors. Nowadays, effective heat generation in doped nanomaterials is studied, mainly with the use of rare earth ions. However, in this case a limiting factor could be encountered which is relatively low absorption cross-section of lanthanides. For this reason, a new approach that involves the Cr^{3+} ions as light-to heat converters was proposed in presented research. The main motivation of this study was the fact that these ions are characterized by a larger absorption cross-section than lanthanide ions. The absorption spectral range of Cr^{3+} ions and the probability of the nonradiative release of the absorbed energy is strongly dependent on the concentration of dopant and the host matrix stoichiometry, thus appropriate modification of these parameters will provide enhancement of the light induced heat generation. The main aim of this research is to investigate the influence of the selected matrix stoichiometry on the efficiency of light-to-heat conversion by Cr^{3+} ions with different concentrations. Acknowledgements The financial support from National Science Center under grant PRELUDIUM 18 2019/35/N/ST5/00553.

10:50 AM

(PACRIM-183-2021) Nanostructured Fluidic Devices for Point of Need Diagnostics (Invited)

S. Mahshid*[1]

1. McGill University, Bioengineering, Canada

Nanostructures display remarkable and unexplored surface characteristics and can act as highly effective sensors and biosensors for specific features of biomolecules. This potential for ultrasensitive analysis of nucleic acids, proteins, and small molecules may be synergistically combined with fluidic sample delivery systems, and lead to development of ideal point of care diagnostic devices with clinically relevant sensitivity and response time in a portable fashion. Such hybrid nanosurface fluidic devices are capable of working with very small sample volumes and precise dosing of reagents, enabling the transition to a portable diagnostic tool. We investigate 1) fabrication of novel nanostructured platforms based on 2D and 3D materials, 2) integration of nanostructures with micro/nanofluidic sample delivery and automated read-out systems and 3) implementation of the device for detection of small molecules, pathogenic disease, and cancer biomarkers. In this regard, we address fundamental questions including optimal interface of nanostructures with fluidic devices, target isolation, preparation and concentration in fluidic devices. We have successfully implemented the nanosurface fluidic devices for rapid and quantitative detection of bacteria and small molecules in combination with optical and electrical read-outs and molecular profiling of extracellular vesicles (EVs) in combination with Raman spectroscopy.

11:20 AM

(PACRIM-184-2021) Synthesis and characterization of SiO2 coated YAG:V nanocrystals and their potential applications

K. Kniec*[1]; L. Marciniak[1]

1. Institute of Low Temperature and Structure Research, Polish Academy of Sciences, Poland

The obtaining of V-doped YAG nanocrystals characterized by narrow size distribution and high stability in polar media is still challenging, but being crucial from the application's perspective. The luminescence of V ions in inorganic lattices, especially the NIR located 3T2→3T1g emission of V3+, is strongly affected by the crystallographic structure of the lattice and the external conditions like temperature. Thus, different synthesis and SiO2 coating approaches were implemented to obtain water-dispersed V-doped nanophosphors. The influence of synthesis method and silica coating on the particles on the spectroscopic properties of V ions, including the emission intensity, the spectral position of emission band and luminescence lifetime was investigated. The applicability potential of the V3+ luminescence in SiO2 coated YAG nanomaterials was examined, particularly considering the development of the contactless luminescence pH sensor, since the variations of pH parameter may give the feedback about the environment conditions in wide range of applications. The accurate detection of pH value is particularly important in terms of the medicinal aspect. Moreover, the properties of YAG host materials provide wide operating range and credible measurement even in harsh conditions. Acknowledgements This work was supported by National Science Center Poland (NCN) under UMO2019/33/N/ST5/00029.

PACRIM Symposium 4: Novel, Green, and Strategic Processing and Manufacturing Technologies

Novel, Green, and Strategic Processing III

Room: Georgia A

Session Chair: Tatsuki Ohji, National Institute of Advanced Industrial Science and Technology (AIST)

1:30 PM

(PACRIM-185-2021) Low Temperature Densification of tantalum carbide ceramics without additives (Invited)

W. Wang*[1]; c. liu[1]; Z. Fu[1]; H. Wang[1]

1. Wuhan University of Technology, China

The dense tantalum carbide ceramics were prepared by spark plasma sintering without any sintering aids using TaC nanopowders as the starting powders. The impacts of temperature and pressure on the densification and grain growth behavior were investigated. The results showed that high temperature and high pressure was prone to promote densification, while fast grain growth of TaC occurs. High density (~97.19%) TaC ceramics with a fine grain size of 2.67 μm were obtained at 1800 °C for 10 min under 80 MPa. Furthermore, the densification kinetics was determined using a creep deformation model. At the initial sintering stage, creep deformation which was controlled by diffusion and grain boundary sliding contributed to the densification mechanism. The apparent activation energy (Q_d) of the mechanism controlling the densification was evaluated at 252.94 KJ/mol at temperatures ranging from 1650 °C to 1750 °C. The dominant mechanism appeared to be grain boundary diffusion and grain boundary slip at the final stage of sintering.

2:00 PM

(PACRIM-186-2021) Boron-Rich Boron Carbide-Titanium Boride Based Composites Fabricated by Reactive Hot Pressing Sintering

Q. He*[1]; A. Wang[1]; T. Tian[1]; L. Hu[1]; W. Guo[1]; W. Wang[1]; Z. Fu[1]

1. Wuhan University of Technology, The State Key Laboratory of Advanced Technology for Materials Synthesis and Processing, China

In this study, $B_{6.5}C$-TiB_2-SiC-BN composite was fabricated via reactive hot pressing from a powder mixture of TiCN, B, and Si. The reaction process of the powder mixture was investigated by heat treatment at elevated temperatures. The as-reacted powder products possessed a core-shell structure of TiB_2 surrounded by h-BN. During densification, the h-BN in this structure acts as a diffusion channel and lubricant, which promoted the densification process. The extremely small size, fresh surface of in-situ formed phases, and uniform dispersion in synthesized powders also contributed to sintering. The designed reactive hot-pressing sintering effectively resolves the difficulty of obtaining uniformly-dispersed BN. The obtained bulk $B_{6.5}C$-TiB_2-SiC-BN composite ceramic shows high relative density (98.8%) and good comprehensive mechanical properties including hardness, bending strength, and fracture toughness of 19.6 GPa, 801 MPa, and 5.31 MPa•m$^{1/2}$, respectively.

2:20 PM

(PACRIM-187-2021) Reactive sintering of nanoceramics using low temperature processes: High Pressure Spark Plasma Sintering and Cold Sintering Process

C. Estournes*[1]; T. Herisson de Beauvoir[1]; A. Flaureau[1]; A. Fregeac[1]; A. Weibel[1]; U. Chung[2]; M. Josse[2]; M. Suchomel[2]; J. Majimel[2]; G. Goglio[2]; C. Elissalde[2]

1. CNRS/University Toulouse, CIRIMAT, France
2. CNRS/University of Bordeaux, ICMCB, France

In the last decades, the development of more energy efficient and environmentally friendly manufacturing processes has encouraged intensive research on low-temperature sintering. These efforts have led to the elaboration of new materials, multi-materials and composites for applications in different socio-economics fields. The presentation will be focused on two recent studies: – Using High Pressure SPS (600MPa) at very low temperature (350°C) pure zirconia ceramics were densified by reactive spark plasma sintering of an amorphous zirconium hydroxide powder. Consolidated nanoceramics made up of monoclinic grains 20nm in size were obtained. Due to their high density of grain boundaries, ceramics exhibit higher Vickers hardness than corresponding materials consolidated by classical SPS or others conventional techniques. – The Cold Sintering Process has been applied to aluminum hydroxide precursors to promote in situ reactive sintering at low temperature (< 400 °C). Partially crystallized translucent boehmite γ- AlOOH was first obtained. Then, an additional heat treatment in an oven at 500 °C induces the transformation of the latter into translucent nanostructured γ-Al_2O_3 ceramics. Here we demonstrate the possibility to produce nanostructured oxides ceramics using a low temperature process (HP-SPS or CSP) by reactive sintering of hydroxides precursors.

2:40 PM

(PACRIM-188-2021) Drucker-Prager-Cap Modelling of Boron Carbide Powder during Spark Plasma Sintering

W. Liu*[1]; W. BAI[1]; W. Wang[1]; C. Peng[1]

1. Wuhan University of Technology, China

The densification behavior of ceramic powder during Spark Plasma Sintering (SPS) can be investigated by Finite Element Analysis (FEA) based on the continuum hypothesis, and an accurate constitutive model of powder material is demanded for FEA. The Drucker-Prager-Cap (DPC) model has been proved for precise continuum modelling. However, the conventional methods of DPC parameter determination are very complex, which is not beneficial to be adopted for FEA, especially for SPS due to the elevated temperature and pressure. Therefore, an inverse identification procedure was proposed to calibrate the DPC parameters of boron carbide powder during SPS. The SPS tests of small cylinder specimen were carried out, and the two-dimensional axisymmetric finite element models were simultaneously presented. The parameters of DPC model were subsequently determined by minimizing the discrepancy between the simulated and experimental loading curves. With comparing the simulated and experimental density distribution of the specimen, the DPC model was validated for FEA of SPS of boron carbide powder. Finally, the FEA with DPC model was used to investigate the influences of specimen dimension, loading temperature and pressure on the densification behavior of boron carbide powder during SPS.

3:20 PM

(PACRIM-189-2021) Advances in pressureless electric field-assisted sintering ceramic ionic conductors (Invited)

R. Muccillo*[1]

1. IPEN, CCTM, Brazil

Sintering, one of the main experimental steps towards preparing ceramic materials for inclusion in devices, usually requires between 1/2 and 3/4 of the absolute melting point and long times [J.E. Burke, J.H. Rosolowski, in Treatise on Solid State Chemistry 1976, ch. 10, 651-659]. Alternatives for decreasing the temperature and time, looking for energy saving, were proposed consistently in the last decades, one being the application of electric fields. In the last ten years a great effort was put on sintering ceramic compacts, supplying additional heat (Joule) by applying an electric field at temperatures and times shorter (then the name flash sintering) than those used in conventional sintering. Stabilized zirconia, doped ceria, BITIVOX, doped barium cerate and zirconate ceramics were pressureless sintered under electric fields at temperatures below the reported temperatures for their conventional sintering. The main results on (enhanced) electrical properties, microstructural features (grain growth inhibition) as well as proposed mechanisms (defect creation, particle surface melting, dielectric breakdown, space charge deployment) concerning the densification by flash sintering these ionic conductors will be presented.

3:50 PM

(PACRIM-190-2021) Ultra-fast firing: Effect of heating rate on sintering of 3YSZ, with and without an electric field

W. Ji*[1]; Z. Fu[1]; R. I. Todd[2]

1. Wuhan University of Technology, China
2. University of Oxford, United Kingdom

It has recently been reported that ceramics can be sintered in a few seconds with the aid of an electric field ("flash sintering"). This investigation tests the possibility that the accelerated sintering is a consequence of the rapid heating rate involved rather than a direct effect of the electric field on mass transport. The sintering of 3YSZ powder compacts at a temperature of ~1300 °C was compared (i) in flash sintering, (ii) with rapid heating rates produced without the application of an electric field, and (iii) with conventional heating rates. The results show that rapid heating can accelerate sintering by over 2 orders of magnitude compared with heating to the same temperature at conventional rates, even without the application of an electric field. It is concluded that the rapid densification in flash sintering of 3YSZ is at least partly a consequence of the rapid heating involved. Possible explanations are discussed. The related publication won the 2019 Ross Coffin Purdy Award from ACerS.

4:10 PM

(PACRIM-191-2021) Preparation and characterization of rGO/B_4C composites by self-assembly polymerization and spark plasma sintering

L. Hu*[1]; Q. He[1]; A. Wang[1]; T. Tian[1]; W. Wang[1]; Z. Fu[1]

1. Wuhan University of Technology, China

A self-assembly polymerization process was used to prepare graphene oxide/boron carbide (GO/B_4C) composite powders, spark plasma sintering (SPS) was used to fabricate reduced graphene oxide/boron carbide (rGO/B_4C) composites at 1800 °C and 30 MPa with a soaking time of 5 min. The effects of rGO addition on mechanical properties of the composites, such as Vickers hardness, flexural strength and fracture toughness were investigated. The results showed that GO/B_4C composite powders were successfully self-assembled and a network structure was formed at high GO contents. The flexural strength and fracture toughness of rGO/B_4C composites were 643.64 MPa and 5.56 MPa•$m^{1/2}$, respectively, at 1 and 2.5 wt.% rGO content, corresponding to an increase of 99.11% and 71.6% when compared to B_4C ceramics. Uniformly dispersed rGO in rGO/B_4C composites played an important role in improving their strength and toughness. The toughening mechanisms of rGO/B_4C composites were explained by graphene pull-out, crack deflection and bridging.

4:30 PM

(PACRIM-192-2021) High-temperature plastic flow of TZP during flexural deformation under a strong electric field

H. Motomura*[1]; Y. Sasaki[2]; H. Masuda[1]; H. Yoshida[1]

1. The University of Tokyo, Materials Science and Engineering, Japan
2. Tokyo University of Science, Materials Science and Engineering, Japan

We have recently demonstrated that yttria-stabilized tetragonal zirconia polycrystal (Y-TZP) with the grain size of about 0.4 μm can exhibit superplastic deformation in a tensile manner at temperatures of <1000 °C and at strain rates of $>10^{-3}$ s^{-1} by applying a strong DC electric field. It would seem that the occurrence of a flash event enhanced diffusion of Zr^{4+} cations, which is the rate-controlling mechanism of the high temperature plastic flow in Y-TZP. In the high temperature tensile tests under the strong field, Joule heating had considerable effect on the enhancement of plastic flow. In the present study, three-point flexural tests of dense Y-TZP with the average grain size of 0.4 μm under the occurrence of flash events were performed; in the three-point flexural test, Y-TZP can deform at relatively low current densities in comparison to the tensile test. As a result, substantial plastic deformation was achieved at a furnace temperature of 1000 °C and initial strain rates in a range of 10^{-4} - 10^{-2} s^{-1} under the current density of 67 mA/mm^2. The specimen temperature during the flexural test was 1250 °C due to Joule heating, whereas flow stress level was comparable to that of 1400 °C without a field. This result suggests that an electric field and/or current had an effect to accelerate the plastic flow in TZP in addition to Joule heating.

PACRIM Symposium 10: Sol-Gel Processing and Related Liquid-Phase Synthesis of Ceramics

Powders, Fibers, Films, Monoliths, and Gels

Room: Regency E
Session Chairs: Andrew Lange, Lawrence Livermore National Laboratory; John Kieffer, University Of Michigan

1:30 PM

(PACRIM-193-2021) Biphasic Sol-gel Synthesis and Microstructural Control of Nanocrystalline Phosphor Materials as Potential Fluorescent Sensors (Invited)

S. Fujihara*[1]

1. Keio University, Japan

This study focuses on the synthesis of nanocrystalline Eu^{3+}-doped YVO_4 materials through a biphasic sol-gel method and their use as fluorescent sensors for detecting H_2O_2. The biphasic sol-gel method is based on immiscible organic and aqueous solutions systems. There are two major advantages of using such biphasic systems: one is that the metal alkoxides can be used without solubility or reactivity problems, and the other is that a reaction can be controlled under mild conditions to achieve new morphologies of target materials. In the present work, while vanadium alkoxide was dissolved in a mixed ethanol/xylene solvent, yttrium and europium chloride were dissolved in water with the addition of sodium carboxylates. The YVO_4:Eu^{3+} materials were then obtained as powders and films, which were characterized as containing micrometer-sized particles comprising primary nanoparticles with high specific surface areas. The photoluminescence intensity of the YVO_4:Eu^{3+} materials was found to be sensitive to the presence of H_2O_2 due to its quenching effect. Furthermore, the H_2O_2 sensitivity was largely dependent on the surface microstructure of the micrometer-sized YVO_4:Eu^{3+} particles. The quantitative dependence of the sensitivity on the H_2O_2 concentration was also observed in the films, indicating that they would be promising as a new fluorescent H_2O_2 sensor.

2:00 PM

(PACRIM-194-2021) Instability of the in-plane residual stress of sol-gel-derived glass and crystalline oxide thin films at room temperature

Y. Nishimura[1]; H. Kozuka*[2]

1. Kansai University, Graduate School of Science and Engineering, Japan
2. Kansai University, Faculty of Chemistry, Materials and Engineering, Japan

Recently we have found that sol-gel-derived, fired oxide thin films on Si(100) wafers exhibit a decrease in in-plane residual tensile stress with time when stored in humid atmosphere at room temperature. The present work was carried out to clarify the mechanism of such a reduction in stress, where the stress was obtained by measuring the radius of substrate curvature. Alkoxide-derived silica films were deposited on Si(100) wafers by spin-coating, followed by firing at 600°C. The films were stored at room temperature at relative humidity over 80% for various periods of time. The films had residual in-plane tensile stress, which decreased with time during storing. The stress, however, was recovered when heated at 300 °C. Infrared (IR) absorption spectroscopy showed that the O-H stretching vibration increases during storing, while decreases on the heat treatment. The film thickness increased during storing, which was detected by in situ spectroscopic ellipsometry. These experimental facts suggest that the decrease in tensile stress is caused by water adsorption, leading to expansion of the film. However, the Si-OH stretching vibration peak was also generated during storing, suggesting that the hydrolysis of the Si-O-Si bonds. Then the consequent structural relaxation cannot be excluded as a part of the cause of the decrease in stress during storing.

2:20 PM

(PACRIM-195-2021) Solution processing of metal chalcogenide and metal halide semiconductor for high performance large area electronics (Invited)

M. Kim*[1]

1. SungKyunKwan University, School of Advanced, Republic of Korea

There is increasing demands of material developments for novel electronic materials in emerging large area electronic applications, such as flexible electronics, wearable electronics, IoT devices, and photovoltaics, which require high electrical performance, novel mechanical functionality, high electrical stability, and/or exotic chemical stability under harsh condition. Furthermore, the emerging applications should be realized with low cost and high throughput processing methods. Although the conventional vacuum processed oxide semiconductor have been successfully utilized for large area electronic applications, the limited material functionality and high processing cost hinder its further applications in next generation large are electronics. In this talk, I will focus on our approaches to develop the high performance inorganic semiconductor materials, such as chalcogenide and halide, with soft chemical processing strategy. For example, the generalized chalco-gel precursors could achieve high performance TFT with low cost solution processing strategy. The optimized TFT device could exhibit a maximum field-effect mobility over 300 $cm^2V^{-1}s^{-1}$ with an Ion/Ioff > 10^7 and a good operational stability. Finally, I will discuss the possibility of transparent p-type metal halides semiconductor for high performance complementary large area electronic applications.

Nanoparticles, Nanofibers, Nanorods, and Nanosheets

Room: Regency E

3:10 PM

(PACRIM-196-2021) High-Temperature Superhydrophobic Ceramic Coatings from Surface-Functionalized Nanoparticles

A. K. Schmidt-Verma*[1]; A. Renner[1]; T. Fischer[1]; S. Mathur[1]

1. Universität zu Köln, Inorganic Chemistry, Germany

The technology of superhydrophobic ceramic coatings with self-cleaning properties gained strong attention during recent years, especially for industrial applications. Innovative coatings that can shield functional surfaces from environmental and temperature impact are essential to increase the life time of devices and reduce the maintenance costs. In this regard, the control over surface properties and chemical structure is mandatory for the development of new materials which can help to increase eco-friendly materials. Nonetheless, the potential of commercial products is immense and their market truly global, but the wide range of possible application is limited by the low temperature stability of these coatings. In this work, a route for the preparation of a superhydrophobic coating with high temperature stability and additional UV-protection will be presented. $[Mg\{Al(O^iPr)_4\}_2]$ was used as precursor to form spinel nanoparticles through sol-gel method. After subsequent functionalization transparent superhydrophobic films with contact angle for water close to 180° after temperature treatment were preserved.

3:30 PM

(PACRIM-197-2021) Localized surface plasmon resonance enhanced photocatalysis and photovoltaic effect (Invited)

G. Kawamura*[1]

1. Toyohashi University of Technology, Japan

Sunlight reaching to the Earth is a vast amount of energy which cannot be compared with energy required by human activities in the world. Photocatalysts and photovoltaic cells are materials/devices that are designed to utilize sunlight energy; however, the energy conversion efficiency against the cost is not sufficient for large-scale commercialization. In this work, the photocatalysis and photovoltaic effect of the devices are enhanced by localized surface plasmon resonance through electron and energy transfer. For instance, titania nanotube arrays decorated with plasmonic silver nanoparticles were employed as a photoanode of dye-sensitized solar cell. On the other hand, strontium titanate nanotube arrays covered by a reduced strontium titanate layer, which exhibits surface plasmon resonance, were used as a photoelectrode for photoelectrochemical water splitting. With these materials, plasmonic enhancements in photovoltaic effect and photocatalysis were confirmed experimentally. Some other case studies and the detailed mechanisms of the enhancement will be discussed on site.

4:00 PM

(PACRIM-198-2021) Preparation of Janus nanosheets via regioselective interlayer surface modification of potassium hexaniobate and subsequent exfoliation

R. Suzuki[2]; M. Sudo[5]; M. Hirano[4]; N. Idota[3]; M. Kunitake[6]; T. Nishimi[7]; Y. Sugahara*[1]

1. Waseda University, Department of Applied Chemistry and Kagami Memorial Research Institute for Science and Technology, Japan
2. Waseda University, Kagami Memorial Research Institute for Materials Science and Technology, Japan
3. Tokyo Institute of Technology, Laboratory for Advanced Nuclear Energy, Institute of Innovative Research, Japan
4. Kumamoto University, Technical Division, Faculty of Engineering, Japan
5. Waseda University, Department of Applied Chemistry, Japan
6. Kumamoto University, Institute of Industrial Nanomaterials, Jersey
7. Japan Technological Research Association of Artificial Photosynthetic Chemical Process (ARPChem), Japan

Janus materials have attracted increasing attention, since they possess two chemically different areas in the individual components, such as particles, and Janus nanosheets attract particular interests because two physically separated surfaces possess different properties. We have demonstrated that Janus nanosheets can be obtained by using potassium hexanioabte which possesses two types of interlayers with different reactivities. We first intercalate relatively bulky dioctadecyldimethylammonium ions only to the interlayer with high reactivity, interlayer I. Then phenylphosphinic acid was grafted on the surface of interlayer I. The surface of the other interlayer with low reactivity, interlayer II, was further modified with carboxypropylphosphonic acid via intercalation of dodecylammonium ions into interlayer II. The exfoliation of the resulting organic derivative resulted in the formation of Janus nanosheets. Another type of Janus nanosheets was obtained by modifying this strategy. The interlayer I was first modified with organophosphonic acid bearing an initiating group for atom transfer radical polymerization (ATRP). Then polyacrylamide-based polymer chains were grown via ATRP. By ion-exchange of the interlayer potassium ions with tetrabutylammonium ions in water, exfoliation occurred at both interlayers I and II to form Janus nanosheets.

4:20 PM

(PACRIM-199-2021) Soft Chemical Synthesis, Hetero-assembly and Emergent Properties of 2D Functional Nanosheets (Invited)

R. Ma*[1]

1. National Institute for Materials Science, International Center for Materials Nanoarchitectonics, Japan

Single-layer atomic crystals or nanosheets peeled from bulk layered compounds have emerged as a forefront of nanoscale materials because of their fascinating properties and diverse applications. Soft chemical exfoliation, in which parent layered compounds are cleaved into single-layer form upon the modification of certain ions and/or solvents, typically accompanies substantial gallery expansion, i.e., osmotic swelling, which significantly weakens the layer-to-layer attraction force. In this context, the understanding of swelling-exfoliaiton scenario is of critical importance to elucidate the mechanism of exfoliation and to achieve controllable preparation of high-quality nanosheets. In this presentation, I will introduce our recent research progress on soft chemical exfoliation of inorganic layered materials. In particular, the controllable preparation and hetero-assembly of transition-metal hydroxide nanosheets with other 2D counterparts will be elaborated. Taking advantages that these single-layer nanosheets are obtained as a monodispersed colloid suspension, we have developed environmentally friendly wet process that can assemble, stack, and complex them as building blocks. we have also demonstrated a promising molecular-scale modulation of heterointerfaces between different 2D unilamellar nanosheets for advanced energy-related applications.

4:50 PM

(PACRIM-200-2021) Sol-Gel Coatings for Optical Fiber Preforms

A. Lange*[1]

1. Lawrence Livermore National Laboratory, Materials Science Division, USA

Vapor-phase manufacturing techniques for optical fiber preforms such as MCVD, OVD, and PVD were key enablers for long-haul communication and the digital age. These approaches have been immensely successful and are still the dominant technologies today for large volume, high-purity fibers (e.g. fiber lasers, amplifiers, and waveguides). However, they are limited in terms of their chemical flexibility as they necessitate volatile precursors that pyrolyze at temperatures close to that of $SiCl_4$. The need for volatile precursors is often circumvented by using solution doping in concert with MCVD in which liquid phase dopants are impregnated into a porous coating that is deposited from the vapor phase. In the work presented here, a technique analogous to MCVD solution doping is pursued in which coatings are deposited via dip coating with sol-gel-derived SiO_2 colloids. It will be shown that liquid phase coating has many advantages such as simplicity, the ability to tune coating morphology via dip coating parameters, and the ability to introduce a wide range of dopants into the core during preform preparation.

PACRIM Symposium 15: Engineering Ceramics and Ceramic Matrix Composites: Design, Development, and Applications

Polymer-Derived Ceramic Matrix Composites

Room: Regency D
Session Chairs: Dong Liu, University of Bristol; Junichi Tatami, Yokohama National University

1:30 PM

(PACRIM-201-2021) Polymer-Derived Ceramics as an innovative processing route toward TiC/SiC nanocomposites with high potential as solar absorbers (Invited)

M. Balestrat[1]; S. Bernard*[1]
1. CNRS, IRCER, France

Silicon-based non-oxide ceramics (SiC, Si_3N_4) have attracted much attention, primarily due to their good mechanical and chemical properties, and also their reliability at room and elevated temperatures. They have great potential for many industrial uses as engineering components. The addition of a second ceramic (nano) phase to SiC or Si_3N_4 leads to materials which promise applications in many fields and offer solutions for most of the market demands. However, the preparation of these materials is still a challenging task according to the fact that the conventional processes unavoidably lead to size and structure inhomogeneities of the different phases and presence of impurities which affect the properties. The Polymer-Derived Ceramics (PDCs) route offers new preparation opportunities in ceramic sciences. The molecular origin of preceramic polymers and the possibility to design advanced ceramics in particular forms play a major role in the preparation of functional ceramics. In this talk, we will present our last results on the modification of polycarbosilanes and polysilazanes with metal-organic species to offer functionalities to the materials obtained after the thermo-chemical conversion of precursors. The chemistry behind the preparation of materials will be described before investigating the properties and applications of the final ceramics.

2:00 PM

(PACRIM-202-2021) Polymer-Derived (Hf,Ta)C/SiC and (Hf,Ti)C/SiC Ceramic Nanocomposites with Excellent High-Temperature Oxidation Resistance (Invited)

E. Ionescu*[1]
1. Technical University Darmstadt, Materials and Earth Sciences, Germany

In the present work, a synergistic strategy to improve the high-temperature oxidation behavior of HfC/SiC is presented and discussed. The strategy consists of a combination of alloying TaC/TiC to HfC and of applying a passivation step to the ceramic nanocomposites prior to their exposure to high-temperature oxidation conditions. Possible mechanisms involved in the oxidation processes of (Hf,Ta)C/SiC and (Hf,Ti)/SiC ceramic nanocomposites are highlighted and critically assessed. It is shown that the passivation treatment activates the formation of silica, which is sluggish at moderate temperatures; whereas the alloying of HfC with TaC and TiC induces the formation of oxidic phases such as $Hf_6Ta_2O_{17}$ and $HfSiO_4/HfTiO_4$, respectively, which are very beneficial for the high-temperature oxidation behavior of the studied ceramic nanocomposites. The solid-solution effect contributes to a significant improvement of the short-term oxidation resistance of the studied nanocomposites, while the passivation of the materials prior exposure of high-temperature oxidation conditions provides a remarkably improved long-term behavior thereof.

2:30 PM

(PACRIM-203-2021) Development SiC fiber and its applications

Y. Lee*[1]; D. Shin[1]; K. Cho[1]
1. Korea Institute of Ceramic Engineering and Technology (KICET), Republic of Korea

Silicon carbide fiber has been spotlighted to advanced reinforcement for high temperature CMCs such as SiCf/SiC composites. As the performance of next-generation turbine engines is improved, SiC fibers with a quality corresponding thereto are required, and the development of a new concept of fiber are also expected. In this presentation, we would like to introduce the current status of SiC fiber development and some products using the fibers in Korea. Now, the main player in the development of SiC fiber has moved to private enterprise, and they are promoting medium- and long-term investments in order to succeed in mass production of SiC fiber. The fiber currently under development is of appropriate quality for civil products, such as heating elements, burner media, and flame retardant products. Mass production of high-performance SiC fibers for CMC is longer-term business. As is well known, it is important to understand and control the properties of polycarbosilane (PCS) as a precursor in the step of spinning fiber and converting them to SiC; such as viscosity, curing conditions, ceramic yield, elemental content, etc. In the process of developing suitable PCS for producing fiber, crystalline fibers with secondary phases was developed to improve high temperature creep performance.

Sintering and Properties of Engineering Ceramics

Room: Regency D
Session Chair: Csaba Balazsi, ELKH Centre for Energy Research

3:40 PM

(PACRIM-204-2021) Dynamic change in the density distribution of alumina ceramics during sintering estimated by in-situ optical coherence tomography

J. Tatami*[1]; M. Tajima[1]; M. Iijima[1]; T. Takahashi[2]
1. Yokohama National University, Japan
2. Kanagawa Institute of Industrial Science and Technology, Japan

Densification of ceramics occurs at high temperatures. For example, lower density region sometimes becomes a fracture origin to degrade mechanical reliability. Therefore, density distribution in the ceramics should be obtained to make better materials. Furthermore, such density distribution usually develops during firing process depending on the internal structure of the green body. This means that the change in the density distribution during firing process should be understood. In this study, the density distribution of alumina ceramics during firing was estimated based on in-situ observation of their internal structure using the optical coherence tomography (OCT). At first, we obtained a calibration curve on the relationship between density and attenuation rate of the OCT signal. Mapping of attenuation rate in the alumina ceramics were obtained by in-situ and 3-dimensional OCT observation at high temperatures using infrared heating furnace. The density distribution was estimated by the mapping of attenuation rate and the calibration curve. It was shown that the density distribution was developed during sintering and it was narrower in the more homogeneous green body.

4:00 PM

(PACRIM-205-2021) Liquid phase sintering of Y_2O_3 studied by master sintering curve method

A. Najafzadehkhoee*; A. Talimian[1]; P. Hvizdoš[3]; D. Galusek[1]

1. Centre for Functional and Surface Functionalized Glass, Alexander Dubček University of Trenčín, Slovakia
2. (Joint Glass Centre of the IIC SAS, TNUAD, and FChPT STU & FunGlass, Alexander Dubček University of Trenčín, Slovakia
3. Institute of Materials Research, Slovak Academy of Sciences, Slovakia

The impact of Al_2O_3 and SiO_2 addition on the sintering of Y_2O_3 ceramics was investigated. The shrinkage data of pure yttria and samples doped with Al_2O_3/SiO_2 upon sintering at various heating rates were used to construct the master sintering curve method. The apparent activation energies of sintering of pure Y_2O_3 and Y_2O_3: $SiO_2+Al_2O_3$ samples were found to be 820 kJ/mol and 430 kJ/mol, respectively. The results indicated that the production of liquid phases from the reaction between Y_2O_3 and the additives decreases the energy barrier of densification, and promotes sintering. Microstructural analyses revealed pockets of glass present at the grain boundary junctions, underpinning the liquid phase assisted sintering hypothesis.

4:20 PM

(PACRIM-206-2021) Microstructure Design and Mechanical Properties of Ceramic/Graphene Thick Coatings for New Emerging Applications

C. Balazsi*[1]; K. Balazsi[1]

1. ELKH Centre for Energy Research, Hungary

The influence of the various content of the multilayered graphene (MLG) on the structural and mechanical properties of the final bulk porous silicon nitride-zirconia (Si_3N_4-ZrO_2) based ceramics was investigated. The ceramic composites were prepared in the form of the laminated structure with different (5-30-5 wt% and 30-5-30 wt%) MLG content by hot isostatic pressing. ZrO_2 particles were incorporated into the Si_3N_4 matrix by attrition milling to improve the mechanical properties of the final composite. Homogeneous distribution of the MLGs, a completed phase transition from α to β-Si_3N_4 in case of 5 wt% MLG have been observed. The structural examinations revealed that the multilayered graphene and zirconia particles owing to their different sizes and shapes influenced the porous microstructure evolution and the related mechanical properties of the composites. The sandwich structures enhanced the mechanical properties compared to reference ceramic with 30 wt% MLG. The position of the layer with higher graphene content, high ratio of a / b phase of Si_3N_4 and higher porosity had crucial effect on the final mechanical properties.

4:40 PM

(PACRIM-207-2021) Heat Resistant and Robust Superhydrophobic Coatings fabricated by functionalized Nanoparticles

A. K. Schmidt-Verma*[1]; R. Weißing[1]; T. Fischer[1]; S. Mathur[1]

1. Universität zu Köln, Inorganic Chemistry, Germany

The technology of superhydrophobic ceramic coatings with self-cleaning properties gained strong attention, especially for industrial applications. Impurities on treated materials are easier removed compared to untreated materials which leads to reduced maintenance costs and offers a green alternative to the usage of aggressive cleaning agents. In this regard, control over surface properties and chemical structure is mandatory for the development of new eco-friendly materials. Although the potential of commercial products is immense and their market truly global, the wide range of possible application is limited by the low temperature and mechanical stability of employed coatings. In this work, a route for the preparation of a superhydrophobic coating with mechanical and thermal stability was fabricated on glass substrates by spin-, and spray coating method. The coating was formed in a solution containing various nanoparticles such as SiO_2, TiO_2 and ZrO_2 and silicic acid, in which the ratio of silicic acid was varied to tune the roughness of the coating. For receiving superhydrophobicity, subsequent and in-situ functionalization with low surface energy material have been performed. After surface functionalization, robust thermally stable films with contact angle close to 180° was preserved even after temperature treatments up to 350 °C and mechanical tests.

5:00 PM

(PACRIM-208-2021) Effects of dopant on electrical and thermal properties of porous SiC ceramics

Y. Kim*[1]; S. Kultayeva[1]

1. University of Seoul, Dept. of Materials Science & Engineering, Republic of Korea

The nitrogen-, vanadium-, aluminum-, boron-, and scandium-doped porous SiC ceramics were fabricated for investigating the effects of dopants on electrical and thermal properties of porous SiC ceramics while keeping the porosity constant as ~61%. The electrical resistivity of nitrogen-doped porous SiC ceramic was $2.1 \times 10^{-1}\Omega$ cm which was four orders of magnitude lower than that of undoped porous SiC ceramic ($1.2 \times 10^{3}\Omega$ cm). The B-doped porous SiC ceramic showed the highest thermal conductivity (16.6 W/(m K)) and flexural strength (25.9 MPa), whereas Sc-doped porous SiC ceramics showed the lowest thermal conductivity (7.7 W/(m K)) and flexural strength (10.5 MPa) among the doped porous SiC ceramics. The electrical resistivity was strongly influenced by doping, whereas the thermal and mechanical properties were primarily dependent on necking area between platelet SiC grains. The results suggest that the electrical conductivity of porous SiC ceramics can be successfully tuned independently of thermal conductivity by suitable doping.

PACRIM Symposium 20: Multiferroic Materials, Devices, and Applications

Multiferroic Materials, Devices, and Applications V

Room: Cypress

Session Chair: Liuyan Zhao, University of Michigan

1:30 PM

(PACRIM-209-2021) Terahertz spectroscopic studies of noncentrosymmetric magnets (Invited)

D. Talbayev*[1]

1. Tulane University, Physics and Engineering Physics, USA

Magnetism and multiferroicity is a rich and fascinating subject in condensed matter physics. We will present spectroscopic studies of magnetic insulators using terahertz (THz) frequency light. This frequency range is important in magnetism because it often hosts the spin resonances and spin waves found in many classes of magnetically ordered crystals. Combined with the ability to collect spectroscopic information at low temperatures and high magnetic fields, THz spectroscopy can serve as a tool for the determination of microscopic magnetic interactions that govern a material's magnetic properties. We will also introduce polarimetry of light and nonlinear optics as a way to interrogate the symmetries inherent in magnetic crystals. When magnetism occurs in noncentrosymmetric crystals, new kinds of optical phenomena often emerge, such as nonreciprocity. In nonreciprocal propagation, a crystal can be transparent to a light wave traveling in one direction and be completely opaque to the same wave traveling in the exact opposite direction. We will consider the necessary symmetry conditions and microscopic origins of the nonreciprocity.

2:00 PM

(PACRIM-210-2021) Manipulating topological transformations and charged domain walls in the ferroelectric film (Invited)

K. Du[1]; Z. Liu[1]; H. Tian*[1]

1. Zhejiang University, China

Base on controllable ferroelectric domain configurations, ferroelectric films offers the opportunity to develop microelectronic devices. Despite the increasing experimental and theoretical insights into various domain structures and domain walls (such as polar vortex and charge domain wall), manipulating the polar structures and comprehensively understanding its underlying mechanism remains lacking. We successfully developed a technique to realize domain structures modulation and take an atomic resolution observation simultaneously. Here we investigate the real-time topological transformations of polar structures in PTO/STO multilayers at an atomic level. The procedure of vortex pair splitting and the transformation from polar vortex to polar wave and out-of-plane polarization are observed step by step. Furthermore, the redistribution of charge in various topological structures has been demonstrated under an external bias. Furthermore, we demonstrated a strategy for the controllable creation and manipulation of charged domain walls in the ferroelectric film. Combining in-situ imaging and in-line holography, the whole dynamic process of creating, moving, reversing, and erasing a charged domain wall was present. These findings advance the current understanding of manipulation by the external field and its origination.

2:30 PM

(PACRIM-211-2021) Hybrid improper ferroelectricity in A-site cation ordered $Li_2La_2Ti_3O_{10}$ ceramic with triple-layer Ruddlesden-Popper structure (Invited)

X. Liu*[1]; B. Zhang[1]; X. Chen[1]

1. Zhejiang University, School of Materials Science and Engineering, China

Hybrid Improper ferroelectricity (HIF), in which the ferroelectric polarization is induced by a complex distortion pattern consisting of two oxygen octahedron tilting modes, has been attracted many attentions because of its potential application in creating room temperature multiferroicity with strong magnetoelectric coupling. The HIF should widely exist in the perovskites and layered perovskites since the oxygen octahedron rotations are ubiquitous distortions in the perovskite-related materials. HIFs have been extensively studied in the double-layer Ruddlesden-Popper oxides in recent years. Although the HIF can also exist in the oxides with A-site cation ordered triple-layer Ruddlesden-Popper structures based on the first-principles calculations, no experimental result has been reported yet. In the present work, the room-temperature ferroelectricity has been observed in $Li_2La_2Ti_3O_{10}$ ceramic with A-site cation ordered triple-layer Ruddlesden-Popper structure. The polar phase has been determined by combining the first-principles calculation and powder diffractions at room temperature. Hybrid improper ferroelectricity was induced by the triple-coupled irreps including the A-site cation ordering. The present work sheds light on designing the hybrid improper ferroelectrics in A-site ordered triple-layer Ruddlesden-Popper compounds.

3:00 PM

(PACRIM-212-2021) Effects of B-site substitution on ferroelectric characteristics of Sr2NaNb5O15 tungsten bronze ceramics

Y. Wang*[1]; T. Sun[2]; X. Zhu[1]; L. Liu[1]; X. Chen[1]

1. Zhejiang University, China
2. Zhejiang University of Technology, China

Tetragonal tungsten bronze compounds constitute the second-largest class of dielectric family next to perovskites. The coexistence of cations, their distribution in A1 and A2-sites, and the related structure modulation played a crucial role to dominate the ferroelectric nature. Meanwhile, the effects of B-site substitution on the physical properties have been attracting increasing attention. Herein, effects of Ta^{5+} in B-site on the crystal structure, dielectric and ferroelectric properties were investigated together with the phase transitions in $Sr_2Na(Nb_{1-x}Ta_x)_5O_{15}$ ceramics. With increasing Ta-content, the crystal structure of the present system changed from P4bm to P4/mbm. The frequency- and temperature-dependent dielectric permittivity and dielectric loss curves exhibited typical relaxor features, and the saturated polarization-electric field hysteresis loop changed gradually to a slim one and eventually to a linear one, though a commensurate structure modulation was determined by the selected area electron diffraction patterns for all compositions. Raman spectra indicated changes of the B–O vibrations during the transition, reflecting the different electron configuration of Nb and Ta and the different bonding of Nb and Ta with O. This work deepens the physical understanding of the crystal structure-ferroelectricity relationship in tungsten bronze ceramics.

PACRIM Symposium 21: Crystalline Materials for Electrical, Optical, and Medical Applications

Scintillation Materials

Room: Regency C
Session Chair: Luisa Bausa, Universidad Autonoma de Madrid

1:30 PM

(PACRIM-213-2021) Photoluminescence and radiation response characteristics of gallate glasses with a luminescence center (Invited)

N. Kawano*[1]; K. Shinozaki[2]; H. Kimura[3]; M. Akatsuka[3]; D. Nakauchi[3]; T. Yanagida[3]

1. Akita University, Japan
2. National Institute of Advanced Industrial Science and Technology (AIST), Japan
3. Nara Institute of Science and Technology, Japan

A scintillator is a type of luminescent materials having a function of transforming high energy radiation into low energy photons, and the scintillator has been used in various application fields such as high energy physics and security. So far, single crystal scintillators are often considered for possible scintillator applications owing to their high light yield. Recently, glass scintillators have gained research attentions thanks to their low cost, large volume fabrication possibility and flexible chemical composition. Gallate glasses should be a candidate for glass scintillators because of their low phonon energy and high effective atomic number. The low phonon energy might lead to efficient luminescence because of the suppression of non-radiative decay loss. Furthermore, the high effective atomic number is important for the high detection efficiency against high energy gamma-ray and X-ray. In this study, we prepared gallate glasses with a luminescence center by the conventional melt-quenching method, and evaluated their photoluminescence and scintillation characteristics. The Eu_2O_3-doped gallate glasses showed scintillation due to the 4f-4f transitions of Eu^{3+}. Among the glass investigated, the 0.5% Eu_2O_3-doped gallate glass exhibited the highest scintillation intensity. Moreover, the Eu_2O_3-doped gallate glasses showed afterglow levels in the range of 140–260 ppm.

2:00 PM

(PACRIM-214-2021) Analysis of Excited States Dynamics in Fluoride Crystals Using Transient Absorption Spectroscopy (Invited)

M. Koshimizu*[1]; S. Yamashita[2]; Y. Muroya[3]; Y. Fujimoto[1]; K. Asai[1]

1. Tohoku University, Department of Applied Chemistry, Japan
2. University of Tokyo, Japan
3. Osaka University, Japan

The development of novel phosphors for radiation detection with excellent performance is of considerable interest. Most phosphors for radiation detection are composed of an insulator host with dopants as the luminescence centers. In such phosphors, excited states are initially generated in the insulator host upon exposure to ionizing radiation. Following the charge or energy transfer, luminescence occurs. Among these basic processes, only little is known about the excited states dynamics in the host crystals, despite its importance. In this study, we analyzed the excited states in fluoride crystals to better understand the excited states dynamics, in particular, the quenching process by measuring transient absorption after irradiation of pulsed electron beams. In BaF_2, decay behavior of the transient absorption coincided with the scintillation temporal profile of the self-trapped excitons (STEs) emission. We observed no additional fast decay component attributable to fast quenching of STEs or decay of core holes. On the contrary, in CeF_3, we observed long decay component of transient absorption, which was much slower than the scintillation temporal profile. This component can be attributed to afterglow, which is caused by the trapping of electrons and holes at different sites and subsequent recombination.

2:30 PM

(PACRIM-215-2021) Engineering of $MgAl_2O_4$ Spinel for Optically Stimulated Luminescence Dosimetry

L. Pan[1]; S. Sholom[2]; S. W. McKeever[2]; L. G. Jacobsohn*[1]

1. Clemson University, Department of Materials Science and Engineering, USA
2. Oklahoma State University, Department of Physics, USA

Motivated by the search of new materials for optically stimulated luminescence (OSL) dosimeters, magnesium aluminum spinels with different Mg:Al ratios were investigated. Powders were prepared via the co-precipitation method with Mg:Al ratios 1:2 (stoichiometric), 1.5:2 (MgO-rich) and 1:3 (Al_2O_3-rich) and calcined from 800 °C to 1090 °C in air for 2h. A higher level of structural disorder was found in the powders than in the single crystal as determined by X-ray diffraction and Raman spectroscopy measurements, while the same luminescence centers were present in all materials as shown by X-ray induced radioluminescence measurements. The nature of the band at 385 nm was identified as being related to anti-site defects. Among the different Mg:Al ratios, stoichiometric spinel presented superior OSL dose response, being linear within 0.16 Gy and 2 Gy, presenting 1% reproducibility of the dosimetric response after repeated exposure to the same dose, having the least fading, and with a minimum detectable absorbed dose (MDD) of 0.65 mGy. This material is based upon work supported by the National Science Foundation under Grant No. 1653016 to Clemson University. The work at Oklahoma State University is supported by the State of Oklahoma.

2:50 PM

(PACRIM-216-2021) Comparison between single-crystalline and glassy $Ce:LaB_3O_6$

D. Yuan*[1]; E. G. Víllora[1]; K. Shimamura[1]

1. National Institute for Materials Science (NIMS), Japan

Cerium (Ce^{3+}) is a bright and low-cost activator, which is widely used in scintillators and phosphors for white light-emitting diodes. The Ce^{3+} emission is favored by the parity-allowed electric-dipole 4f-5d transitions, but nevertheless its efficiency is also greatly influenced by the surrounding host material. Ce^{3+}-doped crystals generally show superior optical properties than those of doped glasses in terms of luminescence efficiency and temperature stability, since they benefit from a periodically ordered atomic structure. However, an exception was observed recently when comparing the single-crystalline and glassy forms of $Ce:LaB_3O_6$ at room temperature (RT). The current study investigates in detail the quantum efficiency of $Ce:LaB_3O_6$ forms at RT and evaluates its temperature dependence in terms of two related but independent measurements, namely the photoluminescence intensity and lifetime. While single-crystalline $Ce:LaB_3O_6$ emits weakly and quenches fast, glassy $Ce:LaB_3O_6$ shows a stronger emission and a slow decrease with temperature. In order to gain insight into this phenomenon, the local structure of Ce activators, is analyzed in both $Ce:LaB_3O_6$ forms by X-ray diffraction and TEM, so that the splitting of the Ce^{3+} $5d_j$ states and their positions within the band gap can be evaluated.

PACRIM Symposium 22: Microwave Dielectric Materials and Their Applications

Microwave Dielectric Materials and Applications

Room: Plaza C
Session Chair: Heli Jantunen, University of Oulu

1:30 PM

(PACRIM-217-2021) Computational studies of transitional behavior in dielectrics at mesoscale (Invited)

S. Nakhmanson*[1]

1. University of Connecticut, Materials Science and Engineering, USA

Ferret is an open-source highly scalable real-space finite-element-method (FEM) based code for simulating transitional behavior of material systems with coupled physical properties at mesoscale. This code is built on MOOSE, Multiphysics Object Oriented Simulation Environment, which allows efficient solving of coupled-physics problems, has a proven record of efficient scaling of execution on up to 10K nodes, and is easily extendable to include new physical models. This approach allows treatment of structures possessing complicated morphology and evaluation of property dependencies on the system shape, size, microstructure and applied boundary conditions. In this presentation we provide an overview of Ferret's technical features and highlight its computational capabilities using a variety of examples. These applications include an investigation of relaxation dynamics and dielectric tunability in the $Ba_xSr_{1-x}TiO_3$ perovskite ferroelectric, which has already seen integration into prototype devices for 5G electronics. The effects of variation in Sr concentration, temperature and polar-domain topologies on the dielectric response and losses are also probed.

2:00 PM

(PACRIM-218-2021) Epitaxial Oxides on Silicon via Graphene Oxide (Invited)

M. Spreitzer*[1]; Z. Jovanović[2]; U. Trstenjak[1]; D. Suvorov[1]

1. Jozef Stefan Institute, Advanced Materials, Slovenia
2. Laboratory of Physics, Vinča Institute of Nuclear Sciences, Serbia

Epitaxial integration of oxides with semiconductor substrates is often limited by the lattice mismatch between the two material systems and their dissimilar chemical properties. However, in case of 2D materials the strict requirements of traditional epitaxy can be mediated by the weak van der Waals interactions. As a results, a 2D material, such as graphene, can allow remote epitaxial registry with a substrate at a long distances or act as a template by itself. In the present work, we have investigated the potential of graphene oxide (GO) as epitaxy-enabling agent for the pulsed-laser deposition (PLD) growth of strontium titanate (STO) on silicon native oxide. We have determined that GO can direct the growth of STO to a smooth, compact and pinhole-free layer, with single (100) out-of-plane orientation. Furthermore, our results show that directing growth of STO on GO is less demanding in terms on interface control and vacuum conditions with respect to systems without GO, which are important advantages for development of large area PLD processes.

PACRIM Symposium 24: Solid Oxide Fuel Cells and Hydrogen Technologies

Proton Conducting SOFC III

Room: Georgia B

Session Chairs: Takuto Araki, Yokohama National University; Yuji Okuyama, University of Miyazaki

1:30 PM

(PACRIM-219-2021) Large Protonic Ceramic Cells for H_2-based applications (Invited)

J. Dailly*[1]; C. Notar[1]; D. Schmider[1]

1. European Institue for Energy Research, Germany

The development of research activities on Protonic Ceramic Cells during the last 20 years has led to a recent growing interest of companies and policies on PCCs, considered as the third generation of solid oxide cells. The protonic conductivity of the electrolyte offers unique advantages for various applications, like the production of dry and pure hydrogen in electrolysis mode. Nevertheless, the low maturity of the manufacturing aspect obstructs its deployment at a higher TRL. The objective is to minimize the bottlenecks of the protonic ceramic cell elaboration to prime a larger development of the technology. Ba-based commercially available powder materials and home-made developed processes based on industrial wet chemical routes have been employed to ensure an easily scalable transfer to larger facilities and to fit industrial requirements. Promising performances were already validated in the past on button cells up to 10 cm^2 in various operation modes like micro-cogeneration cycles, electrolysis and reversibility profiles. Recently, large protonic ceramic-based cells (50 cm^2) have been successfully manufactured with the desired microstructure: 10 to 20 μm thick electrolytes in hydrogen electrode supported cells (thickness of the cell < 500 μm). After a description of the manufacturing processes, the results of microstructural and electrochemical characterizations in different Power-to-X applications will be discussed.

2:00 PM

(PACRIM-220-2021) Development activities for realization of large scale proton conducting ceramic cells (Invited)

M. E. Ivanova*[1]; W. Deibert[1]; C. Lenser[1]; N. H. Menzler[1]; O. Guillon[1]

1. Forschungszentrum Juelich GmbH, IEK-1, Germany

The global effort for reaching highly efficient CO_2-neutral energy supply requires innovative technological solutions based on high performance materials. Ceramic proton conductors (PC) with tailored properties gain increasing scientific and industrial interest due to their multifaceted applications at decreased temperatures. By means of proton conducting ceramic devices, energy can be either i) generated or co-generated together with valuable chemicals (fuel cells) or ii) stored into fuel (electrolyzers) or other energy carriers; iii) CO_2 can be utilized in reactions with protons under certain oxygen withdrawal rates to form chemical products (CO_2/H_2O co-electrolyzers and catalytic membrane reactors); iv) various chemical processes can be intensified through equilibria shifts and carried out at beneficial operation conditions resulting in higher overall energy efficiency, product selectivity and yields. Aiming at realizing the lab-to-market technology readiness levels, large area ceramic cells are required at high cost efficiency besides the high material performance and durability. The present talk will give an overview of the development activities for realization of large scale planar cell design based on proton conducting ceramic and composite materials. Some similarities and / or differences to oxygen-ion conductors will be highlighted too.

2:30 PM

(PACRIM-221-2021) Intermediate Temperature SOEC and SOFC Using Proton-Conducting Perovskites (Invited)

H. Matsumoto*[1]; L. Kwati[1]; Y. Fukahori[1]; P. Wiff[2]; L. Prost[2]

1. Kyushu University, International Institute for Carbon-Neutral Energy Research, Japan
2. K. K. Air Liquide Laboratories, Japan

Solid oxide cells consisting of solid oxide electrolytes play an important role in efficient energy conversion. Solid oxide fuel cells (SOFC) have been commercialized and steam electrolysis cells (SOEC) are extensively studied and developed. A typical oxide ion-conducting zirconia electrolyte has been well established and reliable, however, high operation temperature and utilization/concentration of fuel are the challenges. Proton-conducting perovskite-type proton conductors are an alternative candidate that can solve these issues. In this paper, steam electrolysis and solid oxide fuel cell using proton-conducting oxide electrolyte materials will be reported. Choice of electrode and electrolyte materials and SOFC-SOEC performance will be addressed. Part of the results are of our recent efforts in fabricating and scaling up our protonic membrane devices by sequential tape casting to be presented. there are still challenges on the use of proton-conducting oxide and the discussion on these is the other purpose of this paper. Acknowledgement: This work was supported by Cross-ministerial Strategic Innovation Promotion Program (SIP), the JSPS Core-to-Core Program, METI International joint research and development of innovative energy technology, World Premium International Research Center Initiative, MEXT Japan and Project (JPNP20003), commissioned by NEDO.

Proton Conducting SOFC IV

Room: Georgia B

Session Chairs: Yuki Yamaguchi, National Institute of Advanced Industrial Science and Technology (AIST); Hiroshige Matsumoto, Kyushu University

3:20 PM

(PACRIM-222-2021) Evaluation of PCFC cathodic reaction by using patterned thin film model electrodes (Invited)

K. Amezawa*[1]; K. Nishidate[1]; D. Zhuo[1]; T. Yoshioka[1]; Y. Kimura[1]; T. Nakamura[1]; Y. Mikami[2]; T. Kuroha[2]; F. Iguchi[1]; K. Yashiro[1]; T. Kawada[1]

1. Tohoku University, Japan
2. Panasonic Corporation, Japan

Recent years, protonic ceramic fuel cells (PCFCs) have been developing as intermediate temperature fuel cells. For improving the performance of this type of fuel cells, the enhancement of electrode reaction kinetics, especially at the cathode, is one of biggest technological challenges. For this purpose, it is important to understand mechanisms of PCFC electrode reactions. In this work, we aim to reveal mechanisms of PCFC cathodic reactions from electrochemical and spectroscopic measurements of patterned thin film model electrodes with and without triple phase boundaries (gas/electrode/electrolyte interfaces). We will then discuss the dominant reaction paths in typical PCFC cathodes.

3:40 PM

(PACRIM-223-2021) Layered 50 X 50 mm Cathode-supported Protonic Electrolysis cells for Large Scale Efficient Hydrogen Production

L. Kwati*[1]; M. E. Ivanova[2]; W. Deibert[2]; W. Meulenberg[2]; T. Ishihara[3]; H. Matsumoto[1]

1. International Institute for Carbon-Neutral Energy Research (WPI-I2CNER), Kyushu University, Advanced Energy Conversion Systems Thrust, Japan
2. Institute of Energy and Climate Research, Forschungszentrum Jülich GmbH, Materials Synthesis and Processing (IEK-1), Germany
3. International Institute for Carbon-Neutral Energy Research (WPI-I2CNER), Kyushu University, Applied Chemistry, Japan

Low-temperature steam electrolysis operating below 600 °C with protonic electrolytes constitutes a prospective technology for industrial-scale hydrogen production. A significant challenge, however, is upscaling robust planar-type devices. The fabrication of such multilayered devices, usually via a tape casting process, requires careful control of individual layers' shrinkages to prevent warping and cracks during sintering. In the present contribution, 50X50 mm planar cathode-supported protonic electrolysis half-cell consisting of $Ba(Zr_{0.5}Ce_{0.4})_{8/9}Y_{0.2}O_{2.9}$ electrolyte, NiO-$SrZr_{0.5}Ce_{0.4}Y_{0.1}O_{3-\delta}$ cathode substrate was successfully processed via an inverse tape casting route. The sintering parameters of the half-cells were analyzed and adjusted to obtain defect-free half-cells with diminished warping. The smooth tri-layered green tapes produced yielded suitably dense and gas-tight electrolyte layers after co-sintering at 1350 °C /5h. Current-voltage characteristics and hydrogen evolution rates measured in the temperature range 500-600 °C, using $Ba_{0.5}La_{0.5}CoO_{3-\delta}$ as the steam electrode, demonstrate excellent performance and durability. Our work features a facile approach for realizing the low-cost fabrication of large-sized proton-conducting electrolysis cells, satisfying both high performance and durability.

4:00 PM

(PACRIM-224-2021) Thermal and chemical expansions of $Ba(Ce,Zr)O_3$-based electrolyte materials for PCFCs

K. Nomura*[1]; H. Shimada[1]; Y. Yamaguchi[1]; Y. Mizutani[1]

1. National Institute of Advanced Industrial Science and Technology (AIST), Japan

Protonic ceramic fuel cells (PCFCs), have attracted much attention due to their high energy-conversion efficiency. For the development of highly reliable cells, elucidation of the thermal and chemical expansion behavior of cell constituent materials is indispensable. In this study, we systematically investigated the thermal and chemical expansions of $Ba(Ce,Zr)O_3$-based proton conductors, $Ba(Ce_{0.7}Zr_{0.1}Y_{0.1}Yb_{0.1})O_{3-\delta}$, etc. on the basis of Rietveld refinement results of high temperature X-ray diffraction data between 298 and 1173 K under dry and wet N_2 atmospheres.

4:20 PM

(PACRIM-225-2021) Highly Efficient Cell Design and Cost Analysis of Protonic Ceramic Fuel Cells

J. Otomo*[1]; H. Matsuo[1]; O. C. Andres[1]; S. Yamate[1]

1. The University of Tokyo, Department of Environment Systems, Graduate School of Frontier Sciences, Japan

Protonic ceramic fuel cells (PCFCs) are highly efficient energy conversion devices for next generation fuel cell systems. However, transport properties in PCFCs, including proton, oxide ion, hole, and electron, can induce leakage current in electrolyte membranes and result in the reduction of energy conversion efficiency. Therefore, the control of the transport properties and relevant careful cell design are important issues for developing highly efficient PCFCs. On the other hand, economic assessment is also necessary for social implementation of PCFC systems in future society. In this study, we discuss innovation pathways with cost reduction of PCFCs. We investigate new design of PCFCs with multi-layered electrolyte membranes consisting of $La_{28-x}W_{4+x}O_{54+3x/2v2-3x/2}$ (LWO) and $BaZr_{0.8}Y_{0.2}O_{3-\delta}$ (BZY). The LWO/BZY multi-layered cell showed very high efficiency because LWO hindered hole conduction in the electrolyte membrane adequately. The material costs of the LWO/BZY multi-layered cell was also estimated. The total material cost of the multi-layered cell was 26USD/W. The result is mostly affected by the raw material costs of ZrO_2 and Ni. The production costs will be discussed as well. The current techno-economic approach provides a technological scenario for the cost reduction of PCFCs as well as proposing new cell design with controlling transport properties in multi-layered electrolyte membranes.

4:40 PM

(PACRIM-226-2021) Effect of air utilization on current efficiency distributions in a protonic ceramic fuel cell (Invited)

T. Araki*[1]; F. Suito[1]; R. Takatera[1]; K. Li[1]; A. Ota[1]; M. MORI[2]; H. Shimada[3]

1. Yokohama National University, Faculty of Engineering, Japan
2. Central Research Institute of Electric Power Industry, Japan
3. National Institute of Advanced Industrial Science and Technology, Japan

Protonic ceramic fuel cells (PCFCs) are expected to be super-efficient with higher fuel utilization at intermediate temperatures. On the other hands, current efficiency of PCFCs decrease due to electron hole leakage through the electrolyte. The current efficiency distribution strongly depends on oxygen concentration and the other parameters. In this study two-dimensional Nernst-Planck-Poisson (NPP) system has been conducted for discussing the effect of gas concentration distribution in both of anode and cathode.

PACRIM Symposium 25: Direct Thermal to Electrical Energy Conversion Materials, Applications, and Thermal Energy Harnessing Challenges

Computation and Theories

Room: Oxford

Session Chairs: Michitaka Ohtaki, Kyushu University; Xun Shi, Shanghai Institute of Ceramics, Chinese Academy of Sciences

1:30 PM

(PACRIM-227-2021) First Principles Assessment of Materials for Direct Energy Conversion (Invited)

J. C. Goldsby*[1]; T. L. Benyo[1]

1. NASA Glenn Research Center, Chemistry and Physics, USA

Solid-state energy conversion devices offer the possibility for highly efficient and reliable operations. In a thermionic material, heat to electric potential conversion results from the thermal gradient and the work functions of the electrode collector relative to the emitter electrode. Combined with a thermoelectric, high efficiencies are possible. Computational methods offer an efficient and systematic manner to design new materials and guide their development. Density functional theory allows us to determine the suitability of various materials as thermoelectric and thermionic converters. To this end, low work function materials are highly sought after as thermionic materials, while high values of ZT work best for thermoelectrics. In this presentation, we studied mixed oxide-pyrochlore ceramics for their potential as a thermoelectric material. We also evaluate several oxides and intermetallic emitter candidate materials as a thermionic convertor through their respective work functions. These calculations used a projector augmented wave (PAW) method. A commercial code (Materials Design Inc.) MedeA incorporated the Vienna Ab-initio Simulation Package (VASP) as the computational engine. This study makes predictions and comparisons between experimental and theoretical properties. In addition, the calculated phonon dispersion curves and elastic constants allowed a mechanical stability prediction.

2:00 PM

(PACRIM-228-2021) Exergetic Efficiency Aspect of Direct Energy Conversion and The Role of Ceramics (Invited)

K. Yazawa*[1]; K. Yazawa[2]

1. Purdue University, Birck Nanotechnology Center, USA
2. Atlas Energy Systems, Inc., USA

Thermodynamic efficiency of heat-driven electrical power generation, in terms of societal impact, is the key challenge from the perspective of environmental protection and fossil resource preservation. The direct energy conversion has great potential for on-site or portable power generation. To take full advantage of this heat-energy conversion, there are major challenge in material properties, isotropic or anisotropic thermal conductivity, thermal expansion, and emissivity. Electrical conductivity, and Seebeck coefficient for thermoelectric, must simultaneously be in a good range in design of devices and systems. Temperature dependence of these properties plays a significant role, for example, when having a high temperature energy source above 1000 K. In this talk, Ragone charts will showcase the potential advances and benefits of direct energy conversion devices and systems based on the results of analytical modeling. This shows the characteristics of energy density vs power density. Next, engineering challenges that impedes the technology developments will be discussed. Harvesting large exergy content naturally requires a larger temperature gradient across the device, which also oxidizes the material more. Thermomechanical deformation changes performance depending on the topology. The results of some analytical and numerical hybrid modeling are shown as examples.

2:30 PM

(PACRIM-229-2021) Should we still use ZT as a performance metric?

M. Beekman*[1]

1. California Polytechnic State University, Physics, USA

Since its inception more than 60 years ago, the dimensionless figure of merit, ZT, remains the most commonly used metric for the evaluation of thermoelectric materials. However, a number of recent works have shown that ZT can lead to erroneous conclusions when comparing thermoelectric materials, arising from the fact that the simple model from which ZT originates neglects the temperature dependence of the electrical and thermal transport properties as well as the Thomson effect. In this talk, I will summarize these and various other shortcomings of using ZT as a performance metric for comparing different materials as well as for optimization of the thermoelectric properties of a given material. Alternative approaches to evaluating material thermoelectric performance will also be described, compared, and contrasted.

Oxides, Nitrides, Sulfides

Room: Oxford
Session Chairs: Jon Goldsby, NASA Glenn Research Center; David Berthebaud, CNRS LINK

3:20 PM

(PACRIM-230-2021) Preparation of Ordered Porous Materials Toward the Development of Thermoelectric Materials (Invited)

K. Kuroda*[1]; T. Matsuno[1]; Y. Shimasaki[1]; A. Shimojima[1]; H. Wada[2]

1. Waseda University, Department of Applied Chemistry, Japan
2. Waseda University, Kagami Memorial Research Institute for Materials Science and Technology, Japan

The presence of porosity contributes to the lowering of thermal conductivity, which is one the effective ways to increase the figure of merit of thermoelectric (TE) materials. However, the relation between porosity and thermal conductivity has not yet been fully elucidated. The factors affecting porosity are diverse, including pore arrangement, pore size, pore volume, pore wall thickness, etc. In order to understand the contribution of porosity to the lowering of thermal conductivity requires well designed synthetic methods to correctly understand the contribution of porosity to the TE performance. This paper is related to the preparation of nanoporous transition metal oxides containing ordered mesopores. In particular, the preparation of ordered nanoporous indium tin oxide and nitrogen-doped TiO_2 with ordered mesopores will be presented. Nanoporous metal oxide particles with uniform and ordered spherical pores were synthesized using monodispersed spherical silica nanoparticles as templates and were sintered. The contribution of intra-particle nanopores to the reduction of thermal conductivity was much larger than that of inter-particle pores. This result shows that the precisely controlled porosity suppressed the long-mean-free-path phonon conduction. These results would contribute to the further understanding of phonon transmission and scattering across nano-interfaces.

3:50 PM

(PACRIM-231-2021) Enhanced Thermoelectric Performance of Porous $SrTiO_3$ with Highly Dispersed Exsolved Ni Nanoparticles

M. Ohtaki*[1]; S. Hirata[1]; M. Kimura[1]; K. Suekuni[1]

1. Kyushu University, Interdisciplinary Graduate School of Engineering Sciences, Japan

Here we report an enhanced thermoelectric performance of sintered perovskite-type oxide $SrTiO_3$ (STO) doped with Nb and Ni after a reducing post-treatment resulting in electron doping and precipitation of highly dispersed Ni nanoparticles formed via an exsolution reaction. Samples with a representative composition of $Sr_{0.95}(Ti_{0.8}Nb_{0.2})_{0.95}Ni_{0.05}O_3$ (STNNO) were prepared by calcining mixed raw oxide powders at 1273 K twice to form the STO phase. The resulting powder was mixed with a graphite powder and heated at 1693 K in air, and subsequently reduced at 1623 K under 0% to 20% H_2/N_2. An XRD study confirmed that all the samples have the STO structure even after the reducing post-treatment. SEM/EDS observation of the cross section of the samples revealed discretely dispersed Ni nanoparticles of ca.50 nm in diameter. While the electrical conductivity of the sample reduced in 20% H_2 was significantly higher than that reduced in 5% H_2, the Seebeck coefficient was very insensitive to the H_2 concentration higher than 5%. However, the lattice thermal conductivity was the lowest for the sample reduced in 20% H_2, resulting in ZT = 0.6 at 1073 K as the highest value among the STO-based oxides so far reported.

4:10 PM

(PACRIM-232-2021) Thermoelectric power factor enhancement using ZnO film including nanowires with well-controlled interface

T. Ishibe*[1]; Y. Nakamura[1]

1. Osaka University, Graduate School of Engineering Science, Japan

Thermoelectric materials converting heat to electricity are attracting great interest because a number of heat surround us. The thermoelectric performance is characterized by dimensionless figure-of-merit, $ZT=S^2\sigma T/\kappa$, where S is Seebeck coefficient, σ is electrical conductivity, κ is thermal conductivity, and T is absolute temperature. The crucial dilemma for high ZT is the interdependence among these three parameters (S, σ, κ). Over the several decades, ZT increase has relied on κ reduction. However, ZT value still remains inadequate for the social application. Namely, in addition to κ reduction, the enhancement of power factor ($S^2\sigma$) is demanded. Here, we propose embedded ZnO nanowire structure with dopant-controlled interface. Energetic carriers preferentially go through the nanowire interfaces, leading to S ($S^2\sigma$) enhancement. On the other hand, phonons are scattered effectively at the interface, bringing κ reduction. Thus, the simultaneous realization of $S^2\sigma$ and κ is achieved. This approach can be the universal methodology to obtain high ZT because this simultaneous realization is brought only by introducing nanowires with dopant-controlled interfaces.

4:30 PM

(PACRIM-233-2021) An integrated approach of DFT, machine learning, and experiment toward the development of thermoelectric function of layered complex nitrides (Invited)

I. Ohkubo*[1]; T. Mori[1]

1. National Institute for Materials Science (NIMS), International Center for Materials Nanoarchitectioncs (MANA), Japan

Recent technical advances in the area of density functional theory (DFT) calculation and machine learning have led to unprecedented capabilities for materials discovery and development. Screening and making a selection of materials by DFT calculation and machine learning are accelerating the materials discovery and development. In addition, these have a beneficial effect on the development of unexplored material groups. Among the undeveloped material groups, layered complex nitrides are of interest because remarkable physical properties have been predicted. We have predicted the electronic structures and thermoelectric transport properties of layered complex nitrides, AMN_2 (A and M are s-, p- block metal ions and transition metal ions, respectively.) using DFT and Boltzmann theory calculations. Recently, a closed-loop thin-film growth method utilizing a machine learning approach for the synthesis of transition metal nitrides is developed. High-throughput thin-film growth experiment combined with a machine learning approach will make a significant contribution to the development of layered complex nitrides. Capability of the combination of DFT calculation, machine learning and thin-film growth techniques to facilitate efficient development of layered complex nitrides will be discussed.

PACRIM Symposium 27: Advanced Materials and Technologies for Electrochemical Energy Storage Systems

Advanced Anode and Cathode Materials/ Supercapacitors

Room: Balmoral

1:30 PM

(PACRIM-234-2021) Raising the redox potential in lithiated organic host materials by tailoring the inductive effects in the solid state (Invited)

P. Poizot*[1]

1. University of Nantes, IMN-CNRS, France

Traditional batteries are based on the redox chemistry of inorganic species (mainly metals), of which some are scarce natural resources, often costly (even toxic) and energy greedy at the process level. In this context, redox-active organic compounds could play an important role in the forthcoming battery technologies. Indeed, organic molecules are composed of quite naturally abundant chemical elements (C, H, N, O, S, in particular) and if properly designed, they can exhibit quite interesting electrochemical performance. However, the synthesis of high-voltage lithiated organic cathode materials is rather challenging, so very few examples of all-organic lithium-ion cells currently exist. Herein, we present an inventive chemical approach leading to a significant increase of the redox potential of lithiated organic electrode materials. This is achieved by tuning the electronic effects in the redox-active organic skeleton thanks to the tailoring of inductive effects in the solid state. For example, substituting magnesium (2,5-dilithium-oxy)-terephthalate for lithium (2,5-dilithium-oxy)-terephthalate enables a voltage gain of nearly +800 mV. Other relevant examples will be presented too.

2:00 PM

(PACRIM-235-2021) Chemomechanical strains in MnO_2 nanosheet supercapacitor electrodes during charge/discharge

M. N. Flint*[1]; A. Wallisch[1]; P. Gao[1]; P. Metz[1]; R. Koch[1]; S. T. Misture[1]

1. Alfred University, School of Engineering, USA

We discuss the atomic-scale response of exfoliated 2-D oxides during charge and discharge. Materials of interest are layered MnO_2 and layered perovskites. Mn^{3+} defects in 2-D MnO_2 are Jahn-Teller distorted, causing Mn^{3+} to be displaced out of the plane of the nanosheet, forming surface Frenkel defects. Such defects dramatically improve the electrochemical function, and therefore we extend our study to layered perovskites with aliovalent B-site dopants and A-site vacancies to better understand charge transfer and intercalation mechanisms. High energy X-ray scattering was teamed with X-ray absorption spectroscopy, Raman spectroscopy, and related tools to probe the local atomic structure during electrochemical cycling. For MnO_2, for example, in-plane breathing of the nanosheets occurs during charge/discharge. In bulk materials or in highly restacked variants, we find that the interlayer spacing changes occur as a result of ion intercalation. Raman spectroscopy and XANES confirm the Faradic redox reaction while EXAFS shows that the charging cation is not highly ordered on the nanosheets. Improved cycling stability is therefore attributed to the ability of random and crumpled nanosheets to accommodate strains during charge/discharge in a highly reversible manner.

2:20 PM

(PACRIM-236-2021) New O2-type lithium-rich layered oxides as positive electrode materials for lihtium-ion batteries (Invited)

M. Guignard*[1]; V. Saïbi[1]; L. Castro[2]; C. Delmas[1]

1. ICMCB-CNRS, France
2. Toyota Motor Europe, Belgium

New O2-type lithium-rich layered oxides were synthesized from P2-type sodium layered oxides by ion exchange. The ion exchange reaction was studied by variable temperature in situ X-ray diffraction to optimize the reaction temperature. The structure of the O2-type lithium-ich layered oxides was then studied by high resolution X-ray diffraction. The ion exchange from the sodium phases to the lithium phases results in the gliding of the $(Li,M)O_2$ layers and the structure of the resulting lithium phases presents stacking faults. The occurence of these stacking faults was confirmed by high resolution transmission electron microscopy. The stacking faults were modeled using FAULTS program and the simulated X-ray diffraction patterns were in good agreement with the experimental ones. Finally, the electrochemical performances obtained for batteries made with the new O2-type lithium-rich layered oxides at the positive electrode and compared to those made with the O3-type lithium-rich layered oxides of the same composition.

Na/K/Mg Batteries

Room: Balmoral

Session Chair: Palani Balaya, National University of Singapore

3:00 PM

(PACRIM-237-2021) Revisiting the Structure–Property Relationships in NaSICON Electrode and Electrolytes (Invited)

P. Canepa*[1]

1. National University of Singapore, Materials Science and Engineering, Singapore

Sodium (Na) batteries are promising devices for smart grids and electric vehicles due to cost effectiveness arising from the overall abundance of and its even geographical distribution. The energy density of Na-ion batteries is limited by the positive electrode chemistry. The Natrium Superionic CONductor (NaSICON) is known for its wide range of electrochemical potentials, its structural and thermal stability and its high conductivity. The electrolyte analogue $Na_{1+x}Zr_2Si_xP_{3-x}O_{12}$ ($0 \le x \le 3$) is stable toward other NaSiCON electrodes. Despite the share of research on NaSICONs, the structural and thermodynamic properties of these materials require better understanding. I will analyse the thermodynamic properties of the $Na_{1+x}Zr_2Si_xP_{3-x}O_{12}$ electrolyte by constructing its phase diagram, based on DFT and Monte Carlo simulations. I will discuss the phase diagram as a function of temperature and composition for the rhombohedral phase, which is common in $Na_3Ti_2(PO_4)_3$, $Na_3V_2(PO_4)_3$ and $Na_3Cr_2(PO_4)_3$. I will chart the chemical space of NaSICON electrodes $Na_xMM'(PO_4)_3$ (M, M'= Ti, V, Cr, Mn, Fe, etc.), to analyse their thermodynamic stabilities and the intercalation voltages for Na^+ ions in 28 distinct NaSICON compositions. I will discuss anomalies in previously reported experimental data in this diverse and important chemical space.

3:30 PM

(PACRIM-238-2021) Alloys in Mg-ion systems: From bulky active materials to coating agents (Invited)

R. Berthelot*[1]; C. Pechberty[1]; L. Stievano[1]

1. CNRS / ICGM (France), France

The passivation of magnesium metal with common battery electrolytes is one the main challenge for the development of magnesium-ion batteries as a credible alternative to the Li-ion technology. It involves the design of complex electrolytes that might not be suitable for further large-scale production. Few p-bloc elements (Sn, Bi, Pb) could electrochemically be alloyed with magnesium at low potential, making them possible negative electrode material. Moreover, they seem compatible with more electrolyte, and easy electrode formulation is possible with alloy powders. Here, we propose a systematic benchmark of three alloys (Mg_2Sn, Mg_3Bi_2 and Mg_2Pb) as active material for magnesium-ion systems, including mechanochemical synthesis, optimization of the electrode formulation, preliminary electrochemical tests in half-cell, and extending comparison in full-cell configuration with standard electrodes (Mo_6S_8 and S/C composite). As substituting magnesium by alloys logically induces a loss in final energy density, we are also investigating these alloying-type materials to protect the surface of magnesium electrode. To do so, coating is performed by chemical reduction of a dissolved salt and characterized mainly by X-ray diffraction and X-ray photoemission spectroscopy. The influence of the coating is further checked by plating/stripping tests and full-cell cycling.

PACRIM Symposium 31: Advanced Functional Materials, Devices, and Systems for Environmental Conservation, Pollution Control, and Critical Materials

Automotive Ceramic Sensors / Critical Materials

Room: Plaza A

1:40 PM

(PACRIM-239-2021) Assembly of Barium Titanate Nanocubes by Two Liquids Phase Separation Method under Ultrasonic Irradiation and Their Dielectric Properties (Invited)

S. Ueno[1]; S. Hatakeyama[1]; I. Fujii[1]; S. Wada*[1]

1. University of Yamanashi, Material Science and Technology, Japan

In this study, barium titanate ($BaTiO_3$, BT) nanocubes was prepared by a simple and conventional solvothermal method without surfactants or capping agents. The obtained BT nanocubes were confirmed as ones with a size of around 100 nm. Then, for a dispersion of BT-nanocubes, electric double-layer mechanism was applied and as a result, well dispersed BT nanocubes were clearly observed. Finally, the well-dispersed BT nanocubes were assembled using by Two Liquids Phase Separation (TLPS) method under ultrasonic irradiation at 40 deg. C to measure dielectric properties. For as-assembled samples with relative density of 54 %, dielectric constant was almost 40. This suggested that there is no chemical bonding between attached BT nanocubes. To solve this problem, as-assembled samples were treated under ethanol and water solution with $Ba(OH)_2$ by solvothermal method at 200 deg. C for 20 h. As the results, there were no changes of density and nanocube morphology before and after solvothermal treatment at 200 deg. C, but dielectric constants increased from 40 to 110 by solvothermal treatment.

2:10 PM

(PACRIM-240-2021) Low-temperature Wet Chemical Processes for Barium-Titanate-related Dielectric Ceramics and Dielectric Composites (Invited)

S. Ueno*[1]; I. Fujii[1]; S. Wada[1]

1. University of Yamanashi, Graduate School Department of Interdisciplinary Research, Japan

A Manufacturing of dielectric ceramics generally includes multiple high-temperature heating processes, and thus requires high energy. In contrast, wet chemical processes, which utilize reactions proceeding in a mild condition, are commonly known as eco-friendly processes but are usually employed for preparing nanoparticles, thin films, and porous materials. Here, we focus on the representative dielectric material, $BaTiO_3$, and introduce our wet chemical approaches for fabrication of nanostructured $BaTiO_3$ dielectric ceramics and $BaTiO_3$-related dielectric composites. A solvothermal solidification process is capable of preparing the dense $BaTiO_3$ ceramics below 300°C. Porous TiO_2 green compacts were reacted with Ba source in solutions, and the resultant $BaTiO_3$ eliminated pores inside the compacts due to a volume expansion. These $BaTiO_3$ ceramics whose relative density reached to around 90% showed the ε_r of over 600, and the higher ε_r (~1500) could be achieved by the post-annealing far below sintering temperature. Such process can be also employed for the fabrication of dielectric composites containing metals with a low heat-proof temperature such as high-capacitance Ti metal/$BaTiO_3$ composites with a boundary layer structure. The solvothermal solidification is one of the promising processes for providing component materials in micro- and nano-electric devices.

OCs / Ion-conducting Ceramics II

Room: Plaza A

3:00 PM

(PACRIM-241-2021) Room Temperature Detection of Organic Pollutants under UV Irradiation By a Wide Gap p-n Heterojunction Diode (Invited)

Y. Nakamura*[1]; S. Fujitsu[2]

1. The University of Tokyo, Department of Applied Chemistry, School of Engineering, Japan
2. Tokyo Institute of Technology, Institute of Innovation Research, Laboratory for Materials and Structures, Japan

How to activate the thermodynamically stable molecules such as aromatic VOCs is a key for designing an organic pollutant sensor with lower operation temperature and for this purpose, we propose a new concept for VOC sensing based on the assistance of the "external" energy. A wide-gap semiconductor pn junction interface was applied for the reaction field of the sensor system and high energy holes in VB(valence band) of a p-type semiconductor could efficiently oxidize VOC molecules adsorbed over the junction interface, then, the junction current would be modified. As one embodiment, we prepared a wide-gap pn heterojunction diode by RF magnetron sputtering or PLD technique and processed it into a chemical sensor element by photolithography pattering and etching. A p-NiO/n-ZnO heterojunction sensor element does not work as a VOC sensor under dark, however, under UV light irradiation, its electric current rapidly decreases with introduction of aromatic VOCs. Deep UV light incidence will induces the excitation of six-membered aromatic ring, and that is the necessary condition for the room temperature aromatic VOC sensing in this system. The origin of sensor response is positively ionized (cation) adsorbates, which would be produced by photo-driven electron transfer from excited states of aromatic VOC adsorbate to VB of p-NiO.

3:30 PM

(PACRIM-242-2021) The electrochemical properties of $LiCoO_2$ formed on oxide-based solid electrolytes by a ceramic process (Invited)

Y. Arachi*[1]; H. Oikawa[1]

1. Kansai University, Chemistry and Materials Engineering, Japan

The all-solid-state Li secondary batteries have been expected as a next generation of the existing lithium ion batteries. Compared to sulfide-based electrolyte, oxide-based electrolytes are more challenging to connect with electrodes. The aim of this research is to find the ceramic process to produce a rechargeable interface between positive electrode and solid electrolyte. As a positive electrode and solid electrolytes, we have focused on $LiCoO_2$ (LCO) and $Li_{0.33}La_{0.5}TiO_3$ or $Li_{6.5}La_3Zr_{1.5}Ta_{0.5}O_{12}$, respectively. By heating the starting materials consisting of Li and Co salts, a LCO film with a thickness of several micro meters was formed on the electrolyte substrates. To ensure the mobility of lithium ions between electrolyte and negative electrode, $Li_{3.25}Ge_{0.25}P_{0.75}S_4$ powder was sandwiched between them. The electrochemical properties were evaluated by a galvanostatic charge-discharge testing under various current densities at 323 K. Heating the molten salts resulted in a rechargeable interface between LCO and Li-conducting oxides. The electrical resistances of the interface were estimated by impedance measurements. We will introduce an electrochemical performance of the LCO in addition to an unique microstructure of those interface by both SEM and TEM observations, and discuss the mechanism of the formation of LCO on the solid electrolyte by this process.

4:00 PM

(PACRIM-243-2021) Development of Proton Conductive Materials and Application for Next-Generation Medium Temperature Fuel Cells (Invited)

A. Matsuda*[1]

1. Toyohashi University of Technology, Electrical and Electronic Information Engineering, Japan

Fast proton conducting materials have attracted much attention because of their potential use in protic devices such as capacitors, sensors and fuel cells. There has been a growing interest in nano-size and interface, so-called "nanoionics effects" on fast ionic transfer. We have been studying the highly proton conductive solid electrolytes based on the concept using pores, interface, and defects in the nano-scale. Mechanochemical (MC) synthesis applies high impact forces to a local area of the solid acids or salts, resulting in the significant distortions and the defects in surface structures to form paths suitable for fast proton conduction. Highly proton conductive ortho-oxosalt-heteropoly acid composites are prepared through the ion exchange and formation of three-dimensional hydrogen networks by the MC process. We have been studying composite electrolyte membranes composed of polybenzimidazole (PBI) and mechanochemically synthesized inorganic solid acid complex for the fuel cell applications. Here we report the materials design of highly conductive solid state proton conductors based on the MC synthesis and their application to the next generation medium temperature fuel cells as the electrolyte membranes for power sources of cogeneration systems and electric vehicles.

4:30 PM

(PACRIM-244-2021) Preparation of Li_4SnS_4 Solid Electrolytes by Aqueous Ion-Exchange Method and Fabrication of All-Solid-State Batteries

R. Matsuda*[1]; T. Kokubo[1]; H. Muto[1]; A. Matsuda[1]

1. Toyohashi University of Technology, Japan

Research and development of all-solid-state lithium ion batteries are blooming because they used inorganic solid electrolytes, a promising alternative to conventional organic solvents containing electrolytes, and they have the potential to improve battery safety and enhance energy density and capacity. Sulfide-based solid electrolytes have been examined as candidates for use in all-solid-state batteries because they show high ion conductivity as same as organic solvent and unique deformation property. However, handling of sulfide solid electrolytes must be done in dry atmosphere because of their vulnerability to moisture in air. In general, hydrolysis of sulfide solid electrolytes causes structural change and generates H_2S gas. To enhance the chemical stability in air, in the present study, we focused on the Li_4SnS_4 system. In previously report, the Li4SnS4 solid electrolyte was prepared by solid-phase synthesis. We developed a new method using aqueous ion-exchange for its preparation and then evaluated the structure, electrochemical and mechanical properties. All-solid-state batteries using the prepared Li_4SnS_4 in the composite cathode, with titanium sulfide as an active material, were fabricated to evaluate the electrochemical performance and stability of Li_4SnS_4

PACRIM Symposium 34: Glass and Ceramics for Nuclear Waste Treatment and Sequestration

Waste Form Matrices-Synthesis and Characterization III

Room: Regency F
Session Chair: Theodore Besmann, University of South Carolina

1:30 PM

(PACRIM-245-2021) Structural dependence of sulfur solubility in borosilicate glasses

R. Saini*[1]; D. R. Neuville[2]; R. Youngman[3]; H. Eckert[4]; A. Goel[1]

1. Rutgers University, Materials Sci. & Engg., USA
2. institut de physique du globe de Paris, Université de Paris, France
3. Science and Technology Division, Corning Incorporated, USA
4. São Carlos Institute of Physics, São Paulo University, Brazil

The US-DOE is considering implementing the direct feed LAW (DFLAW) and direct feed HLW (DFHLW) approach to expedite the vitrification of radioactive waste stored at the Hanford site. If implemented, the increasing sulfate content in direct-feed LAW and HLW poses a challenge for the established models for predicting the sulfur solubility in LAW glasses and is expected to aggravate the complexity of HLW vitrification. Therefore, to design advanced glass formulations with enhanced waste loadings, it is essential to comprehend the fundamental compositional and structural drivers governing the sulfur solubility in borosilicate glasses. In this pursuit, the structural dependence of sulfur solubility in sodium borosilicate glasses is investigated over a broad compositional space. Sulfur is incorporated in the baseline glasses using Na_2SO_4 as a precursor while maintaining a constant Na_2O content (as per baseline glass). The structure and sulfur solubility in these glasses is investigated by XRD, DSC-TGA, ICP-OES, Raman, and ^{11}B MAS-NMR spectroscopies. The focus is to elucidate the impact of (1) structural role of Na^+, (2) degree of network polymerization, and (3) the formation of Si–O–B and B–O–B linkages on the sulfur solubility in the borosilicate glasses.

1:50 PM

(PACRIM-246-2021) Impact of irradiation on the properties of gel layer formed after aqueous corrosion of borosilicate glasses

A. Jan*[1]; N. Krishnan[1]

1. Indian Institute of Technology Delhi, India

Confinement of radionuclides in borosilicate glass matrix is subject to the durability of vitrified nuclear glass in the aqueous medium on the geological time scales under repository conditions. It is now well accepted that under aqueous condition there is a formation of a gel layer. However, the properties of this gel layer are not well understood yet. In this work using atomistic modelling, a series of borosilicate glasses- pristine and irradiated- are prepared and further, gels are prepared by replacing boron and sodium by hydrogen. These gels are then aged at different temperatures - 500K, 1000K, 1500K and 2000K. It is seen that, indeed, there is a difference in the properties gel (connectivity, short and medium range order) formed from pristine and irradiated glass. For example, the mean square displacement of hydrogen is orders of magnitude (exact magnitude depending upon the composition) higher in the gel formed from irradiated glass as compared to pristine glass. Overall, the present study suggests that the alteration layer formed from irradiated and pristine glasses are notably different, which also varies as a function of the glass composition.

2:10 PM

(PACRIM-247-2021) Candidate Waste Forms for the Immobilization of Dehalogenated Electrorefiner Salt (Invited)

L. D. Gardner*[1]

1. Argonne National Lab, Chemical and Fuel Cycle Technologies, USA

The use of electrochemical processing technology has proven effective for the recovery of fissionable actinides from used nuclear fuel. This operation generates salt waste streams containing fission product and actinide halides that must be further processed to produce durable waste forms that ensure radionuclides will be contained in a repository over geologic timescales. Glass-bonded sodalite is currently the baseline waste form for waste salt immobilization, but several alternative disposition paths have been proposed and studied. Several pathways involve halide ion removal from the waste salt (dehalogenation) that generate $HCl_{(g)}$ or similar secondary waste streams. These reactions provide large decreases in solid waste volume. This presentation provides an overview of optional dehalogenation reactions and waste forms for the resulting waste streams, including H-Y zeolite-derived sintered waste forms and iron phosphate glasses. Microstructural characterization, chemical durability, and estimated final storage volumes are used to assess the effectiveness of these candidate waste forms. Improvements in chemical durability and waste volume relative to glass-bonded sodalite waste forms will be highlighted. This work is supported by the U.S. Department of Energy, Office of Nuclear Energy, under contract # DE-AC02-06CH11357 with the U.S. Department of Energy.

Waste Form Matrices-Synthesis and Characterization IV

Room: Regency F

Session Chair: Theodore Besmann, University of South Carolina

3:20 PM

(PACRIM-248-2021) A Predictive Model for Offgas Composition in Waste Glass Melters

A. Barron[1]; D. P. Guillen*[1]; M. Hall[2]; A. Abboud[1]; S. Davidson[2]; A. A. Kruger[3]

1. Idaho National Laboratory, Materials Science and Engineering, USA
2. Pacific Northwest National Lab, USA
3. U.S. Department of Energy, USA

During the batch-to-glass conversion of legacy tank waste at the Waste Treatment and Immobilization Plant, water vapor, air, aerosols, particulate matter, and other gaseous species are evolved in the melter plenum by evaporation, chemical reactions, and influx air. To predict the offgas composition for different feeds, an Aspen Plus model was developed. Currently, the offgas prediction is limited to nitrous oxides and carboxides. The model employs a constant stirred tank reactor block and user-defined reaction chemistry to determine the offgas composition. The offgas compositions predicted by the model were validated by available test data for ten different feed compositions. The results indicate that Aspen Plus can reliably predict the evolution of nitrous oxides and carboxides in the melter plenum for both high level waste and low activity waste feeds. Future work will expand the suite of gases to predict include other plenum components of interest, such as SOx, and products of incomplete combustion. Other improvements include the addition of reaction chemistry for side reactions through development of reaction stoichiometry consistent with experimental results and improved reaction kinetics by developing a method to predict the Arrhenius constant or by developing a new kinetic model through experimentation.

3:40 PM

(PACRIM-249-2021) Redox effect on the structures of iron containing boro-aluminosilicate glasses from Molecular Dynamics Simulations

M. I. Tuheen*[1]; W. Sun[1]; J. Du[1]

1. University of North Texas, Materials Science and Engineering, USA

Iron oxide is found in many industrial or commercial glass compositions. It is also one of the most common constituents in glasses to immobilize nuclear waste. The interaction of the multiple glass formers with each other and also with different glass modifiers is still not fully understood. This work aims to focus on the preference of different glass formers such as B, Al and Fe to be charge compensated by constant Na_2O content in three glass systems with four glass compositions using molecular dynamics (MD) simulations. It was observed that, there is indeed competition among the glass formers for modifiers ions. Al is always preferred by Na before B but between Al and Fe, the competition is still debatable. Moreover, the ratio of any two glass formers affect the structural properties of these glasses. The coordination numbers and other detailed structural analyses were performed to investigate their roles and the competitions between the B, Al, and Fe for charge compensation by glass modifiers cations.

4:00 PM

(PACRIM-250-2021) Thermodynamic assessment of Zn-containing hollandite as a robust waste form for alkaline-earth metals

J. Schorne-Pinto*[1]; A. Mofrad[1]; K. Brinkman[2]; J. Amoroso[3]; S. T. Misture[4]; T. M. Besmann[1]

1. UofSC, Mechanical engineering, USA
2. Clemson University, Materials Science and Engineering, USA
3. Savannah River National Lab, USA
4. Alfred University, USA

The hollandite phase is a potential nuclear waste form for alkaline-earth metals such as Cs and Ba by chemical incorporation into the crystalline lattice. The general hollandite formula is $A_xB_8O_{16}$ with the A-site habitually occupied by large mono- or divalent cations (e.g., Ba, Sr, Na, K, Rb, Cs, Pb), and the B-site by small di-, tri-, or tetravalent cations (e.g., Ti, Al, Fe, Cr, Sc) and it crystallizes in a tetragonal or monoclinic structure depending on the composition. Recently, our group has developed a thermodynamic database to describe the BaO-Cs_2O-TiO_2-Cr_2O_3-Al_2O_3-Fe_2O_3-FeO-Ga_2O_3 system based on the Calphad method, where the hollandite solid-solution was modeled using the Compound Energy Formalism (CEF). Recent findings show the impact of varying A-site Cs stoichiometry of Zn-containing hollandite compositions on relative thermodynamic stability. Utilization of this understanding allowed improved performance of the hollandite phase in the multi-phase waste form termed

SYNROC. We here report on our efforts to include the Zn in our waste form thermodynamic database, modeling the phase equilibria of the complex hollandite-forming oxide waste system to allow more comprehensive optimization of the system compositions.

4:20 PM

(PACRIM-251-2021) Dissolution kinetics and mechanisms of zirconolite based wasteforms for UK plutonium disposal

C. Gausse[1]; E. Ormerod[1]; L. Blackburn[1]; L. J. Gardner[1]; S. Sun[1]; N. C. Hyatt[1]; M. C. Stennett[1]; C. L. Corkhill*[1]

1. University of Sheffield, United Kingdom

A small proportion of the 140 tons of UK civil separated plutonium has been regarded as unsuitable for reuse as mixed-oxide fuel (MOX). Several different wasteforms, including zirconolite ceramics and zirconolite glass-ceramics produced by Hot Isostatic Pressing (HIP), have been investigated as suitable candidates for immobilising plutonium. These materials have demonstrated excellent suitability thanks to their good radiation stability, high chemical flexibility and promising chemical durability [1-3]. However, further understanding of the long-term behaviour of these wasteforms within a geological disposal setting is required. We present here a systematic dissolution study, with the aim of understanding the dissolution kinetics and mechanisms of a zirconolite ceramic ($Ca_{0.8}Ce_{0.2}Zr_{0.90}(Ce, U)_{0.10}Ti_{1.6}Fe_{0.4}O_7$) and glass-ceramic produced via HIP ($CaZr_{0.9}Ce_{0.1}Ti_2O_7$: $Na_2Al_2Si_6O_{16}$). The chemical durability of these materials was determined using a range of surface analytical techniques, e.g. Vertical Scanning Interferometry (VSI) and aqueous geochemical techniques, complemented by thermodynamic modelling.

PACRIM Symposium 36: Advanced Multifunctional Bioceramics and Clinical Applications

Biological Cell Response / Clinical Research / Calcium Phosphates

Room: English Bay

Session Chair: Qiong Wang, University of British Columbia

1:30 PM

(PACRIM-252-2021) Improvement of surface wettability and cellular interactions by bioceramic electrets (Invited)

M. Nakamura*[1]; L. Bergara Muguruza[2]; U. Sarwar[2]; K. Yamashita[3]

1. University of Turku/Tokyo Medical and Dental University, Finland
2. University of Turku, Finland
3. Tokyo Medical and Dental University, Finland

The surface modification of ceramic biomaterials used for medical devices is expected to improve the osteoconductivity through control of the interfaces between the materials and living tissues. The purpose of this study was to investigate a mechanism through which the surface wettability of biomaterials can be improved and determine the effects of biomaterial surface characteristics on cellular behaviors. Polarization treatment induced surface charges on hydroxyapatite and carbonate-substituted hydroxyapatite. Characterization of the surfaces revealed that the wettability of the polarized ceramic biomaterials was improved through the increase in the surface free energies compared with conventional ceramic surfaces. The effects of polarization were maintained after sterilization and cell culture. The polarization accelerated the cell adhesion and differentiation.

2:00 PM

(PACRIM-253-2021) Synergistic effect of Na/K contents, polarization and external electrical stimulation on biocompatibility of $Na_xK_{1-x}NbO_3$ (x = 0.2 - 0.8) piezoceramics (Invited)

D. Khare*[1]; A. K. Dubey[1]

1. Indian Institute of Technology(BHU), Department of Ceramic Engineering, India

Endogenous bioelectrical signals play a crucial role in hard tissue repair and regeneration. The piezoelectric charges developed on the surface of stressed human bone activate intracellular as well as extracellular biological factors responsible for cell growth. In this perspective, the present article combines the synergistic effect of surface charge polarization on the piezoelectric sodium potassium substrate and direct external electrical stimulation to achieve enhanced biological response. Apart from bioelectrical factors, the effect of compositional variation of Na and K on the biocompatibility of $Na_xK_{1-x}NbO_3$ (x = 0.2 to 0.8) were also investigated as both of elements play an important role in the metabolism of human body. The result outcome reveals that the cell viability is higher on the electrically stimulated - polarized samples preferably on the sodium rich NKN i.e., $Na_{0.8}K_{0.2}NbO_3$ and potassium rich NKN i.e., $Na_{0.2}K_{0.8}NbO_3$ samples. Therefore, the synergistic utilization of surface charge polarization, external electrical stimulation and optimized compositional modification provides a novel way to improve the biological response of NKN piezobioceramics.

2:30 PM

(PACRIM-254-2021) Mechanism of bone regeneration in extraction sockets grafted with SCPC

A. El-Ghannam*[1]; M. Nakamura[2]; U. Sarwar[2]; L. Bergara Muguruza[2]; R. Al Fotawi[3]; M. Hassan[1]; R. Horowitz[1]

1. University of North Carolina at Charlotte, USA
2. University of Turku, Finland
3. King Abdulazeez University, Saudi Arabia
4. The NYU College of Dentistry, USA

Resorption of graft material and stimulation of new vascularized bone formation are critical for maximizing vital bone regeneration. The dissolution products of the graft should have dual function; first activate the new bone formation pathway and second, inhibit cellular activities that compromise the mineralization of new bone matrix. Silica-calcium phosphate composite (SCPC) was grafted in extraction sockets and core bone biopsy taken at the time of implant insertion was analyzed using histology. Immunohistochemistry showed presence of specific markers of mature bone and absence of osteoclasts at the interface with the remnants of SCPC graft. In vitro cell culture studies showed the release of Si, P and Na ions from SCPC in tissue culture medium. In conjunction with the ceramic ion release, the bone resorptive activity of osteoclasts was inhibited as indicated by the significant decrease in multinucleated TRAP stained cells and the volume of resorption pits on bone slices in vitro. In contrast, addition of SCPC to hBMSC cultured in conventional medium promoted higher RUNX 2 ($p< 0.05$), OSC ($p< 0.01$) and BSP ($p<0.01$) than that expressed by control cells grown in the absence of SCPC. Mathematical modelling showed that the predicted dissolution time of porous SCPC particles is 6.9 months. Promoted osteoblasts are responsible for the rapid bone formation and maturation inside grafted defects.

PACRIM Symposium 37: Material and Technology Needs for Medical Devices, Sensors, and Tissue Regeneration

Device, Sensor, and Tissue Regeneration III

Room: Stanley
Session Chairs: Rizhi Wang, University of British Columbia; Kohei Soga, Tokyo University of Science

1:30 PM

(PACRIM-255-2021) 3D Printing of Porous Biphasic Calcium Phosphate Scaffolds for Bone Regeneration via Digital Light Processing

Y. Wang[3]; J. Bai[2]; M. Wang*[1]

1. The University of Hong Kong, Department of Mechanical Engineering, Hong Kong
2. Southern University of Science and Technology, Department of Mechanical and Energy Engineering, China
3. Department of Mechanical and Energy Engineering, Southern University of Science and Technology, Department of Mechanical Engineering, The University of Hong Kong, Hong Kong, China

Biphasic calcium phosphate (BCP), a mixture of hydroxyapatite (HA) and beta-tricalcium phosphate (beta-TCP), is a promising bioceramic for bone tissue regeneration. Digital light processing (DLP) is a fast 3D printing technique that can fabricate ceramics with complex architectures. But there are rare reports on the fabrication of bioceramic implants via DLP. A comprehensive investigation was therefore conducted to study and optimize the DLP process and parameters to make high-performance porous BCP scaffolds. First, nano-sized HA and β-TCP powders were thoroughly mixed in a planetary mill. They were then formulated with optimized photosensitive resins to prepare BCP slurries. Four dispersants were compared for BCP DLP, aiming to decrease bioceramic nanoparticle agglomeration and achieve BCP slurry with high solid loading ($\geq$70 wt.%) and low viscosity. After DLP printing, debinding and sintering of printed green parts were conducted, obtaining crack-free BCP with desirable biological and mechanical properties. Finally, porous BCP scaffolds with different designed pore morphologies, pore sizes and porosities were obtained under optimized parameters. DLP-formed BCP scaffolds showed good cell viability, improved osteoconductivity and controlled degradation rates. This study has demonstrated the enormous potential of DLP for constructing complex bioceramics scaffolds.

1:50 PM

(PACRIM-256-2021) In-situ TEM Studies of Biomineralization (Invited)

R. Shahbazian Yassar*[1]; T. Shokuhfar[1]

1. University of Illinois at Chicago, USA

This talk will provide an overview of the PIs' efforts to understand the dynamics of biomineralization via in-situ transmission electron microscopy. We demonstrate our success to observe the biomineralization of hydroxyapatite (HA) crystals. Our results show that HA crystals follow classical and non-classical nucleation theories to from within a supersaturated solution. We also studied the effect of molecular modifiers on the effect of calcium oxalate crystals that are the primary constituent of kidney stones. We show that the addition of citrate can affect the crystallization pathway of these minerals. Interestingly, the citrate molecules affect the pre-nucleation stage of CaOx crystals making them thermodynamically stable. In addition, we will showcase some examples of biomineralization of iron oxide core in ferritin proteins and demonstrate the ability to monitor the biomineralization of these crystals using graphene liquid cells in TEM. We will show that the ratio of L and H subunits in the ferritin protein shells can affect the nucleation and growth of iron oxide cores. We also will present our latest results on the biomineralization of magnetosomes in magnetotactic bacteria grown in iron-rich media using in situ GLC-TEM studies. We observed that such bacteria can remain alive and intact during TEM imaging and follows the classical nucleation theory for the biomineralization of magnetosomes.

2:20 PM

(PACRIM-257-2021) Novel borosilicate bioactive glass material for bone implants

A. Szczodra*[1]

1. Tampere University, Finland

Annually, 12–18 million surgeries for dental implants and over two million surgeries for bone grafting are performed. The current dental implants mainly use Titanium 6Al4V alloy, despite many problems like early implant rejection, leaching of Ti over time and allergic inflammatory reaction. The aim of the present study was to develop borosilicate bioglass scaffolds using technological progress in 3D printing. Borosilicate glasses were found efficient in producing glasses with fast conversion into hydroxyapatite and with thermal properties allowing sintering into 3D scaffolds. Moreover, boron substitution for SiO2 showed strong ability to stimulate osteogenic commitment and upregulate endothelial markers. In this study, borosilicate glasses were processed into 3D scaffolds using the porogen burn-off technique and 3D prototyping. As the overall porosity is known to effect glass dissolution both scaffold types were processed with similar porosity. The impact of preincubation in TRIS buffer solution and αMEM culture medium on ions release profiles was studied by ICP-OES. Furthermore, the cytotoxicity of scaffolds to MC3T3 pre-osteoblast cell line are studied in direct and indirect cell culture and compared to a dense sintered glass body. Focus was on cell viability, proliferation and morphology. The aim is to assess the impact of the pore structure and borosilicate glass composition on the cells behaviour.

Device, Sensor, and Tissue Regeneration IV

Room: Stanley
Session Chairs: Min Wang, The University of Hong Kong; Reza Yassar, University of Illinois

3:20 PM

(PACRIM-258-2021) Composition-Property Relationships of Multicomponent Glasses for Transarterial Embolization: Modelling and Optimizing for Imaging and Density

K. O'Connell*[1]; E. Tonkopi[3]; A. Headley[1]; D. Dobrowski[1]; D. Boyd[2]; S. Kehoe[1]

1. ABK Biomedical Inc., Canada
2. Dalhousie University, Applied Oral Sciences and School of Biomedical Engineering, Canada
3. Nova Scotia Health, Diagnostic Imaging, Canada

27 glass microspheres were developed using an I-optimal quadratic design. Mixture components (mol%) were constrained to: SiO_2: 55% to 75% / Y_2O_3: 0 to 25 / BaO: 0 to 20 / Na_2O: 0 to 20/B_2O_3: 0 to 20 / Ta_2O_5: 0 to 10. Formulations were examined by X-ray diffraction, differential scanning calorimetry, dynamic image analysis, helium pycnometry and in vitro cytotoxicity (L929 fibroblasts). Axial computed tomography (CT) and region of interest analysis was completed using a Siemens Somatom Definition AS+ scanner at 70 and 120kVp. Second-order Scheffé polynomial mixture equations were fitted for each property. Polynomial mixture regression models and associated ANOVAs along with their coefficients were determined. An optimized formulation (minimum density, maximum radiopacity) was identified, synthesized, and verified. Significant model terms were developed for each property. Densities ranged 3.3g/cm^3 to 3.9 g/cm^3, with CT measurements between 6000 and 22000 Hounsfield Units (HU). Optimization yielded a microsphere

with a density about 3.3 g/cm^3 and radiopacity of 16,000 HU. An optimized microsphere with minimized density and maximized radiopacity has been identified based on statistical and experimental modelling and is suited to advanced intra and post procedural imaging for transarterial embolization.

3:40 PM

(PACRIM-259-2021) Processing of Calcium Phosphate Nanostructure as a Potential Assay for Studying Osteoporosis (Invited)

S. Chen[1]; Q. Wang[1]; F. Eltit[1]; Y. Guo[1]; M. Cox[1]; R. Wang*[1]

1. University of British Columbia, Canada

Osteoclastic resorption of bone plays a central role in both osteoporosis. A reliable in vitro assay that simulates osteoclastic resorption in vivo would significantly speed up the process of developing effective therapeutic solutions for those diseases. A challenge in the past has been to establish a standard calcium phosphate substrate with well-controlled structure. In a series of laboratory studies, we have developed a robust nano-structured calcium phosphate coating with unique functions on the track-etched porous membrane by using a novel ammonia-induced mineralization (AiM) technique. This report presents unique nanostructure of the processed coating together with formation mechanisms. We will then demonstrate the applications of the processed calcium phosphate coating as a drug screening device by using alendronate, a widely used drug for osteoporosis. The application will be supported by in-depth analyses of drug release and osteoclast activities using various microscopy techniques. Finally, applications of this novel assay in other diseases such as prostate cancer bone metastasis will be discussed.

4:10 PM

(PACRIM-260-2021) Rare-Earth Doped Ceramics for Near Infrared Bio and Medical Photonics (Invited)

K. Soga*[1]; K. OKUBO[1]; M. Umezawa[3]; D. DUNG[2]; M. Kamimura[1]

1. Tokyo University of Science, Dept Mater Sci & Tech, Japan
2. Tokyo University of Science, RIBS, Japan
3. Tokyo University of Science, RIST, Japan

Near infrared light has been known to be transparent for biological tissues because of the suppressed reflection, refraction and scattering. The authors have been developing the nanostructured materials with rare-earth doped ceramics nanoparticles for bio and medical sensing and imaging. Not only to these, the nanostructure with the ceramic nanoparticle and biopolymer with certain biologically or photoactive molecules can also be used for therapy such as photo dynamic therapy with near infrared excitation. Recent imaging system development is enabling 3D temperature imaging by combining fluorescence thermometry and computer tomography. The use of the near infrared expands the possibility of the optics and photonics to be applied for the thickness with centimeter order. The paper will review the materials and system development for those near infrared bio and medical photonics.

PACRIM Symposium 40: 6th International Richard M. Fulrath Symposium, "Frontiers of Ceramics for a Sustainable Society"

Global Human Health Challenges

Room: Plaza B
Session Chair: Sheikh Akbar, Ohio State University

1:30 PM

(PACRIM-261-2021) Two Photon Polymerization of Microstructured and Nanostructured Surfaces for Medical Applications (Invited)

R. Narayan*[1]

1. NC State University, USA

Two photon polymerization has been used for 3D printing of surfaces with microscale and nanoscale features for biological applications. For example, photon polymerization utilizes a titanium:sapphire laser pulses to excite photoinitiator molecules for polymerization and hardening within a photosensitive material. Polymerization occurs at locations in which excitation threshold of the photoinitiator is exceeded. Processing of structures with microscale and nanoscale features is possible since the two photon absorption process exhibits a nonlinear relationship with the incident light intensity; structures below the diffraction limit can be obtained. Two photon polymerization has been used to create surfaces with microscale and nanoscale features out of ceramic-containing materials such as zirconium oxide hybrid materials and other organically-modified ceramic materials. Several types of medically-relevant surfaces, such as scaffolds for tissue engineering as well as microneedles for drug delivery and biosensing, have been created using two photon polymerization. In this talk, biological and functional testing of two photon polymerization-created surfaces will be considered.

2:00 PM

(PACRIM-262-2021) Ferroelectrics with a controlled oxygen-vacancy distribution (Invited)

Y. Noguchi*[1]

1. Kumamoto University, Faculty of Advanced Science and Technology, Japan

Controlling point defects in transiton metal oxides provides an additional degree of freedom not only for increasing physical properties but also for inducing additional functionalities. In ferroelectric oxides, an accumulation of point defects at specific interfaces often deteriorates a polarization-switching capability, but yields interface-driven phenomena. Despite of intensive researches over several decades, it remains challenging to manipulating oxygen vacancies to achieve a desirable defect structure. Here, we report a practical route to controlling oxygen-vacancy distributions by exploiting the interaction with transition-metal dopants. Our thin-film experiments combined with DFT theoretical calculations for $BiFeO_3$ demonstrate that isovalent dopants such as Mn^{3+} with a partly or fully electron-occupied e_g state can trap oxygen vacancies, leading to a robust polarization switching. Our approach to controlling oxygen vacancy distributions by harnessing the oxygen-vacancy-trapping capability of isovalent transition-metal cations has a possibility for realizing full potential of switchable polarization in ferroelectric oxides.

Emerging Ceramic Materials and Technologies III

Room: Plaza B

Session Chair: Roger Narayan, North Carolina State University

3:20 PM

(PACRIM-263-2021) AC Poling Treatment in Grain-oriented BT-BNT Piezoceramics (Invited)

S. Kim[1]; I. Fujii[1]; S. Ueno[1]; S. Wada*[1]

1. University of Yamanashi, Material Science and Technology, Japan

The engineered domain technique is one of the well-known domain engineering techniques for piezoelectric enhancement, which is composed of two parts, (1) domain wall fixing technique on the basis of crystallographic anisotropy and (2) increasing domain wall density on the basis of grain size control. This is because domain wall itself is very soft and possess very high dielectric and piezoelectric properties. For tetragonal perovskite oxide materials, crystallographic directions are <111> and <110> directions to fix domain walls into the materials. We previously reported the preparation of the <111> oriented barium titanate ($BaTiO_3$, BT) ceramics using by electrophoresis deposition (EPD) method under high magnetic field (HM-EPD) of 12 T, and piezoelectric enhancement with d_{33} of 500pm/V. On the other hand, we also reported the preparation of the <110> oriented BT ceramics using by template grain growth (TGG) method, and piezoelectric enhancement with d_{33} of 800pm/V. However, the Curie temperature (T_C) of BT is 132 °C, and this is very low for the most of piezoelectric applications, which need the higher TC over 200°C. Thus, in this study, we focus bismuth sodium titanate ($(Bi_{0.5}Na_{0.5})TiO_3$, BNT) – BT (BNT-BT) system ceramics with high T_C of around 250°C. Using tetragonal 0.85BNT-0.15BT chemical compositions, two kind of grain-oriented ceramics along <111> and <110> directions were prepared.

3:50 PM

(PACRIM-264-2021) Ceramic Nano-heterostructures by Materials Design: Platforms for Sensing and Biomedical Applications (Invited)

S. A. Akbar*[1]

1. Ohio State University, Materials Science and Engineering, USA

This talk summarizes R&D efforts in the author's laboratory on the fabrication of oxide nano-heterostructures, exploiting intrinsic material properties, that are highly scalable and do not require use of lithography. One such process creates crystallographically oriented nanofiber arrays of TiO_2 via nano-carving and on Au-coated SnO_2, showing directional growth on crystal facets in H_2/N_2 environment. Another process creates nanofibers of TiO_2 on Ti metal/alloys via oxidation under a limited supply of oxygen. In another process, SnO_2 nanowires grown from commercial FTO slides using the VLS method were placed in a MW-assisted hydrothermal chamber where TiO_2 nanorods nucleated radially from the SnO_2 nanowire cores. We have developed yet other interesting nano-structures (nanoislands and nanobars) during thermal annealing of an oxide (GDC) on top of another oxide (YSZ) that self-assembles along the softest elastic direction of the substrate. All these structures are fabricated without the use of lithographic techniques and involve simple gas-phase reactions and stress-driven processes. These nano-structures can be used as platforms for chemical sensing, catalysis, photocatalysis, photovoltaics and biomedical applications. Opportunities and challenges in gas sensing are presented including an Open access Database Of Resistive type gas Sensors (ODORS) that is under development.

4:20 PM

(PACRIM-265-2021) Advancement of electro-optic effect in ferroelectric thin films: controlling classical ferroelectrics and exploring emerging ferroelectrics (Invited)

T. Yamada*[1]; S. Kondo[2]; H. Funakubo[3]; T. Nagasaki[1]

1. Nagoya University, Japan
2. Okayama University, Japan
3. Tokyo Institute of Technology, Japan

In recent years, novel thin-film-based electro-optic (EO) modulators, which are compact, energy-saving, and high-speed, have attracted a great deal of attention. In such devices, the use of ferroelectric materials, which show large EO coefficients, is preferable compared to that of non-linear polymers from the points of view of long-term stability. However, it is widely observed that the EO coefficients of thin, ferroelectric films are often different from those of the bulks, and these origins have not been deeply understood, unlike the ferroelectric and piezoelectric effects. Therefore, the strategy to enhance the EO effect in ferroelectric thin films has not been established yet. In this presentation, we first theoretically and experimentally demonstrate the influence of strain on the EO effect in classical $(Ba, Sr)TiO_3$ thin films, and discuss the extrinsic contribution to the total EO response in classical $Pb(Zr, Ti)O_3$ thin films. Then, we explore the EO response in emerging HfO_2-based ferroelectric thin films as a candidate material compatible with Si-based CMOS technology.

4:50 PM

(PACRIM-266-2021) Growth and characterization novel single crystals for electro-optical applications (Invited)

K. Shimamura*[1]; E. G. Víllora[1]

1. National Institute for Materials Science, Japan

Here, recent activities of our group related with novel single crystals with advantageous characteristics will be reviewed. A new concept of high-brightness white LEDs based on Ce:YAG single crystal phosphor plates (SCPPs), which can overcome the conventional temperature- and photo-degradation problems, is proposed. SCPPs demonstrated excellent thermal stability with no temperature quenching, high values of luminous efficacy and increased quantum efficiency. TSLAG single crystals have been designed and grown for high-power laser machinery. They showed a higher visible-UV transparency and a larger Faraday rotation than TGG. They are therefore very promising material in particular for new magneto-optical isolator applications. KNN lead-free single crystals are attracting much attention for various piezoelectric applications due to their relatively high piezoelectric coefficient and high Curie temperature. KNN single crystals were successfully grown, and the properties were systematically studied. Authors would like to thank to Koha Co., Ltd., and Fujikura Ltd., for the collaboration.

PACRIM Symposium 42: Young Investigator Forum - Next-Generation Materials for Multifunctional Applications and Sustainable Development, and Concurrent Societal Challenges in the New Millennium

Health: Diagnostics and Therapy Towards Multifunctional Theranostics II

Room: Prince of Wales
Session Chair: Artiom Skripka, Institut National de la Recherche Scientifique

1:30 PM

(PACRIM-267-2021) The pathway to optimized lanthanide-based luminescent thermometers (Invited)

M. Suta*[1]; A. Meijerink[1]

1. Universiteit Utrecht, Chemistry, Netherlands

Luminescence thermometry with lanthanide ions is a novel emerging technique to perform non-invasive and remote temperature sensing with various applications. Despite many potential suggestions for material candidates, a theoretical approach allowing for clear user-friendly guidelines aiming at a removal of the usually encountered trial-and-error principles in that field was lacking up to now. In this talk, I will present and explain those guidelines and what requirements a robust and insightful model has to fulfill in this research area. The strength of such a combined feedback between experiment and theory will be explicitly demonstrated on some examples. It will be shown how the performance of the NIR-emitting Nd^{3+} ion for reliable physiological temperature sensing based on Boltzmann thermometry could be optimized, while Tm^{3+} and Ho^{3+} are examples, for which both modelling measurements of both the temperature-dependent multiphonon transitions and energy transfer allows reliable predictions on optimum doping concentrations for highest precision thermometry for any temperature range of interest.

2:00 PM

(PACRIM-268-2021) Nanocrystalline luminescent thermometer based on transition metal ions (Invited)

L. Marciniak*[1]; K. Kniec[1]; K. Elzbieciak-Piecka[1]; K. M. Trejgis[1]; J. Drabik[1]

1. Institute of Low Temperature and Structure Research Polish Academy of Sciences, Poland

The growing interest in the use of luminescence thermometry (LT) for noncontact temperature reading in very specific conditions imposes the need to develop an approach allowing modification of the luminescence parameters of the thermometer accordingly to the requirements. Therefore, in response to these expectations, novel luminescent thermometers based on the luminescence of transition metal (TM) ions are developed. Spectroscopic properties of the TM including the spectral position of the emission band, a lifetime of the excited state and their susceptibility to luminescence thermal quenching can be easily modified by the crystal field strength and host material composition. This feature of TM is especially desired from the thermometric perspective Therefore during this lecture different strategies of the optimization of the thermometric properties of TM (Cr3+,Mn4+,Mn3+, V3+, V4+, V5+, Ni2+, Ti3+, Co2+) based LT will be presented. The method of knowledgeable modulation of the relative sensitivity of TM-based luminescent thermometers by the intentional modification of the crystal field strength of the host material will be discussed in detail. Acknowledgments: The, High sensitive thermal imaging for biomedical and microelectronic application' project is carried out within the First Team program of the Foundation for Polish Science co-financed by the European Union under the European Regional Development Fund

Health: Diagnostics and Therapy Towards Multifunctional Theranostics III

Room: Prince of Wales
Session Chair: Artiom Skripka, Institut National de la Recherche Scientifique

3:20 PM

(PACRIM-269-2021) Towards controlled photothermal therapy - nanoparticles combining the luminescent nanothermometer and nanoheater functionalities

K. R. Maciejewska*[1]; A. Pasciak[1]; L. Marciniak[1]

1. Institute of Low Temperature and Structure Research, Division of Optical Spectroscopy, Poland

One of the most desirable applications of luminescent nanoparticles (LNPs) is their use in photothermal therapy. In this technique light absorbed by the LNPs is converted into heat through non-radiative processes, in order to overheat the cancer cell to generate its apoptosis. Nevertheless, a side effect of this kind of treatment is related to the risk of overheating of healthy cell neighboring the tumor. The solution of this issue is a real time temperature sensing of the nanoparticles provided by the use of luminescent thermometry- a technique that enables for non-contact temperature reading by using LNPs. Therefore, combining light-to-heat conversion and luminescence thermometry into a single system should present significant advantages for targeting therapies. However, the combination of these two techniques is a great challenge both in terms of synthesis and instrumentation. Therefore, the present study will focus on obtaining biocompatible LNPs comprising in one system the dual role of nanoheaters and luminescent nanothermometers. Hence, the novel multifunctional of NaYF:Cr,Nd nanocrystals for light-induced heating and in-real time temperature monitoring will be presented and their applicative potential will be discussed. This work was supported by National Science Center Poland (NCN) under project no. 2020/37/N/ST5/00536.

3:40 PM

(PACRIM-270-2021) A SBR-based luminescent thermometer exploiting GSA and ESA in Nd^{3+} doped fluorides

K. M. Trejgis*[1]; L. Marciniak[1]

1. Institute of Low Temperatures and Structural Research PAS, Division of Optical Spectroscopy, Poland

Due to its unique properties, such as a noncontact working mode and the possibility of its application even in the most demanding environments, luminescence thermometry has become a rapidly developing and desirable field of science. One of the most popular luminescence thermometers (LTs) were those based on the band-shape, however, they are characterized by certain limitations in the form of the difficulties of spectral separation of bands that are often in close proximity, which limits the obtained temperature resolution of measurements. Moreover, due to the dispersive dependence of the extinction coefficient, each of the analyzed emission bands occurring in different spectral regions may be disturbed in various ways by the medium in which the LT is located, which reduces the reliability of measurement. Recently, a novel single-band ratiometric concept using two excitation beams matched to the absorption from the ground (GSA) and the excited state (ESA) to overcome these limitations has been proposed. Due to the dense ladder of energy levels and the possibility of cross-relaxation, which is favorable for the ESA, Nd^{3+} ions are ideal candidates for this type of LT. In this work, an highly sensitive SBR based LT in Nd^{3+}-doped $NaYF_4$ and $NaGdF_4$ synthesized via co-precipitation method will be presented. This work was supported by National Science Center Poland under project No.UMO-2019/33/N/ST5/00011

4:00 PM

(PACRIM-271-2021) Sensitization of the Ln^{3+}-based thermometers operating in NIR spectral region through Cr^{3+} ions

W. Piotrowski*[1]; L. Marciniak[1]

1. Institute of Low Temperature and Structure Research PAS, Division of Optical Spectroscopy, Poland

The possibility of the subcutaneous temperature determination with the luminescent thermometers (LTs) is especially attractive from the biological perspective since it provides information about thermal gradients in the biological samples in-real time in a non-invasive way. In the case of LTs, the accuracy is strongly dependent on the sensitivity to temperature changes and its emission intensity. Therefore in this work, a novel approach of the sensitization of the near-infrared emitting Ln^{3+}-based LTs with Cr^{3+} ion is discussed. One of the main advantages of using Cr^{3+} as a sensitizer is its high absorption cross section. Additionally due to the energy mismatch between excited states of Cr^{3+} and Ln^{3+} the probability of the $Cr^{3+} \rightarrow Ln^{3+}$ energy transfer is strongly dependent on the temperature. Therefore the enhancement of the emission intensity and the relative sensitivity of the LTs can be achieved. The proposed approach is evaluated on the YAG nanocrystals doped with 1%Nd^{3+}, 1%Er^{3+} and x%Cr^{3+} (where x = 0,1,2,5,10,20). The significant effect of the sensitization process was observed above 5% Cr^{3+} and the effect gradually increases with the sensitizer concentration up to 20% Cr^{3+} for which, S_R = 1.5%/°C was obtained with respect to the 0.2%/°C for Cr^{3+}-unco-doped YAG:Nd^{3+}, Er^{3+}. Acknowledgments: This work was supported by the NCN Poland under project no. DEC-UMO-2018/31/G/ST5/03258.

4:20 PM

(PACRIM-272-2021) Magnetic nanostructures for biological separation, sensing and imaging applications (Invited)

I. Gessner*[1]; S. Mathur[2]; R. Weissleder[1]

1. Massachusetts General Hospital / Harvard Medical School, Center for Systems Biology, USA
2. University of Cologne, Inorganic Chemistry, Germany

Magnetic nanomaterials, most commonly prepared from iron oxides, have become valuable materials for a host of biomedical applications. These materials can be tailored to achieve high magnetic moments, biocompatibility, non-fouling and conjugatability for surface modification. Depending on the biological application, different synthetic pathways have been developed that result in unique magnetic materials with uniform morphology, size, and composition. Besides simple core-only structures, more complex core-shell structures can be produced to obtain dual-functional e.g., plasmonic-magnetic materials desired for sensing applications at the nanoscale. Moreover, the controlled attachment of functional ligands such as antibodies, peptides, small molecules and oligonucleotides can deliver specificity for biomarkers and cellular components. Here, we present on the controlled synthesis of new magnetic nanomaterials based on iron oxide, spinel ferrites and their composites with other materials (SiO_2, Au, polymers). We discuss their use for next generation biosensing and imaging applications.

4:50 PM

(PACRIM-273-2021) Exploiting the biofilm microenvironment to prevent an oral disease via nanocatalysts (Invited)

Y. Huang*[1]

1. National Institute of Health, USA

Human dental caries is an intractable biofilm-associated disease caused by microbial interactions and dietary sugars on the host's teeth. Commensal bacteria help control opportunistic pathogens via bioactive products such as hydrogen peroxide (H_2O_2). However, high-sugar consumption disrupts homeostasis and promotes pathogen accumulation in acidic biofilms that cause tooth-decay. Here, we exploit the pathological (sugar-rich/acidic) conditions using nanocatalysts to increase intrinsic H_2O_2 production and trigger pH-dependent reactive oxygen species (ROS) generation for efficient biofilm virulence targeting. The nanosystem contains glucose-oxidase that catalyzes glucose present in biofilms to increase intrinsic H_2O_2, which is converted by iron oxide nanoparticles with peroxidase-like activity into ROS in acidic pH. Notably, it selectively kills Streptococcus mutans (pathogen) without affecting Streptococcus oralis (commensal) via preferential pathogen-binding and in situ ROS generation. Furthermore, the treatments based on sequential nanocatalysts potently reduced dental caries in a rodent model. The data reveal therapeutic precision of nanocatalysts against a biofilm-related disease in a controlled-manner activated by pathological conditions.

Wednesday, December 15, 2021

PACRIM Symposium 2: Frontier of Modeling and Design of Ceramics and Composites

Modeling and Design of Ceramics and Composites

Room: Regency B

Session Chair: Yu Song, University of California, Los Angeles

8:30 AM

(PACRIM-274-2021) High-entropy rare earth silicates for multifunctional thermal and environmental barrier coating materials (Invited)

J. Wang*[1]

1. Shenyang National Laboratory for Materials Science, Institute of Metal Research, Chinese Academy of Sciences, Advanced Ceramics and Composites Division, China

Rare earth (RE) silicates are promising candidates of multifunctional thermal and environmental barrier coating (TEBC) for CMCs against high temperature corrosions in aviation engines. Current challenges address to the balance of the critical mechanical, thermal and chemical properties of silicates containing single or simple RE species. High entropy or multicomponent design of REs may stimulate new breakthroughs on property modifications for harsh engine environments. This lecture shows our recently progresses on the advancement of outstanding phase stability, tailoring of thermal expansion, and enhancement of CMAS corrosion resistances of high entropy rare earth silicates. The present works provide new perspective on the design of TEBC materials through engineering of complex multicomponent rare earth silicates.

9:00 AM

(PACRIM-275-2021) Ceramics and glasses from atomistic simulations: Enhanced understanding of materials for energy technology

P. Kowalski*[1]; M. Sun[2]; S. C. Finkeldei[3]; Z. He[1]; J. Stackhouse[4]; N. Huittinen[5]; K. Kvashnina[6]; T. Wang[2]; G. Murphy[1]; Z. Zhang[7]; B. Kennedy[8]; E. di Napoli[1]

1. Forschungszentrum Juelich, Germany
2. Lanzhou University, China
3. University of California, Irvine, USA
4. University of California, Berkeley, USA
5. HZDR, Germany
6. ESRF, France
7. ANSTO, Australia
8. University of Sydney, Australia

We will discuss some important success studies of using atomistic modeling methods for prediction of properties of some ceramic and glassy energy materials. We will demonstrate how a careful selection of the computational methodology and an effective combination of theoretical and experimental efforts leads to a superior characterization of materials properties. In particular we will discuss: (1) prediction of redox state and electronic structure of mixed uranium oxides, (2) structural incorporation of radionuclide/Li ions into pyrochlore and phosphate-type ceramics, (3) computational methodologies for d- and f-elements and (4) investigation of radiation damage in ceramics and glasses. We will demonstrate an efficient both-ways beneficial interaction between the experimental and modeling efforts. Last but not least, we will emphasize importance of modern AI and ML algorithms for better understanding of the data, including key structure-property relationships.

9:20 AM

(PACRIM-276-2021) De Novo Inverse Design of Nanoporous Ceramics by Machine Learning

H. Liu[1]; Y. Liu[1]; M. Bauchy*[1]

1. University of California, Los Angeles, Civil and Environmental Engineering Department, USA

Numerical simulations have revolutionized the design of ceramic materials. However, although simulations excel at mapping an input material to its output property, their direct application to inverse design (i.e., mapping an input property to an optimal output material) has traditionally been limited by their high computing cost and lack of differentiability—so that simulations are often replaced by surrogate machine learning models in inverse design problems. Here, taking the example of the inverse design of a porous ceramic matrix featuring targeted sorption isotherm, we introduce a computational inverse design framework that addresses these challenges. We reformulate a lattice density functional theory of sorption in terms of a convolutional neural network with fixed hard-coded weights that leverages automated end-to-end differentiation. Thanks to its differentiability, the simulation is used to directly train a deep generative model, which outputs an optimal porous matrix based on an arbitrary input sorption isotherm curve. Importantly, this pipeline leverages for the first time the power of tensor processing units (TPU)—an emerging family of dedicated chips, which, although they are specialized in deep learning, are flexible enough for intensive scientific simulations. This approach holds promise to accelerate inverse materials design.

10:00 AM

(PACRIM-277-2021) Atomistic simulations of tellurite glasses: Effect of quenching rate, system size and ensemble

A. Jan*[1]; N. Krishnan[1]

1. Indian Institute of Technology Delhi, India

Tellurium oxide-based glasses are among the most promising candidates for integration in non-linear optical devices, such as up-conversion frequency systems and high speed optical- switches. However, these glasses are known to be prone to fast devitrification and hence, it is necessary to understand their formation and properties to be able to take full advantage of their peculiarity. Though a lot of experimental work has been carried out in the field of tellurite glasses but these glasses have not been explored using atomistic modelling. In this work Tellurium oxide glass has been simulated, using the interatomic potentials based on the framework of Born model of ionic solids and further, core shell model. This work explores the effect of system size, quenching rate, and ensemble (NPT and NVT) in terms of connectivity, short-range and medium range order. The glass is estimated to be sensitive to the quenching rate and ensemble choice but not much to the system size and further, this glass is concluded to be weakly connected with the presence of large number of terminal oxygen. Overall, this work bolsters the confidence in the atomistic simulations and presents its importance in providing key insights in understanding the structure and properties of tellurite glasses.

10:20 AM

(PACRIM-278-2021) Fluoride releasing borate glasses for oral health applications

K. N. MacDonald-Parsons[1]; G. E. Boudreau[1]; D. Boyd*[1]

1. IR Scientific, Canada

The potential of fluoridated silicate free borate glasses to release F^- and support tooth mineralization was investigated. A design of mixtures approach was used to model the composition, structure, behaviour relationships of glasses comprising 45-95% B_2O_3, 1-50% MgO, CaO, or Na_2O and 1-30% NaF or KF. Melt quench glasses were produced and characterized through XRD, He pycnometry, and DSC. Mass loss and F^- release was evaluated in a 1:100 mass to volume ratio of Tris buffered saline over 1, 12 and 24 hours. A lead composition comprising 47.5% B_2O_3, 9.5% Na_2O, 14.3% CaO, 19.1% MgO and 9.5% NaF was incorporated into a toothpaste at 5wt% and evaluated for enamel fluoride uptake and enamel microhardness in vitro. Glass formation tendency was favoured by increasing NaF or KF, with weak effects for other components. T_g was increased though CaO addition, and decreased with addition of Na_2O, NaF or KF. Mass loss was favoured by high B_2O_3, Na_2O or KF contents, and decreased with MgO or CaO addition. F^- release was favoured both by K addition as well as Na_2O. Simulated toothpaste application resulted in a F^- uptake of 9.0 $\mu g/cm^2$ and 58.8% recovery of surface microhardness after acid attack, relative to F^- uptake of 24.2 $\mu g/cm^2$ and a 23.3% increase in microhardness with a NaF paste. Silicate free borate bioactive glasses show potential for use as oral care additives capable of releasing F^- and increasing enamel hardness.

10:40 AM

(PACRIM-279-2021) Deciphering the viscosity of glass materials with symbolic regression

Y. Song*[1]; M. Bauchy[1]

1. University of California, Los Angeles, Civil and Environmental Engineering, USA

Viscosity is one of the most investigated properties of glass materials during the past decades. Despite the large number of studies involved with glass viscosity, the huge design space of glass leaves the conventional experimental and simulation methods extremely inefficient for advancing the understanding of how the chemical compositions of glasses affect their viscosity performance. In recent years, machine learning methods have been shown to provide promising solutions for mapping the oxide composition of unknown glasses to their viscosities. However, the commonly used machine learning approaches are usually not easily interpretable and do not work well for extrapolation. Here, we investigated the use of symbolic regression—a machine learning method that yields its model as an analytical expression—to predict the viscosity of different glass systems as a function of their oxide fractions and temperature. Our symbolic regression models for estimating glass viscosity broadly agree with the classic equations that are obtained from established studies. More importantly, this approach greatly facilitates explaining the explicit relationship between each individual model input and the glass viscosity.

PACRIM Symposium 4: Novel, Green, and Strategic Processing and Manufacturing Technologies

Novel, Green, and Strategic Processing IV

Room: Georgia A

Session Chair: Tatsuki Ohji, National Institute of Advanced Industrial Science and Technology (AIST)

8:30 AM

(PACRIM-280-2021) Robocasting (RC) of Ceramic-based Composites (Invited)

L. Wahl*[1]; N. Travitzky[1]

1. University of Erlangen-Nuremberg, Germany

The wider use of advanced ceramic-based composites depends on the availability of industrial processing routes to fabricate parts with required geometries. Since current technology-based processes are not capable of producing complex-shaped ceramic parts with the desired microstructures and properties, Additive Manufacturing (AM) is becoming increasingly important. The mechanical properties of the materials fabricated by AM techniques are in many cases similar to the corresponding properties of commercially available ceramic-based materials fabricated by other methods. In this talk, the emphasis is placed on the fabrication of ceramic-based materials with relatively complex geometries by RC process. Robocasting offers the possibility to fabricate oxide and non-oxide ceramics. The results of robocasted single and multi-material specimens will be shown and the microstructures, physical and mechanical properties of these fabricated materials will be presented. The associated problems, such as the development of residual stresses in the fabricated composites, will be discussed.

9:00 AM

(PACRIM-281-2021) Low-temperature approach for fabricating α-Al_2O_3 composite coating as tritium permeation barrier (Invited)

H. Li*[1]

1. Huazhong University of Science and Technology, China

Over the decades, fusion device has been designed to prove the feasibility of fusion as a large-scale and carbon-free source of energy. In the fusion device, the fusion reaction between two hydrogen isotopes, deuterium (D) and tritium (T), release tremendous amounts of energy in the process. In fusion reactor, structural materials generally have high permeability of hydrogen isotopes. Lessening the tritium permeability of structural materials has been a critical issue in the D-T fusion reactor systems. So far, depositing ceramic coatings on the structural materials has been recognized as the most effective method to address this challenge. Ceramic coatings can function as tritium permeation barrier (TPB) to effectively reduce tritium penetration. Among the emerging materials, α-Al_2O_3 based coatings have been recognized as the most promising TPB due to their ultra high permeation reduction factor and excellent chemical stability. However, α-Al_2O_3 phase was generally attainable at a high temperature above 1000 °C, that can not be borne by the steel substrate. To overcome this issue, new approaches including thermal-chemical reaction method and multi-element-penetration embedding technique were developed, which enabled α-Al_2O_3 composite coatings to be formed at a relatively low temperature, with no deterioration in performance of the steel substrate.

9:50 AM

(PACRIM-282-2021) Coupling precursor chemistry and rapid hot pressing: Toward the design of functional polymer-derived carbide and nitride ceramics (Invited)

M. Balestrat[1]; A. Lale[1]; O. Hanzel[2]; Z. Lences[2]; P. Sajgalik[2]; S. Bernard*[1]

1. CNRS, IRCER, France
2. Slovak Academy of Sciences, Institute of Inorganic Chemistry, Slovakia

Silicon-based non-oxide ceramics (SiC, Si_3N_4) have attracted much attention, primarily due to their good mechanical and chemical properties, and also their reliability at room and elevated temperatures. They have great potential for many industrial uses as engineering components. The addition of a second ceramic (nano) phase to SiC or Si_3N_4 leads to materials which promise applications in many fields and offer solutions for most of the market demands. However, the preparation of these materials is still a challenging task according to the fact that the conventional processes unavoidably lead to size and structure inhomogeneities of the different phases and presence of impurities which affect the properties. Here, we modified polycarbosilanes and polysilazanes with boron and titanium, respectively. Then, the precursors were pyrolyzed into amorphous powders to be sintered without sintering additives to form dense pellets in the Si-B-C and Si-Ti-N systems. In this presentation, the chemistry behind the preparation of powders will be briefly described before investigating the properties of the final ceramics. The latter have been fully characterized and in particular their electrical conductivity, affected by the chemistry behind the preparation of powders, will be discussed.

10:20 AM

(PACRIM-283-2021) Near Net Size Fabrication of Porous Ceramics Through Controlling Shrinkage during drying and sintering (Invited)

C. Wang*[1]

1. Tsinghua University, School of Materials Science and Engineering, China

In the preparation process of porous ceramics, especially in the drying and sintering stages, large volume shrinkage is often accompanied, which affects the high porosity and comprehensive properties of porous ceramics, and increases the subsequent processing costs. Therefore, it is very important to control and reduce the shrinkage in the preparation process and realize near net size preparation for high performance porous ceramics. This report focuses on some specific measures to control and reduce the shrinkage in the preparation process of porous ceramics, so as to realize the near net size preparation of porous ceramics. First, by selecting the solvent with high saturated vapor pressure and low surface tension, the shrinkage of porous ceramic green body in molding and drying is reduced by using gelcasting process. Secondly, by adding shrinkage compensating agent, the shrinkage of porous ceramics can be reduced effectively, and even zero shrinkage can be achieved, so as to realize the near net size preparation of porous ceramics. Finally, in the process of reactive sintering of porous ceramics, low-density materials are formed to generate volume expansion, which counteracts the sintering shrinkage, thus realizing the preparation of near net size porous ceramics.

PACRIM Symposium 7: Synthesis, Processing, and Micro-structural Control of Materials using Electric Currents, Magnetic fields and/or Pressures

Synthesis, Processing, and Micro-structural Control of Materials using Electric Currents, Magnetic fields and/or Pressures

Room: Regency D
Session Chair: Javier Garay, University of California, San Diego

8:30 AM

(PACRIM-284-2021) Creating anisotropic texture in porous ceramics (Invited)

H. Le Ferrand*[1]

1. Nanyang Technological University, Singapore

Texture in ceramics can enhance their functional and structural properties. In porous bodies, it may be used for applications as implants, batteries, filters, etc. Current methods for texturation in dense ceramics employ plate-like particles or microplatelets as starting powders and processes that drive their orientation in specific directions, such as freeze-casting, tape-casting, pressing, or external fields. However, anisotropy in porous bodies is generally created by aligning the pores in specific directions, using anisotropic porogens, sacrificial templates, or freeze-drying. There is thus lesser control of the grain orientation and texture in porous ceramics. Here, we use rotating low magnetic fields to create alumina ceramics with anisotropic grains, having controllable orientations and texture, grain length of 3 to 7 um, open porosity between 30-40% and interesting cracking behaviour.

9:00 AM

(PACRIM-285-2021) Stabilized zirconia ceramics Spark Plasma Sintered: From the study of mechanisms to the control of their microstructure and mechanical properties (Invited)

C. Estournes*[1]; A. Flaureau[1]; A. Fregeac[1]; M. Rousselle[1]; T. Herisson de Beauvoir[1]; G. Chevallier[1]; A. Weibel[1]; F. Ansart[1]; G. Fradet[2]; S. Selezneff[2]; C. Elissalde[3]; F. Mauvy[3]

1. CNRS/University Toulouse, CIRIMAT, France
2. Safran Aircraft Engines, France
3. CNRS/University Bordeaux, ICMCB, France

Ceramics exhibit interesting mechanical properties (hardness, stiffness, wear resistance...) but are penalized by their low toughness (K_{Ic}). However, zirconia based ceramics (ZrO_2) described as "ceramic steel" because, while retaining the usual properties of ceramics, exhibit exceptionally high toughness which paved the way for many applications. SPS allows to densify materials, at lower temperature and shortest time, leading to nanoscale microstructures that are hardly achieved by others methods. Few years ago, we have densified 200 nm grain size yttria stabilized zirconia (3Y-ZrO_2) ceramics exhibiting excellent mechanical properties (s_f = 692 MPa and K_{Ic} = 10.3 MPa.m$^{1/2}$). The presentation will be focused on: – A study of the sintering mechanisms (densification and grain growth) of two stabilized zirconia powders (3- and 8Y-ZrO_2). – In- and ex-situ impedance spectroscopy measurements on 3- and 8Y-ZrO_2, the results are discussed to assess any influence of the sintering parameters and/or specific effects of the SPS process on their electrical conductivity. – The effect of the nature, size and morphology of the grains of YSZ starting powders and SPS conditions on the microstructure of the ceramics obtained and how their mechanical properties are impacted.

9:50 AM

(PACRIM-286-2021) Crack Healing in Zirconia Ceramics under DC Current (Invited)

K. Morita*[1]

1. National Institute for Materials Science (NIMS), Japan

Self-healing techniques were reported in several material systems. For Al_2O_3 dispersed with SiC particles as the healing agent, for example, the cracks were healed by filling with the SiO_2 phases transformed from the healing agent SiC. Since the self-healing takes place automatically at high temperatures, it is a large advantage for the industrial applications. In most of the self-healing techniques, however, it is necessary to disperse second phases as the healing agent. Thus, it is important to develop a new healing technique that does not require any healing agents that form glassy phases. Recently, since current effects are reported to accelerate high temperature phenomena of ceramic materials, it might have a potential to be an effective healing technique. In this work, flash event caused by DC electric field/current was applied to the crack healing of 8mol% Y_2O_3 stabilized cubic ZrO_2 (8Y-CSZ). For 8Y-CSZ, the flash event occurred by applying the DC power higher than a critical value of W_c » 100 mW/mm^3 at 800 °C. For above the W_c value, the micro cracks can successfully be healed under the flash condition within several minutes. The flash healing cannot be explained only by the temperature effect and were accelerated about 2 times through the current-enhanced diffusional processes.

10:20 AM

(PACRIM-287-2021) Fields Matter: Phase and Morphology Control in Iron Oxides Through Magnetic Field Assisted CVD (Invited)

S. Mathur*[1]

1. University of Cologne, Institute of Inorganic Chemistry, Germany

This talk will highlight the importance of field-matter interactions in the synthesis of nanomaterials to control their microstructure, defects and to achieve phase selective synthesis. Application of external magnetic fields during chemical vapor deposition (CVD) offers an independent and additional processing parameter to manipulate the evolution of microstructure and phase composition by influencing atomistic processes involved in the thin film growth. We report here on the role of applied magnetic fields (0.5 T) in altering the mass transport of the iron precursor ($[Fe^{III}(O^tBu)_3]_2$) during a cold-wall CVD process that affected the crystal growth and phase composition. Thin films grown in the presence or absence of external magnetic fields showed striking differences in compositions and preferred growth directions of crystallites. The application of external magnetic field (0.5 T) produced larger particulates with enhanced crystallinity and out-of-plane magnetic anisotropy, as confirmed by microstructural studies (SEM, AFM and XRD) and magnetization (VSM) data. Comparison of the coercive field (H_c) in films obtained in magnetic fields (H_c=11mT) and without external field (H_c=60mT) indicated lower defect concentration in the former case. X ray photoemission electron microscopy for both the field-assisted and zero-field samples confirmed the selective formation of magnetite.

PACRIM Symposium 8: Porous Ceramics: Innovative Processing and Advanced Applications

Additive Manufacturing for Porous Ceramics

Room: English Bay

8:30 AM

(PACRIM-288-2021) 3D printing hierarchical porous ceramics with pastes containing aqueous suspension-oil formulations (Invited)

G. V. Franks*[1]; S. S. Chan[1]; M. L. Sesso[2]; S. Y. Kim[2]

1. University of Melbourne, Chemical Engineering, Australia
2. LaTrobe University, Engineering, Australia

Multiscale porous ceramics are produced by 3D printing (Direct Ink Writing) ceramic pastes. Millimeter scale porosity is created by the 3D printed scaffold filaments. 20 micron scale porosity is produced within the scaffold filaments by adding oil to aqueous suspensions to create pores via either particle stabilized emulsions or capillary suspensions. Micron scale porosity is developed by partial sintering. The rheological (flow) properties of the emulsion or capillary suspension pastes (storage modulus and yield stress) must be carefully controlled to produce paste inks suitable for printing by extrusion through the print nozzle. Control of the internal filament microstructure between particle stabilised emulsions and capillary suspensions is possible by controlling the amount of oil, surfactant and dispersant concentration. The objects become strengthened by sintering at high temperature. Formulations have been developed for alumina, ultra-high temperature ceramics and bioceramic materials. Complex shaped objects can be printed and sintered into crack free components, but distortion during drying and sintering lead to poor shape and tolerance control. X-ray tomography is used to characterize the internal structure of the printed components. 4 point bend strength measurements demonstrate high strength to density ratio.

9:00 AM

(PACRIM-289-2021) Injection Molded Lattice Structures with Variable Poisson Ratio - Influence on Mechanical and Piezoelectric Properties

D. Köllner*[1]; T. Fey[1]

1. University of Erlangen-Nuremberg, Germany

The combination of additive manufacturing and ceramic injection molding allows the production of monolithic lattices with tunable lattice parameters. In particular, honeycomb and auxetic structures, whose Poisson's ratio were set between -2 and +2, were produced in this work. Single hexagons, defined as unit cells, were varied by their angle between -35° and +35°, with negative angles representing an auxetic structure and positive angles an hexagonal. First, unit cells were characterized, followed by lattices with a 2x2 arrangement. Furthermore, different unit cells were combined to obtain lattices with a zero Poisson's ratio. The negative molds of the structures were produced using fused deposition modeling (FDM) and stereolithography (SLA). These molds were filled with Al_2O_3, $BaTiO_3$ and $(Ba_{0.85}Ca_{0.15})(Zr_{0.1}Ti_{0.9})O_3$ by ceramic injection molding. The influence of the lattice parameters on the compresive strength, Young's modulus and the d_{33} were investigated. Piezoelectric properties were determined by direct excitation via the Berlincourt method and indirectly by P-E loops and Digital Image Correlation. Additionally, Finite element methods (FEM) were performed to cover a wider range of parameters and to compare the results with the experiments. The simulations showed an maximum structural Young's modulus between -15° and -10°, independent of the material.

Catalytic Performances of Porous Ceramics

Room: English Bay

9:40 AM

(PACRIM-290-2021) Porous composites containing non-noble metal clusters for catalytic hydrogen production (Invited)

H. Inokawa*[1]

1. Sojo University, Division of Applied Chemistry, Japan

Metal clusters smaller than 5 nm have a great potential for catalytic application. In this presentation, 2 types of porous composites containing non-noble metal clusters are introduced. The one is nickel clusters with an excellent thermal stability supported in and/or on pores of zeolite Y. The Ni cluster-zeolite composite showed 5-10 times higher activity for ammonia decomposition at 500 °C than Ni particles supported on zeolite Y and alumina by a conventional impregnation method. The Ni clusters in/on zeolite pores maintained their small size (< 5 nm) and high dispersion after the ammonia decomposition, whereas Ni particles supported by the conventional impregnation method formed huge (> 50 nm) particles. The other topic is cobalt metal clusters synthesized in interlayer of Layered Double Hydroxide (LDH). Although LDH is a 2-dimensional non-porous material, the composite had a porous structure because metal clusters had a role of pillars to support the hydroxide layers and form slit shape pores. Hydrogen production rates via sodium borohydride hydrolysis reaction were compared among the Co cluster-LDH composites with different size of the Co clusters. The Co clusters with an optimal size showed an excellent hydrogen generation rate, 4520 mL min^{-1} gCo^{-1} at 25 °C.

10:10 AM

(PACRIM-291-2021) Combined CFD and experimental investigations on transport properties of open cellular structures for catalytic applications (Invited)

G. Groppi*[1]

1. Politecnico di Milano, Energy, Italy

Open cellular structures with random (foams) and periodically ordered (POCS) geometry have attracted growing interest as substrate materials for catalytic applications. In fact, their typical geometry with high void fraction and specific surface offers potential in emission abatement catalytic processes, which demand a trade-off between low pressure drops and high conversion rates under mass transfer controlled conditions. Besides the structures can be manufactured with different materials, which combined with the geometrical features allow tuning other transport properties (e.g. thermal and electrical conductivity) of potential interest in chemical processes. To fully assess the potential of open cellular structures as catalytic substrates, a quantitative evaluation is needed. In this talk the results of combined numerical CFD and experimental investigations on gas-solid momentum and mass transfer properties will be described. Engineering correlations for the calculations of pressure drops and mass transfer coefficients in foams and POCS will be presented, which allow to quantitatively assess the trade-off between pressure drops and conversion rates depending on geometry and operating conditions. The issue of geometrical characterization/reconstruction of random open cellular structures will also be addressed.

10:40 AM

(PACRIM-292-2021) Synthesis and Photocatalytic Properties of Tantalum (Oxy)Nitride with an Inverse Opal Structure (Invited)

T. Moriga*[1]; N. HIRAYAMA[1]; K. YOSHIDA[1]; A. NAKANISHI[1]; K. MURAI[1]; W. CHEN[2]; G. WATERHOUSE[2]

1. Tokushima University, Japan
2. University of Auckland, New Zealand

Certain oxynitrides and nitrides demonstrate potentials as visible-light-driven photocatalysts. However these (oxy)nitrides intrinsically possess an abundance of anion vacancies or defects which act as electron-hole pair recombination centers, thereby limiting photocatalytic performance. In this work, we introduce photonic bandgap (PBG) engineering as a strategy for suppressing the recombination in tantalum nitride thin films to improve the performance. Batches of monodisperse polymethylmethacrylate (PMMA) colloid spheres with diameters ranging from 230 to 430 nm were first synthesized by a surfactant-free emulsion polymerization of methyl methacrylate. PMMA colloidal crystal thin films were then grown on silica glass slides using a flow-controlled vertical deposition technique. Several drops of $TaCl_5$ solution were immersed into the colloidal crystal to make the interstitial voids filled with a tantalum hydroxide sol. Then the films were dried and fired at 500°C to be burnt off the PMMA colloids. After an ammonolysis of the films, the obtained IO Ta_3N_5 films possess a 3-dimensionally ordered macroporous structure with periodicity on the scale of visible-light wavelengths, resulting in pseudo PBGs at visible wavelengths. Strong reflection due to [111] PBG was clearly observed at 543nm for the IO amorphous Ta_2O_5 film and at 581nm for the IO Ta_3N_5 film, respectively.

PACRIM Symposium 10: Sol-Gel Processing and Related Liquid-Phase Synthesis of Ceramics

Hierarchical Structuring Method

Room: Regency E
Session Chair: Carl Brozek, University of Oregon

8:30 AM

(PACRIM-293-2021) Morphological Control of TiO_2-based Materials via Mild Solution Processes

G. Hasegawa*[1]; K. Nakanishi[1]

1. Nagoya University, Japan

Morphological control of porous ceramics is a key to produce functional materials with improved performances. Many synthetic methodologies have been developed to fabricate ceramic materials with complex nanostructures while most of them require multi-step processes and/or strong conditions. For example, TiO2-based materials, which show high chemical stability, remain intact even under hydrothermal conditions in 1 M NaOH. Considerably strong conditions like >150 °C in 10 M NaOH are necessary to change the morphology of TiO2 particles into nanotubes. In this study, we have developed mild solution processes with reaction temperatures below 100 °C which allow morphological change of TiO2-based materials. So far, we have found that two kinds of aqueous solutions can be employed to change the morphologies of TiO2; lithium hydroxide and phosphoric acid. By utilizing this technique, we have successfully prepared porous TiO2 monoliths with flower-like architectures on pore surface. Effects of the reaction conditions on final morphologies have been investigated, which reveals that the condition range effective for the morphological change is limited in both systems. The introduced flower-like structures can be preserved at up to 700 °C.

8:50 AM

(PACRIM-294-2021) Scalable Fabrication of Nanostructured Metal Oxides (Invited)

C. Chang*[1]

1. Oregon State University, Chemical Engineering, USA

Microreactor-Assisted Nanomaterial Deposition (MAND) process combines the merits of microreaction technology and the chemical synthesis of nanomaterials. In synthesis, microreactor technology offers large surface-area-to-volume ratios within microchannel structures to accelerate heat and mass transport. This accelerated transport allows for rapid changes in reaction temperatures and concentrations, leading to more uniform heating and mixing in the deposition process. The possibility of synthesizing nanomaterials in the required volumes at the point of application eliminates the need to store and transport potentially hazardous materials while providing new opportunities for tailoring novel nanostructures. MAND processes can be replicated to produce higher chemical production volumes. This critical feature opens a promising avenue in developing scalable nanomanufacturing. Furthermore, the continuous flow microreactor opens up the opportunity to conveniently assemble unique nanostructures and nanostructured thin films. Results-to-date demonstrates the possibility to control the reacting flux, including small intermediate-reaction molecules, nanoclusters, nanoparticles, and structured assembly of nanomaterials. In this presentation, I will discuss our recent progress in using MAND to enable scalable fabrication of nanostructured metal oxides for functional applications.

Porous Low-density Materials (Aerogels)

Room: Regency E
Session Chair: Chih-hung Chang, Oregon State University

9:30 AM

(PACRIM-295-2021) Combining sol-gel-derived thin films with metasurfaces to tailor the optical functionalities (Invited)

S. Murai*[1]

1. Kyoto University, Material Chemistry, Japan

Localized surface plasmon resonances (LSPRs) are plasma oscillations in the metallic nanoparticles that resonate with lightwaves. Large scattering cross section accompanying LSPRs makes metallic nanoparticles particularly attractive optical components in modern nanophotonics. Especially, precise spatial arrangement of nanoparticles on a substrate creates metasurfaces that can control the scattering wavefront to modulate the light, allowing for designing next-generation optical devices. We have been working on metasurfaces in combination with optically functional materials to modulate their optical performances. Typical example includes photoluminescence and upconversion enhancement. In this talk, we will present the combination of metasurface with sol-gel-derived mesoporous silica and nano- Layered double hydroxides (LDHs).

10:00 AM

(PACRIM-296-2021) Dynamic Bonding in Mircroporous Materials (Invited)

C. K. Brozek*[1]; A. Andreeva[1]; J. McKenzie[1]; K. Fabrizio[1]

1. University of Oregon, Chemistry and Biochemistry, USA

Microporous materials attract widespread interest due to their high surface areas, tunable transport properties, and diverse compositions. Zeolites, porous carbon, and metal-organic frameworks (MOFs) represent prominent classes of microporous materials with applications ranging from battery electrodes and separators to heterogeneous catalysts and purification membranes. As solid-state substances, their structures are commonly assumed to be static and rigid, but useful properties often derive from structural dynamics, such as the spontaneous surface reconstruction of metal catalysts and lithium ion diffusion through battery electrode pores. Here, we present general evidence for dynamic bonding in MOF materials long assumed to be static, and explain how the existence of fluctuating structures helps to explain important MOF phenomena. Additionally, we describe how these types of dynamic bonds can be leveraged to design a novel class of phase change materials with transient microporosity. While these topics involve dynamic covalent bonding, we will also describe how dynamic electrostatic interactions between ions and charges dictates the fundamental electrical properties of microporous materials and how ion-coupled charge transport impacts wide ranging technologies.

10:30 AM

(PACRIM-297-2021) Preparation of Polymethylsilsesquioxane/ TEMPO-Oxidized Cellulose Nanofiber Composite Aerogels

D. Kazaoka[1]; G. Hasegawa[1]; K. Kanamori[2]; K. Nakanishi*[1]

1. Nagoya University, Japan
2. Kyoto University, Graduate School of Science, Japan

Based on the poly(methylsilsesquioxane), PMSQ, network, composite aerogels were prepared by a two-step acid-base sol-gel reaction of methyltrimethoxysilane, MTMT, in the co-presence of TEMPO-oxidized cellulose nanofiber, TOCN. Pluronic surfactant was used to surppresses macroscopic phase separation. Wet gels thus prepared, after washing and solvent exchange with 2-propanol, were subjected to supercritical CO_2 drying to obtain PMSQ/TOCN composite aerogels. The PMSQ/TOCN composite aerogels with low bulk density (ca. 0.14 g cm^{-3}) were obtained by optimizing the starting composition and synthetic conditions. The visible light transmittance was found to be as high (ca. 80 %; wavelength 550 nm; 10 mm thickness equivalent) as that of the pristine PMSQ aerogels. The SEM examination confirmed that the fibrous structures were dispersed in the porous structure of PMSQ without showing remarkable aggregation. PMSQ/TOCN aerogels were capable of 50 % uniaxial compression without cracking, and showed high compressive deformation resistance; the Young's modulus is about 1.2-2.5 times higher (ca. 2.5 MPa) than that of the PMSQ/CNF aerogels of the previous study. The uniform dispersion of TOCN having a smaller fiber diameter into the porous structure of PMSQ reinforces the PMSQ network more strongly and increases the Young's modulus of the entire material.

PACRIM Symposium 12: Specific Reaction Field and Material Fabrication Design

Specific Reaction Field and Material Fabrication Design

Room: Plaza B

8:30 AM

(PACRIM-298-2021) Development of Nanoparticle-loaded Organic Scintillators (Invited)

M. Koshimizu*[1]; A. Magi[1]; A. Watanabe[1]; Y. Fujimoto[1]; A. Yoko[2]; G. Seong[3]; T. Tomai[4]; T. Adschiri[2]; R. Haruki[5]; F. Nishikido[6]; S. Kishimoto[5]; K. Asai[1]

1. Tohoku University, Department of Applied Chemistry, Japan
2. Tohoku University, WPI-AIMR, Japan
3. Tohoku University, NICHe, Japan
4. Tohoku University, Institute of Multidisciplinary Research for Advanced Materials, Japan
5. High-Energy Accelerator Research Organization, Japan
6. National Institute for Quantum and Radiological Science and Technology, National Institute of Radiological Sciences, Japan

There is an increasing demand for radiation detectors that can detect high-energy photons at a high counting rate. One such detector is a scintillation detector with a fast scintillator. Plastic scintillators are suitable for this application, but they suffer from the disadvantage of low efficiency for detecting high-energy photons. Therefore, we attempted to improve the efficiency of a plastic scintillator for detecting high-energy photons by sufficiently increasing the amount of high-atomic-number elements in the active region of the detector. This approach overcomes the limitations of plastic scintillators such as their low density and low atomic numbers of constituent elements. In this study, we added nanoparticles of heavy-metal oxides to plastic scintillators to increase the probability of interaction with high-energy photons. An important aspect of this approach is the addition of nanoparticles to organic media with high dispersibility. Surface modification of heavy-metal oxide nanoparticles with organic ligands improves their dispersibility in the organic medium, thus rendering them suitable for this approach. We have successfully developed nanoparticle-loaded plastic scintillators having higher efficiencies for detecting high-energy photons and higher light yields than those of a commercially available Pb-loaded plastic scintillator, EJ-256.

9:00 AM

(PACRIM-299-2021) Hydrothermal Synthesis and Thermochromic Properties of Fluorine Doped Vanadium Oxides for Smart Window Application

S. Yin*[1]

1. IMRAM, Tohoku University, Japan

The monoclinic phase of Vanadium dioxide (VO_2 (M)) shows a reversible metal-to-semiconductor transition (MST) at a critical temperature of 68 °C. Until now, VO_2(M) is the most promising material for thermochromic applications. However, its high critical temperature, low transparency, and undesirable color hindrance the practical application. Metal ions and non-metal anions doping has been shown as an effective strategy to enhance the performance of VO_2. In this research, fluorine-doped VO_2(M) was successfully synthesized to decrease the critical temperature and improve its thermochromic performance. The synthesis was carried out by hydrothermal reaction to obtain fine particles. The XRD patterns of the produced VO_2 particles matched to VO_2 (monoclinic) without any impurity. The actual concentration of the fluorine dopant was measured by XPS analysis, within the range from 0.10 to 0.62 at.%. The fluorine doping decreased the critical temperature of the VO_2. The SEM and TEM images of the obtained samples display a spherical-like morphology with 100-200 nm range diameter. All undoped and doped VO_2 thin films were tested at a low temperature of 25°C and high temperature above the critical temperature at 95°C. The prepared thin films of fluorine-doped VO_2 possessed excellent thermochromic performance, indicating its potential applications as smart window materials.

9:20 AM

(PACRIM-300-2021) Sorption of Co^{2+} on seaweed-like sodium titanate mats synthesized by template-free hydrothermal method (Invited)

T. Goto*[1]; Y. Kondo[2]; T. Sekino[1]

1. Osaka University, ISIR, Japan
2. Osaka University, Graduate School of Engineering, Japan

The contamination to water resources by heavy metals or radionuclides is serious problem for our health and environment fields. Layered sodium titanates, which is one of the cation exchangers, have been widely used for water purification. In this study, we synthesized a sodium titanate mat (SST) using a template-free alkaline hydrothermal method at 200 °C. SST has a unique seaweed-like structure composed of randomly distributed nanofibres of layered sodium titanate. For the sorption test, cobalt(II) nitrate hexahydrate solution of 0.2-4.0 mmol/dm^3 were prepared without pH adjustment. SST sample were shaken in Co^{2+} test solution with various concentration at 25 °C for 1 day. To compare the sorption density, chemical reagent of sodium metatitanate was also tested under the same conditions. The Co^{2+} sorption isotherm of SST was fitted with the Langmuir isotherm model and maximum sorption density was 1.85 mmol/g, and sorption density of SST was higher than that of chemical reagent of sodium metatitanate. The sorption mechanism of Co^{2+} of SST showed the mainly ion-exchange, while that of sodium metatitanate showed the ion-exchange and the precipitation of cobalt hydroxide. These findings indicated that the design of the crystal structure and morphology of cation exchanger is important for the control of sorption properties of Co^{2+} from waste water.

9:50 AM

(PACRIM-301-2021) Ecomomical and Ecological Nanomaterial Fabrication for Sustainable Development Goals

Y. Hayashi*[1]

1. Tohoku University, School of Engineering, Japan

Nanomaterial is useful of energy saving and environmental purification for Sustainable Development Goals. However the practical use of the nanomaterial did not advance very much, because of nanomaterial price. In general, nanomaterial synthesis is low yield and high environment impact, high cost in traditional processing. The innovative processing with new concept is necessary for the promotion in practical use of the nanomaterial. We developed the nanomaterial processing of the new concept to solve a conventional problem. It is our process using liquid-solid reaction which raw materials do not dissolve in a solvent. Liquid-solid reaction enable high concentrated synthesis. Our process uses oxide and metal to solid raw materials as non toxic starting material. Therefore harmful counter anion such as the nitric acid does not exist. Ultrasound and Microwave is used as reaction energy. Everyone can use it as ultrasonic cleaner and microwave oven in laboratory. By these combinations, our process can realize Sustainable Development Goals in the processing of the nanomaterial. In presentation, various nanomaterials (metal, ceramics, nanocomosite) are introduced by application of this concepts for SDGs.

PACRIM Symposium 14: Functional Nanomaterials for Energy Harvesting and Solar Fuels

Functional Metal Oxide Nano- and Heterostructures for Photocatalysis and Solar Fuels

Room: Plaza A

8:30 AM

(PACRIM-302-2021) Bottom-up chemical synthesis of visible-light responsible peroxo-modified titanate nanotubes and their photochemical properties (Invited)

T. Sekino*[1]; Y. Kondo[2]; H. Park[1]; S. Chou[1]; T. Goto[3]; M. Kakihana[1]

1. Osaka University, SANKEN (The Institute of Scientific and Industrial Research), Japan
2. Osaka University, Graduate School of Engineering, Japan
3. Osaka University, Institute for Advanced Co-Creation Studies, Japan

Visible-light responsible titanate nanotubes (TNTs) modified by peroxo functional groups was prepared via facile solution chemical routes by hydrogen peroxide (H_2O_2) treatment of pristine TNT and one-step bottom-up synthesis from peroxo titanium complex ion precursor. Obtained peroxo modified TNT (PTNT) had the cylindrical-tubular morphology with an outer diameter of approximately 10 nm. The peroxo functional group (-O-O-) was formed between the interlayers of the lepidocrocite-type titanate crystal that was the base structure of TNTs, with the interlayer distance estimated at approximately 10 Å. The formation of the peroxo groups exhibited a reduced band gap of 2.6 eV, which is lower than that of pure TNTs (typically 3.3 eV). In addition, the peroxo titanium functional group increased the chemical adsorption performances with positively charged Rhodamine B molecules in water. The photocatalytic performance of the PTNTs was enhanced under the visible light irradiating in comparison with TNTs. It was speculated from the active species detection experiment using various radical trap agencies that the photo-exited electron played a major role for visible-light driven photocatalytic degradation of organic dye molecules. In this paper, detailed synthesis processes, formation mechanisms, physical, chemical and photochemical properties will be discussed.

9:00 AM

(PACRIM-303-2021) Multifunctional Materials for Emerging Technologies (Invited)

F. Rosei*[1]

1. INRS, Canada

This presentation focuses on structure property/relationships in advanced materials, emphasizing multifunctional systems that exhibit multiple functionalities. Such systems are then used as building blocks for the fabrication of various emerging technologies. In particular, nanostructured materials synthesized via the bottom–up approach present an opportunity for future generation low cost manufacturing of devices. We focus in particular on recent developments in solar technologies that aim to address the energy challenge, including third generation photovoltaics, solar hydrogen production, luminescent solar concentrators and other optoelectronic devices.

9:50 AM

(PACRIM-304-2021) Transient absorption studies of photocatalytic water oxidation and reduction at TiO_2 and Pt-TiO_2 surface (Invited)

Y. Tachibana*[1]

1. RMIT University, School of Engineering, Australia

Solar water splitting is one of the most attractive energy generation reactions to utilize solar energy. Light generated hydrogen is a clean energy source, since their consumption for energy generation produces only water. Moreover, generated hydrogen can be stored and used whenever required. Semiconductor TiO_2 is known as a photocatalyst and has been intensively studied for solar water splitting reactions over the last several decades, however has still been employed to assess the capability of photocatalysis reactions. Despite their efforts, mechanisms of the primary water oxidation and reduction reactions at the TiO_2 surface remains to be clarified. In this presentation, we will demonstrate quantitative assessment of primary photocatalytic water oxidation and reduction mechanisms at the TiO_2 or Pt- TiO_2 surface by employing a series of transient absorption spectroscopies covering from 150 fs to 100 s over UV-VIS-NIR wavelength ranges. The reaction paths and their reaction dynamics will be discussed. This work was partly supported by ARC DP fund (DP180103815), ARC LIEF fund (LE170100235) and ARC LIEF fund (LE200100051), Australia. We also acknowledge supports from JSPS KAKENHI Grant Number 19H02813, Japan, and Project Research Center for Fundamental Sciences, Faculty of Science, Osaka University.

10:20 AM

(PACRIM-305-2021) Nanotwins Enable Efficient Photocatalytic Hydrogen Production (Invited)

M. Liu*[1]

1. Xi'an Jiaotong University, International Research Center for Renewable Energy, State Key Laboratory of Multiphase Flow, China

Photocatalysis has long relied on a photosensitive photocatalyst to facilitate the chemical and energy transformations. However, the fact is that available photocatalysts usually have suitable band gap for photo absorption but quite low activity for the conversion. In a recent job of our group, we successfully synthesized a visible-light-active $Cd_{0.5}Zn_{0.5}S$ solid-solution photocatalyst with novel one-dimensional ordered twinned nanostructures. We demonstrated that the nanotwins can effectively separate photo-generated charges and facilitate the transfer of charge carriers and reacting species by coupling twin-induced junction and spatial nanowall effects. Compared to traditional hetero-junction, we define such junction as homo-junction considering that no other components are introduced for its formation. This superiority permits us to tune the band structures as well as the spatial configuration of semiconductors to the specific application in a more precise way. The quantum efficiency of solar hydrogen evolution over this photocatalyst, without noble metal loading, reaches 62%. Our results highlight the power of forming long-range ordered nanotwins at the nanoscale for large-scale activation of the transportation of energy and mass involved in photocatalytic and photoelectrochemical reactions.

PACRIM Symposium 17: Multifunctional Coatings for Structural, Energy, and Environmental Applications

Multifunctional Coatings

Room: Regency A

8:30 AM

(PACRIM-306-2021) Hexavalent chromium-free functional plating alternative technology using aerosol deposition (AD) method (Invited)

J. Akedo*[1]; I. Ohno[2]

1. AIST, ACT, Japan
2. Hojitsu Seiko Co. Ltd., Japan

Plating has generally been used as a rust preventive treatment on metal surfaces. In particular, in hard chrome plating widely used for mechanical parts, a hexavalent chromium solution is often used in order to obtain mechanical strength. However, hexavalent chromium has become a social problem in terms of adverse effects on workers in the processing process and is subject to many regulations as an environmentally hazardous substance in Europe and elsewhere. Even in Japan, the industry has longed for a hard chrome plating alternative technology that does not use hexavalent chromium. On the other hand, the aerosol deposition (AD) method, which utilizes the room temperature impact consolidation (RTIC) phenomenon, is known as a technology that can form a dense ceramic film on a metal substrate at room temperature. It is very promising as an alternative technology to these conventional rust prevention and wear resistance. In this presentation, by examining in detail the conditions such as the surface finish of the base material coated by the AD method and the spray angle of the ceramic particles, we have realized a rust-proof and abrasion-resistant three-dimensional coating without pinholes. The hexavalent chromium is not used at all from the work process at a mass production level. The results of a comparative study with the conventional method will be reported.

9:00 AM

(PACRIM-307-2021) Measurement of in-plane coefficient of thermal expansion for ceramic coatings from room temperature to 1400°C

H. Kakisawa*[1]; T. Nishimura[1]; T. Yokoi[2]; N. Yamaguchi[2]; S. Kitaoka[2]

1. National Institute for Materials Science (NIMS), Japan
2. Japan Fine Ceramics Center, Japan

A measurement method of in-plane coefficient of thermal expansion (CTE) was proposed using optical microscopy. The surface of a mullite coating as an environmental barrier coating for ceramic matrix composites was observed by a blue-laser scanning microscope having a pin-hole during the heating from 25°C to 1400°C. We obtained images of the coating surface with a constant contrast independent of temperature, i.e., regardless of the effect of thermal radiation at high temperature, due to the combination of the blue laser illumination and pin-hole system. Digital image correlation (DIC) was applied to the series of obtained images to measure the thermal strains, and the CTE of the coating in the in-plane direction was determined. Accuracy and reliability of the measurement were also evaluated using a c-plane sapphire plate of which CTE was well established. Surfaces of the coatings adhered on the substrate with various thickness were also observed as well as self-standing coating. The effects of constraint by the substrate and bending of the coating/substrate laminates due to the thermal expansion mismatch on the measured CTEs were discussed. It was suggested that we could estimate the in-plane CTE of the coating by applying a proper laminar model to the analysis of the obtained data, even when we were not able to prepare a self-standing coating.

9:20 AM

(PACRIM-308-2021) Preparation of nanostructured HfO_2–Al_2O_3 coatings using chemical vapor deposition

S. Matsumoto*[1]; A. Ito[1]

1. Yokohama National University, Graduate School of Environment and Information Sciences, Japan

ZrO_2 have been widely studied and used as structural ceramics and thermal barrier coatings due to its thermal insulation properties and high fracture toughness. The mechanical properties of ZrO_2 can be improved by synthesizing composites with alumina (α-Al_2O_3). These composites have been widely reported in the form of bulk specimen prepared using sintering and melt-solidification techniques. HfO_2 has similar properties and crystallographic features as ZrO_2. Because HfO_2 has a higher melting point (3100 K) and higher monoclinic–tetragonal transition point (2200 K) than ZrO_2, HfO_2-based ceramics are candidate materials for thermal barrier coatings. In addition to HfO_2-based ceramics, HfO_2–Al_2O_3 eutectic composites have also been prepared using melt-solidification methods. To make use of the excellent mechanical and thermal properties of HfO_2–Al_2O_3 composites in the form of coatings, a feasible deposition technique must be established. We have previously studied on chemical vapor deposition (CVD) of ZrO_2–Al_2O_3 nanocomposite films, which exhibited high hardness and high thermal insulating properties. In addition, high speed deposition of self-oriented HfO_2 has also been achieved. In the present study, we demonstrated the preparation of HfO_2–Al_2O_3 nanocomposite films, the effects of deposition conditions on phase, and the microstructure of the films was investigated.

9:40 AM

(PACRIM-309-2021) In-situ formation of nanostructures in chemical vapor deposition of MgO–$MgAl_2O_4$–Al_2O_3 system coatings

A. Ito*[1]; M. Ikai[1]

1. Yokohama National University, Environment and Information Sciences, Japan

MgO–$MgAl_2O_4$–Al_2O_3 pseudoternary system has excellent chemical stability and mechanical, electrical, and optical properties, and solid-state reaction of MgAl2O4 solid solution formation between MgO and Al_2O_3 has been widely studied. We employed a laser-assisted chemical vapor deposition technique to demonstrate a surface modification by formation of nanostructures in MgO–$MgAl_2O_4$–Al_2O_3 coatings, and solid solution formation in the system will be discussed through morphological and crystallographic studies. Both MgO rock salt and $MgAl_2O_4$ spinel films showed self-oriented growth with (111) preferred orientations. (111) MgO films had a solid columnar structure with triangular pyramidal caps, while (111) $MgAl_2O_4$ films consisted of nanocrystalline $MgAl_2O_4$ flakes. Mg-rich $MgAl_2O_4$ solid solution films exhibited a similar columnar structure to the (111) MgO film, whereas surface of the columnar grain was modified by formation of nanostructures. The lattice parameters of $MgAl_2O_4$ solid solution decreased as the increase of Al content in the film, and the trendline was lied on the lattice parameter changes between MgO rock salt and γ-Al_2O_3 spinel structures. In-situ topotactic reaction between MgO and $MgAl_2O_4$ during chemical vapor deposition resulted in the surface modification by formation of nanostructures.

10:00 AM

(PACRIM-310-2021) Influence of filler materials on the internal stresses and thermal annealing behavior of ceramic films formed by Powder Aerosol Deposition

D. Paulus*[1]; J. Exner[1]; J. Kita[1]; R. Moos[1]

1. University of Bayreuth, Functional Materials, Germany

The Powder Aerosol Deposition (PAD) method is a novel method to produce ceramic films at room temperature. During the coating process, a ceramic powder forms an aerosol and is transferred to a vacuum deposition chamber. Particles are accelerated by a nozzle and subsequently impact on a substrate where a dense film is formed. PAD films typically bear a high internal compressive stress up to 2.5 GPa as an intrinsic feature. Firstly, this stress may lead to a film delamination in case of high film thicknesses, if the interface of film and substrate cannot withstand the occurring shear stress. Secondly, film stresses cause a significant deterioration of electric properties like the electronic or ionic conductivity. Thermal annealing diminishes the internal stress and regains high electrical properties, yet this treatment opposes to the advantage of PAD of ceramic room temperature processing. In this work, the influence of a filler material on the formation of the internal stress and the annealing behavior of Yttrium stabilized Zirconia (8YSZ) films is investigated. Here, different fillers with varying properties were used to deposit composite films by Aerosol Co-Deposition using powder mixtures. Results describe the change of the internal stress (microstrain) and electrical annealing behavior in dependence of the filler content.

10:20 AM

(PACRIM-311-2021) Enhanced CMAS Resistance and Mechanical Durability of a Next-Gen Thermal Barrier Coating

D. E. Wolfe*[1]; M. Schmitt[2]; A. K. Rai[3]; P. Lauer[1]; P. Albert[1]; R. Spangler[1]

1. Pennsylvania State University, USA
2. HAMR, USA
3. UES, Inc., USA

Thermal barrier coatings (TBC) are commonly applied to gas turbine engine blades to increase operating temperature, efficiency, component lifetime. Modern TBCs use yttria stabilized zirconia (YSZ) top coats, but this material system is susceptible to calcium-magnesia-aluminosilicate (CMAS) spallation. A proposed material system to replace YSZ for improved CMAS resistance is gadolinium zirconate (GZO). However, it is susceptible to erosion, reducing its performance. Novel material systems such as gadolinium aluminum perovskite (GAP) has shown enhanced system performance. GAP has shown improved erosion susceptibility and high CMAS resistance. This work details the thermochemical response of GZO and GAP when exposed to CMAS at varying temperatures and heating times. Reaction mechanisms were analyzed using XRD, SEM, TEM/EDS. Mechanical properties of both materials were evaluated: elastic modulus, fracture toughness, and erosion resistance were determined with resonance ultrasound spectroscopy, indentation fracture, and erosion testing, respectively. When compared to GZO, GAP displayed improved CMAS resistance. GAP samples displayed increased elastic modulus, fracture toughness, and reduced erosion susceptibility. The phase stability at operating temperatures and the thermochemical and mechanical material properties observed suggest that GAP-GZO composite systems could replace YSZ and GZO TBCs.

PACRIM Symposium 19: Geopolymers: Low Energy and Environmentally Friendly Ceramics

Geopolymers

Room: Stanley

Session Chair: Nishant Garg, University of Illinois Urbana-Champaign

8:30 AM

(PACRIM-312-2021) Processing and characterization of fly ash geopolymers for encapsulation of nuclear waste containing cesium (Invited)

S. Jain*[1]; T. Troczynski[1]; N. Banthia[2]

1. The University of British Columbia, Materials Engineering, Canada
2. The University of British Columbia, Civil Engineering, Canada

Improper disposal of nuclear waste can lead to long-term adverse environmental impacts, harming millions of lives by triggering cancerous growths or genetic problems for many generations. In low and intermediate level nuclear waste management, conditioning is a crucial step, which ensures safe embedding of the radioactive waste into a solid matrix, preventing the leakage of radionuclides into the environment. Radioisotopes ^{137}Cs and ^{135}Cs are of paramount concern due to their long half-lives, decay by high-energy pathways, high solubility and chemical reactivity. Traditional low-cost inert binders such as Portland cement-based materials, bitumen, or cement polymer composites for cesium encapsulation are porous, relatively easily leached, and have low thermal stability and poor acid corrosion resistance. In this work, microstructural transformations of NaOH based fly ash geopolymers (NaOH-FAGP) were investigated in terms of the processing parameters (composition, curing conditions), and linked to their ability to encapsulate and retain non-radioactive equivalent of cesium. The Cs leaching and adsorption studies demonstrated NaOH-FAGP reliability as a conditioning matrix for cesium, significantly exceeding that of Portland cement. These results could aid in a wider acceptance of this relatively new class of environmentally-friendly materials for such critical applications.

9:00 AM

(PACRIM-313-2021) Municipal Solid Waste Incineration Ashes – A Precursor for Sustainable Ceramics (Invited)

V. Kumar[1]; N. Garg*[1]

1. University of Illinois Urbana-Champaign, Civil and Environmental Engineering, USA

Municipal solid waste (MSW) management involves three primary practices: landfilling, recycling, and incineration for energy recovery (waste-to-energy or WTE). WTE is a potentially sustainable method of MSW management because it reduces landfilling and generates energy. Incineration reduces input waste mass by 70%. The remaining 30%—in the form of bottom and fly ashes—has to be discarded or landfilled. A main barrier to beneficial use of these ashes is the variability in their composition, which renders them as an unreliable byproduct. In this work, we pursue new characterization approaches to determine the chemical and phase composition of these complex ashes. On one hand, we find that these ashes are similar to other supplementary cementitious materials such as coal fly ashes and blast furnace slags. On the other hand, however, these ashes have non-trivial concentrations of Cl, Pb, S, Zn, and Ti which may affect their performance while substituting Portland cement or formulating geopolymers. Most importantly, we report and define a new parameter based on chemical composition which can accurately predict the mechanical performance of ceramics developed from these ashes. These investigations pave the way for utilizing these materials in various structural applications, opening doors for a truly circular economy.

9:30 AM

(PACRIM-314-2021) Porosity control of acid or basic geopolymer foam (Invited)

S. Rossignol*[1]

1. IRCER, France

This study focuses on the porosity control of geopolymers in order to implement targeted applications. These materials can be synthesized in basic or acid media from aluminum silicate sources in the presence of alkaline or phosphoric solutions respectively. To form foam-like materials, one of the particular points is the synthesis method used to generate porosity. In addition, the various parameters, such as the reactivity of precursors, the introduction of additives or the manufacturing process, governing the formation of the binder are also decisive. The objective of this work is to define the key parameters to be controlled in order to obtain the best porosity possible. Indeed, it is one of the determining factors of thermal property as well as mechanical property. In addition, in order to optimize the structure and microstructure, certain additives may be introduced into the formulations. It should be noted that composition, in particular molar ratios (Al/P and Al/Si and Si/K) appear as the first points. All this data influences the reactive mixture and changes the viscosity and therefore the porosity. For example, a conductivity value of 40-50 mW/K is reached whatever basic or acid geopolymer foamsamue. Finally, a wide range of geopolymer materials has been synthesized up to the TRL4 scale.

PACRIM Symposium 21: Crystalline Materials for Electrical, Optical, and Medical Applications

Optical Materials I

Room: Regency C
Session Chair: Mariola Ramirez, Universidad Autonoma de Madrid

8:40 AM

(PACRIM-315-2021) High-Throughput Synthesis of nano-phosphors by novel water-assisted solid-state reaction method (Invited)

K. Toda*[1]

1. Niigata University, Japan

Many people believe that an ionic-diffusion in an ionic crystal is very slow at room temperature. Therefore, a conventional solid-state reaction (SSR) method requires a high temperature to synthesize the ceramic materials. In order to perform rapid screening for new phosphor materials, we have applied novel soft chemical synthesis technique, water-assisted solid-state reaction (WASSR) method. The WASSR method is very simple and can synthesize nano-materials just by storing or mixing raw materials added a small amount (typically 10wt%) of water at low temperature below 500 K. Typical particle sizes (10-100 nm) of the samples prepared by the WASSR method were smaller than that (2 - 10 micrometer size) of the sample prepared by the SSR method. The most probable mechanism is new type water accelerated solid acid-base reaction on the surface. We successfully synthesized various phosphor materials ($(Cs,Rb)VO_3$, YVO_4:Eu, Ba_2SiO_4:Eu, $LaPO_4$:Ce,Tb, $CsPbX_3$) by the WASSR method. This synthesis technique is a powerful tool for rapid screening of nano-phosphor materials. Acknowledgment This research was partly supported by the Ministry of Trade, Industry and Energy(MOTIE) and Korea Institute for Advancement of Technology(KIAT) through the International Cooperative R&D program (P0006844_Development of color conversion nanocrystal luminescence materials for next generation display).

9:10 AM

(PACRIM-316-2021) Impact of glass crystallization method on the strain, defect formation, and thermoluminescence of YAG: Ce (Invited)

K. Shinozaki*[1]; G. Okada[3]; K. Sato[2]; M. Affatigato[4]

1. AIST, Nanomaterials Research Institute, Japan
2. Tokyo Gakugei University, Department of Environmental Sciences, Japan
3. Kanazawa Institute of Technology, Japan
4. Coe College, USA

The effects of the crystallization path of YAG: Ce on the local structure of its emission site and defect formation were investigated. Ce-doped YAG crystals were prepared in three ways: (1) solid-state reaction, (2) heat-treatment of YAG glass, and (3) crystallization from a supercooled YAG melt by a controlled cooling process from a deeply supercooled state. For the sample obtained by the crystallization of a supercooled melt, the photoluminescence peak red-shifted by 3 nm relative to those of the other samples, and the thermoluminescence increased by a factor of 50–5000. As the thermoluminescence intensity has been associated with the defects formed, positron annihilation lifetime spectroscopy was performed to investigate the defects. The positron lifetimes of all crystallized samples exhibited two components, and the crystallized sample obtained by melt-cooling showed the longest lifetime among the samples.

Optical Materials II

Room: Regency C
Session Chair: Kenji Toda, Niigata University

10:00 AM

(PACRIM-317-2021) Partial air reduction of $Eu^{3+} \rightarrow Eu^{2+}$ in Ca_2SiO_4:Eu phosphors derived from agro-food wastes

M. Kaur[1]; K. Singh*[1]

1. Thapar Institute of Engineering and Technology, School of Physics and Materials Science, India

Rice husk ash and egg shell powder are used to synthesize self-activated, europium doped calcium silicate ($Ca_{2-x}SiO_4$:xEu^{3+} x(%)= 0.25, 0.50, 0.75, and 1.00) phosphors by solid-state reaction method. X-ray diffraction data confirm stabilization of β-Ca_2SiO_4 phase at room temperature. X-ray photoelectron spectroscopy (XPS) results indicate that Eu^{3+} ions undergo partial reduction to Eu^{2+} in the air atmosphere during the processing of the samples. XPS also confirms the presence of titanium ions which are inherently present in agro-food wastes. The photoluminescence emission spectrum contains highly intense peaks of titanium ions which are found in trace amounts in agro-food wastes. In comparison, the emission due to europium ions has very low intensity. The CIE coordinates of the waste-derived samples lie in the green region. This approach presents an alternate use of the widely available agro-food wastes to synthesize phosphors for solid-state lighting applications.

10:20 AM

(PACRIM-318-2021) Plasmon induced coherence and directionality in the spontaneous emission of Rare Earth ions in a solid-state platform (Invited)

J. Fernández-Martínez[3]; S. Carretero-Palacios[3]; L. Sanchez-García[3]; j. Bravo-Abad[1]; P. Molina[3]; N. van Hoof[2]; M. Ramirez[3]; J. Gómez Rivas[2]; L. E. Bausa*[3]

1. Universidad Autonoma de Madrid, Física de la Materia Condensada, Spain
2. Eindhoven University of Technology, Netherlands
3. Universidad Autonoma de Madrid, Spain

In the last years, optically active Rare Earth (RE) ions have been the subject of an intense research activity due to their wide-range applicability in a vast diversity of relevant technological fields. However, a precise control of the emission properties of RE ions at the nanoscale is still a challenge to fulfil the current requirements of emerging fields such as quantum information or optical nanocircuits. In this context, the association of plasmonic structures with RE doped crystals appears as useful approach. Here, we show the ability of a single plasmonic chain formed by closely-spaced Ag nanoparticles to induced extended spatial coherence and directional outcoupling in the emission of Nd^{3+} ions spatially located in the proximity of that chain. The plasmon induced coherence has been experimentally evidenced by the appearance of an interference pattern in the Nd^{3+} emission, and corroborated by numerical simulations, which evidenced the need of a near-field coherent coupling of Nd^{3+} emitters to the plasmon mode supported by the chain. The results open new routes for the manipulation of RE emission and to control the spatial coherence between quantum emitters at room temperature, which could be of interest for the development of novel solid-state platforms for imaging, quantum technologies or sensing applications.

10:50 AM

(PACRIM-319-2021) Synthesis of boron carbon oxynitride films by laser chemical vapor deposition (Invited)

H. Katsui*[1]; K. Harada[2]; M. Hotta[1]

1. National Institute of Advanced Industrial Science and Technology (AIST), Multi-Material Research Institute, Japan
2. Nagoya University, Institute of Materials and System for Sustainability, Japan

Materials related to hexagonal boron nitride (h-BN) and graphite, such as boron carbonitride (BCN) and boron carbon oxynitride (BCNO), are promising for a variety of applications in electronic and biomedical luminescent devices. While the structural similarities between h-BN and graphite have attracted a tremendous amount of attention regarding the development of BCN films with a variable energy band gap, there are few reports on the thin-film synthesis of BCNO which is a novel-rare-earth-metal-free phosphor emitting visible light with the high quantum efficiency and tunable spectra. Chemical vapor deposition (CVD) techniques utilizing laser beams enable the high-speed and conformal deposition of functional thin films with excellent controllability for the structure and compositions of constituent light elements. CVD often suffers from the harmful natures of precursors and by-products; plural species of hazardous gaseous precursors, such as toxic and corrosive halides or diborane as a boron source, ammonia as a nitrogen source, and hydrocarbons as a carbon source, are used for the deposition of B–C–N–O compounds. In the presentation, the vapor-phase growth of BCN and BCNO phosphor films with a turbostratic BN structure will be demonstrated by a laser CVD technique using an alkyl-amino precursor as a single source.

11:20 AM

(PACRIM-320-2021) Luminescence Thermometry - Striving a Breakthrough

E. Zych*[1]; P. Bolek[1]; M. Sójka[1]; D. Kulesza[1]; J. Trojan-Piegza[1]

1. University of Wroclaw, Poland

For the last dozen or so years, the number of publications on luminescence thermometry has been growing steadily and rapidly. In the meantime, this field has split into several parallel research directions. Medical and biological applications set far different requirements for a luminescent thermometer than, for example, space research, aviation, or thermal imaging of electronics. In the former applications, sensitivity is a prime concern, but within a limited temperature range. The latter uses require a wider measuring range and can be satisfied with a lower sensitivity. In our presentation, we will critically discuss possible expectations from luminescent thermometers. We will pay special attention to the issue of luminescent thermometers, which could measure the temperature in a wide range, offering relatively high sensitivity. Physics puts clear limitations here. Can we take a kind of game here that will turn limitations into opportunities? We will be discussing this challenge during our presentation. Supported by the National Science Centre (#UMO-2018/29/B/ST5/00420)

11:40 AM

(PACRIM-321-2021) Loss reduction via annealing and lattice Kerker effect in silicon metasurfaces

L. Liu*[1]; F. Zhang[1]; S. Murai[1]; K. Tanaka[1]

1. Kyoto University, Japan

The resonant phenomena of metasurfaces highly depend on the scattering strength of each component and their interferences. The losses modify the phase and reduce the amplitude of all multipoles, thus loss control is vital for obtaining the designed properties. Amorphous (α-)Si has a refractive index comparable to that of crystalline form while its loss is much higher, limiting its optical application. We found a simple rapid thermal annealing (RTA) path to refine the α-Si metasurfaces. We applied it to the sputtering-made α-Si metasurface comprising a square array of nanodisks. While the large loss smears out the resonances for the as-made metasurface, the sharp and near-zero reflection with near-perfect absorption is achieved after the RTA, satisfying the lattice Kerker condition via the interference of excited magnetic and electric dipoles. The numerical results are all found to be in good agreement with the experimental results, and the multipole expansion analysis for the single nanodisk gives the physical background of this observation. This refinement of α-Si metasurfaces by RTA treatment paves the simple and robust way of realizing thriving optical and optoelectrical applications, such as detectors and sensors.

PACRIM Symposium 24: Solid Oxide Fuel Cells and Hydrogen Technologies

Electrode Materials

Room: Georgia B
Session Chair: Fatih Dogan, Missouri University of Science and Technology

8:40 AM

(PACRIM-322-2021) Nanostructured Composite Electrodes for Redox Stable Solid Oxide Fuel Cells (Invited)

F. Dogan*[1]

1. Missouri University of Science and Technology, Dept. of Materials Science and Engineering, USA

Recent results on development of SOFC electrodes prepared by co-sintering of starting powders and infiltration methods will be reviewed. While co-sintering of powders leads to achieving of relatively high power densities, redox stable SOFCs are obtained by electrode infiltration techniques. Infiltration of a polymeric nickel oxide precursor into a sintered porous yttria–stabilized zirconia (YSZ) skeleton is a promising approach to achieve redox stable cells. In order to ensure that the porous YSZ skeleton was mechanically strong enough to withstand the stresses caused by the volumetric expansion of the Ni phase upon oxidation, a polymeric YSZ precursor was infiltrated into the porous YSZ skeleton, prior to NiO infiltration. It was shown that infiltration of YSZ precursor strengthened the porous YSZ skeleton without compromising from the porosity or initial YSZ particle size significantly. The relationship between the processing, microstructural development and electrochemical performance of SOFC electrodes will be discussed.

9:10 AM

(PACRIM-323-2021) Sandvik Surface Technology: Nanotechnology in large volumes for the rapid industrialization of Solid Oxide Cells (Invited)

C. Bernuy-Lopez*[1]

1. Sandvik Materials Technology, Sweden

Sandvik Surface Technology is a production unit of Sandvik Materials Technology (SMT). It is specialized in the mass-manufacturing of coated stainless-steel strips for various applications such as batteries or fuel cells. Stainless steel-based interconnects for Solid Oxide Cells (SOC) are essential parts in obtaining competitive devices. Pre-coated stainless steels can be produced in large volumes and roll-to-roll processes which brings cost savings in the fabrication of the interconnect. The cost-effective AISI 441 ferritic stainless steel grade can be used with a Ce/Co nanocoating in order to form a protective spinel oxide layer. This material is already manufactured by SMT under the name Sandvik Sanergy® HT 441. This material has shown to form a protective $(Co,Mn)_3O_4$ spinel layer that improves corrosion resistance and heavily reduces Cr evaporation. Therefore, this work will show the solutions that SMT put in the market to industrialize fuel cells. Finally, the progress of the R&D activities regarding SOC interconnects in topics such as Co-free coatings and the self-healing mechanism will be presented.

Electrode Materials and Interconnects

Room: Georgia B
Session Chair: Sebastian Molin, Gdansk University of Technology

10:00 AM

(PACRIM-324-2021) Development of environmentally friendly interconnect coatings - modifications of Mn_2CuO_4

S. Molin*[1]

1. Gdansk University of Technology, Laboratory of Functional Materials, Faculty of Electronics, Telecommunications and Informatics, Poland

The current state of the art interconnect protective coating material is the $(Mn,Co)_3O_4$ spinel. Cobalt is regarded as a carcinogenic element, and is quite costly. Therefore an interest in developing environmentally friendly coatings, which are based on $(Mn,Cu)_3O_4$ spinels. These materials offer high electronic conductivity and good compatibility with typical stainless steels used, but their long-term properties are much less documented. Our group is researching Mn-Cu-oxide-based spinels, including their compositional modifications for phase stabilization. This work will give an overview of our activities.

10:20 AM

(PACRIM-325-2021) Colloidal processing approach toward high performance ceramic fuel cells

K. Sato*[1]; H. Abe[2]

1. Gunma University, Division of Environmental Engineering Science, Japan
2. Osaka University, Japan

This talk offers the current status of colloidal processing based bottom-up technology for better performance of ceramic fuel cells at lower temperature. This technology is based on the synthesis of nanoparticles and nanocomposite with well controlled size and composition. The approach is composed of the following two steps, including growth of aqueous dispersible oxide nanocrystals such as yttria-stabilized zirconia (YSZ) and gadolinium doped ceria (GDC), and subsequent fabrication of nanocomposite particles with the nanocrystals. The nanocrystals play as seed, and facilitate heterogeneous nucleation of other phases in aqueous medium. Solid oxide fuel cells with the nano-composite electrodes fabricated through the approach showed extraordinary high performance at intermediate temperature range. The approach may also be a powerful tool for development of higher performance protonic ceramic fuel cells.

10:40 AM

(PACRIM-326-2021) New approach: The effect of La-doped ceria buffer layers fabricated by ultrasonic spray for electrodes of LSGM electrolyte cells

S. Lee*[1]; T. Shin[1]

1. Korea Institute of Ceramic Engineering and Technology (KICET), Energy & Environmental Division, Republic of Korea

Solid oxide fuel cells (SOFCs) have attracted much attention for energy conversion, due to their highly efficient electricity generation. However, the high operation temperature of SOFC has caused limitations and instability of the materials. Recently, $La_{0.8}Sr_{0.2}Ga_{0.8}Mg_{0.2}O_3$ (LSGM) electrolytes have been studied due to high ionic conductivity. In electrolyte-supported SOFCs, the cell using Ni-based anode and LSGM electrolyte showed degradation in the cell performance because of Ni diffusion causing formation of secondary phase. Thus, inserting of buffer layer between the electrolyte and electrode is considered as an alternative. In this study, a new process, ultrasonic spray was used to form a uniform and thin La doped ceria (LDC), buffer layer and the optimal thickness was investigated through cell performance evaluation. The ideal microstructure of the fabricated cell was performed through SEM analysis. In terms of performance, the cell with buffer layer of 10 μm exhibits the performance of 1.13 W cm^{-2} at 1073 K. In addition, the electrochemical behavior was confirmed through electrochemical impedance spectrum and DRT analysis, in detail.

11:00 AM

(PACRIM-327-2021) Surface passivation enables fabrication of $La_{0.6}Sr_{0.4}Co_{0.2}Fe_{0.8}O_3$ nano-structured electrodes

N. Tsvetkov*[1]; H. Seo[1]; W. Jung[1]; J. Kang[1]

1. Korea Advanced Institute of Science and Engineering (KAIST), Republic of Korea

The development of the stable nanocrystalline electrodes with high surface area that can boost electrode catalytic activity is one of the major goals to for lowering down the operation temperature in the energy conversion systems such as solid oxide fuel/electrolysis cells (SOEFCs). However, perovskite oxides, materials that are typically used as SOEFC electrodes, especially nano-structured materials, suffer from Sr segregation as well as from the particle agglomeration that leads to the decrease of the active surface area and related electrode activity. In this study the effect of the surface passivation with 0.5 nm-thick Hf-based monolayer on the surface and bulk stability of nano-structured $La_{0.6}Sr_{0.4}Co_{0.2}Fe_{0.8}O_3$ (LSCF) electrodes was investigated. We have demonstrated how surface passivation enables the improved electrochemical stability of the passivated LSCF electrodes during operation in humidified air at 650 °C compared to pristine LSCF electrodes. Moreover, for the first time we have demonstrated that the ultra-thin surface passivation layer can prevent agglomeration of LSCF particles at temperatures as high as 1300 °C. Thus, the stabilization of the perovskite oxide surface through passivation can open a new route for designing nanocrystalline electrodes with high surface area for enabling low-temperature SOFCs.

11:20 AM

(PACRIM-328-2021) Optimization of electrophoretic deposition technique to control doping and densification of protective spinel coatings for SOC interconnects

E. Zanchi*[1]; S. Molin[2]; J. Ignaczak[2]; B. Kamecki[2]; G. Cempura[3]; A. R. Boccaccini[4]; F. Smeacetto[1]

1. Politecnico di Torino, Applied Science and Technology, Italy
2. Gdansk University of Technology, Faculty of Electronics, Telecommunications and Informatics, Poland
3. AGH University of Science and Technology, Department of Metallurgy and Powder Metallurgy, Poland
4. University of Erlangen-Nuremberg, Department of Materials Science and Engineering, Germany

Manganese cobaltite spinel coatings have been reported to limit oxidation and Cr-evaporation from ferritic stainless steel interconnects in solid oxide cell stacks; however, the implementation of the functional properties of the base Mn–Co spinel coating and compatibility with the substrate can be pursued through the optimisation of the coating composition, as well as the deposition method and sintering profile. Electrophoretic deposition (EPD) allows to deposit homogeneous layers in few seconds on complexly shaped steel components; it also offers the possibility to produce in-situ doped coatings, avoiding time and energy consuming multi-step processes. In this work, various EPD suspensions are optimised to achieve a single step co-deposition of CuO, Fe_2O_3 and $Mn_{1.5}Co_{1.5}O_4$ on Crofer 22 APU. Different Fe-Cu doped Mn–Co spinel are successfully obtained by controlling the precursors amount in the EPD suspension and subsequent reactive sintering, as proved by detailed SEM and TEM analyses. Improved functional properties of produced coatings are evaluated in terms of oxidation kinetics and area specific resistance. Both the iron and copper amount in the coating and the sintering process significantly influence the coating densification, with benefits to the protective properties and thermomechanical compatibility with the interconnect.

PACRIM Symposium 25: Direct Thermal to Electrical Energy Conversion Materials, Applications, and Thermal Energy Harnessing Challenges

Tellurides

Room: Oxford

Session Chairs: Matt Beekman, California Polytechnic State University; Emmanuel Guilmeau, CNRS CRISMAT

8:30 AM

(PACRIM-329-2021) Thermoelectric Properties of Various Barium Copper Chalcogenides (Invited)

H. Kleinke*[1]

1. University of Waterloo, Department of Chemistry and Waterloo Inst for Nanotechnology, Canada

Binary copper chalcogenides as well as several ternary and quaternary variants have been investigated because of their low thermal conductivity values, high earth abundancy constituents, and tunable electrical properties. In particular, materials such as $Cu_{2-x}S$ and $Cu_{2-x}Se$ have low thermal conductivity values (< 1 W $m^{-1}K^{-1}$) due to their 'liquid like' structures. Unfortunately, these materials often lack stability due to the migration of Cu ions, and therefore efforts have been undertaken to hinder the Cu ion movement, enhancing the stability. Several Ba containing copper chalcogenides such as $BaCu_2Te_2$, $Ba_3Cu_{14-x}Te_{12}$, $BaCu_{6-x}S_{1-y}Te_{6+y}$ and $BaCu_{6-x}Se_{1-y}Te_{6+y}$, $Ba_3Cu_{16-x}S_{11-y}Te_y$ and $Ba_3Cu_{16-x}Se_{11-y}Te_y$ have been characterized in the last decade or so. These materials exhibit typical copper chalcogenide attributes, such as $Cu(S,Se,Te)_4$ tetrahedral frameworks and p-type semiconductivity within a moderate temperature range, in part with peak zT values of the order or even in excess of 1. In this presentation, these materials will be compared to each other, and their properties presented, as well as efforts to stabilize them despite noticeable Cu ion conductivity.

9:00 AM

(PACRIM-330-2021) Defect engineering to enhance thermoelectric performance of GeTe

T. Mori*[1]

1. National Institute for Materials Science (NIMS), Japan

There is a vital need to develop technologies for energy saving and also dynamically harvest energy from surroundings to power IoT applications. Thermoelectrics is a good candidate, and nanostructuring yield some powerful material enhancement principles. Initially trying to achieve a magnetic enhancement effect, Cr doping was tried in GeTe, and had the serendipitous effect to lower the formation energy of Ge defects. This led to creation of homogenously distributed Ge precipitations and Ge vacancies, coupled with typical band convergence doping to lead to ZT~2. A high entropy approach of AgInTe2 alloying into GeTe, stabilized the cubic phase, thereby enabling enhanced doping of Bi, leading to the first stable n-type conduction in GeTe. The hidden role of rhombohedral distortion degree on the Ge-vacancy formation energy was revealed and utilized leading to high power factor and excellent average ZT.
Acknowledgments: Members of JST Mirai Large-Scale Program (JPMJMI19A1) are thanked.

9:20 AM

(PACRIM-331-2021) First principles calculations of electronic and thermoelectric transport properties of $Pb_2Bi_2Te_5$ with different atoms sequences

W. Ma*[1]; P. Boulet[2]; M. Record[1]

1. Aix-Marseille University, IM2NP, France
2. Aix-Marseille University, MADIREL, France

Chalcogenide compounds such as PbTe, $PbBi_2Te_4$, $PbB_{14}Te_7$ and Bi_2Te_3, belong to the homologous series of ternary phases n(PbTe) m(Bi_2Te_3), exhibit both excellent topological insulators and promising thermoelectric properties. For 2(PbTe)(Bi_2Te_3) also read as $Pb_2Bi_2Te_5$, two different possible atoms sequences have been reported: -Te-Pb-Te-Bi-Te-Bi-Te-Pb-Te- (S1) and -Te-Bi-Te-Pb-Te-Pb-Te-Bi-Te- (S2). In this paper, we performed a series of first principles calculations using density functional theory (DFT) to determine electronic and thermoelectric properties of $Pb_2Bi_2Te_5$ with these 2 different atoms sequences. The related compounds PbTe and Bi_2Te_3 were investigated for comparison. Different exchange-correlation functionals (LDA, PBE, EV, TB-mBJ potentials, and rev-vdW-DF2) were tested, w/o spin-orbit coupling, which has been found to have important effects. The calculated electronic bands indicate that both of the 2 sequences lead to indirect band gap semiconductors. We also calculated elastic moduli, dielectric constants, Born effective charges, and phonon dispersion within the quasi-harmonic approximation. Based on the above-mentioned calculations results, thermal conductivity has been obtained by solving the Boltzmann transport equation. The most interesting compound for thermoelectric applications was found to be $Pb_2Bi_2Te_5$ with the S2 sequence.

Emerging Materials

Room: Oxford

Session Chairs: Holger Kleinke, University of Waterloo; Takao Mori, National Institute for Materials Science (NIMS)

10:00 AM

(PACRIM-332-2021) Intermetallic and Chalcogenide Materials for Mid to High Temperature Thermoelectric Applications (Invited)

D. Berthebaud*[1]

1. CNRS LINK, Japan

This presentation should highlight our recent works focus on the development of thermoelectric materials for mid to high temperature range applications. We will present our recent results on intermetallic materials ($CrSi_2$, β-$FeSi_2$, MnSiγ, FeGeγ) which were selected for their low cost and/or advantageous properties. For example, the related Nowotny-chimney ladder compounds MnSiγ (γ ~ 1.73) and FeGeγ (γ ~ 1.52) were studied through high temperature XRD (synchrotron or lab scale) to assess their crystal structure and thermoelectric properties stability. MnSiγ is particularly interesting as it exhibits sufficiently high ZT values (≥ 0.5 at 500 °C) in a wide temperature range, which is an advantage for stable electricity production. $CrSi_2$ is also an interesting p-type alternative materials with similar structural features to above mentioned ones, and we will present here recent results on the improvement of its thermoelectric properties through a slip-casting process under a strong magnetic field. Chalcogenides materials with attractive thermoelectric figure of merit (zT) at high-temperature regime (823 – 873 K) have also been considered such as GeTe and SnTe. Here we will also focus on the case of Ti/Zr doped SnTe which were found to be efficient dopants for SnTe to realize a stable and improved thermoelectric performance for mid temperature regime (323 – 723 K).

10:30 AM

(PACRIM-333-2021) Texturization and improved thermoelectric properties using a magnetic slip casting process - the illustrative case of $CrSi_2$

S. M. Le Tonquesse*[1]; W. Zhang[2]; T. Mori[2]; J. Halet[1]; D. Berthebaud[1]; T. S. Suzuki[2]

1. National Institute for Materials Science (NIMS), LINK LAb, France
2. National Institute for Materials Science (NIMS), Japan

$CrSi_2$ is a promising thermoelectric for industrial application because it is low cost and toxicity. As a bulk polycrystal, it shows an high power factor but the maximum ZT remains moderate (~0.13) due to its relatively high thermal conductivity. For this reason, most of the researches aims at reducing its thermal conductivity via alloying or nanostructuration. An alternative approach is crystallographic texturization along preferential directions. However, conventional thermomechanical texturization methods are not well adapted due to its purely elastic mechanical behavior. Here, an alternative texturization process will be presented which consists in slip casting a dispersion of $CrSi_2$ particles under a magnetic field. This process takes advantage of the anisotropic susceptibility of diamagnetic materials to orient the particles along their easy magnetization axis. In this presentation, this synthesis method will be described in details. Combined structural and texture analyses realized using XRD shows a strong {001} texture along the magnetic field direction. As a result, the directional thermoelectric properties were found to compete with those measured on single crystals. To conclude, a theoretical study based on DFT calculations and aiming at better understanding the origin of the large anisotropy of the transport properties in $CrSi_2$ will be presented.

10:50 AM

(PACRIM-334-2021) Advances in Thermoelectric Ternary and Quaternary Sulphides (Invited)

E. Guilmeau*[1]

1. CNRS CRISMAT, France

Complex copper sulfides can provide an eco-friendly high-performance low cost alternative by using elements that are abundant in naturally occurring minerals. Most of these materials exhibit low thermal conductivity possibly determined by local structural distortions, rattling phenomena, or strong bond anharmonicity. However, the improvement of the TE performances of these materials remains a challenge. The presence of structural defects, the deviation to stoichiometry, the nature of the chemical bonds, and of the ordering/disordering of cations in these complex structures are still a matter of debate, which is of capital importance for the optimization of their TE properties. Our recent investigations on some thermoelectric Cu-based sulphides derived from the sphalerite-type structure will be presented. The talk will be especially focused on colusite Cu26(V,Cr)2(Ge,Sn)6S32 compounds, and the recenltly discovered Cu5Sn2S7 and Cu22Sn10S32 phases. Mechanical alloying, SPS sintering, as well as structural and microstructural features will be reported, together with electrical and thermal properties. Band structure and vibrational properties from first principles calculations will be discussed.

11:20 AM

(PACRIM-335-2021) Realizing high thermoelectric performance in N Type Mg_3Sb_2 Based zintl compounds through multi-element doping strategy (Invited)

L. Miao*[1]; J. Liang[2]; H. Yang[3]

1. Shibaura Institute of Technology, Material Science and Engineering, Japan
2. Guangxi University, China
3. Huaiyin Normal University, China

Mg_3Sb_2-based thermoelectric materials have been attracted more attention in the thermoelectric fields because of their low-cost, earth abundant and environment-friendly constituent elements. Mg_3Sb_2 is an intrinsic p type semiconductor has poor electrical performance and thermoelectric properties because of its low intrinsic carrier concentration. Some results have shown that n type Mg_3Sb_2-based thermoelectric materials possess good thermoelectric properties, but how to efficiently and cheaply prepare n type Mg_3Sb_2-based thermoelectric materials are less systematically studied. In addition, the carrier transport mechanism is less explicated. Therefore, our work developed a synthesis method, which has the advantage of short synthesis period and low cost. Increasing the carrier concentration of n type Mg_3Sb_2-based thermoelectric materials by doping Bi/Se on Sb site and Mn/Y on Mg site, which enhances the power factor and thermoelectric properties.

PACRIM Symposium 26: Materials for Solar Thermal Energy Conversion and Storage

Materials to Produce Synthetic Fuels

Room: Plaza C

8:30 AM

(PACRIM-336-2021) Efficient Photon-harvesting Technologies for Water Splitting Reactions (Invited)

S. Mathur*[1]

1. University of Cologne, Institute of Inorganic Chemistry, Germany

Metal oxide nanostructures with hetero-contacts and phase boundaries offer unique platform for designing materials architectures for energy harvesting applications. As viable alternative to water electrolysis, photoelectrochemical (PEC) water splitting has emerged as a competitive technology being capable of converting solar energy directly into chemical energy using stable and efficient photocatalysts for solar hydrogen production. Besides the size and surface effects, the modulation of electronic behaviour due to junction properties leads to modified surface states that promote selective decomposition of analytes and adsorbates. The growing possibilities of engineering nanostructures in various compositions and forms has intensified the research on the integration of different functional material units in a single architecture to obtain new photocatalytic materials. Even though the potential of hematite thin films for water splitting applications are widely accepted, researchers are still tackling the 'rust challenge'. We report here on the influence of external magnetic fields applied parallel or perpendicular to the substrate during plasma enhanced chemical vapor deposition of hematite nanostructures. Investigations on the water splitting properties of the hematite films in a photoelectrochemical reactor revealed superior photocurrent values of hematite photoanodes deposited in external magnetic field.

9:00 AM

(PACRIM-337-2021) Redox Materials for Solar Thermochemical Fuel Production

M. Roeb[1]; C. C. Agrafiotis*[1]

1. DLR - German Aerospace Center, Germany

Thermochemical two-step processes for fuel production and solar energy storage can be used to enhance the availability of solar energy in terms of energy transport, of energy demand/supply management and of potential energy carrier related applications. Coupling concentrated sunlight to suitable sequences of thermochemical reaction enables the production of fuels like hydrogen or syngas by water- and/or CO_2-splitting as well as the storage of solar energy in suitable reversible reactions. One barrier to technological success of many of those processes is the identification of redox materials exhibiting satisfactory reactivity and durability. It is investigated how key materials will be developed and shaped for the intended use as water and CO2 splitting material on solar tower applications. One part will concentrate on the development of redox materials capable of working in a cyclic mode. A second part will elaborate on methods to fabricate and shape the respective redox materials to the required form, in the form of granules to serve moving bed solar interfaces, and in the form of porous monolithic bodies to enable the realization of volumetric receivers. And in a third part the role of materials in enhancing solar fuels production by thermochemical oxygen pumping is analyzed. The experimental results are complemented with an approach to correlate this performance to material properties, e.g. to the tolerance factor.

9:20 AM

(PACRIM-338-2021) Materials for Solar Thermal Energy Conversion and Long-term Storage in Solid Sulphur

C. C. Agrafiotis*[1]; V. Thanda[1]; D. Thomey[1]; L. de Oliveira[1]; L. Mevißen[1]; H. Noguchi[2]; M. Roeb[1]; C. Sattler[1]

1. DLR - German Aerospace Center, Institute of Solar Research, Germany
2. Japan Atomic Energy Agency, Japan

Specific advanced ceramic materials have been established as the configuration of choice in a variety of solar thermal energy conversion and storage applications, due to their inherent combination of properties like capability for concentrated solar irradiation absorbance, refractoriness and thermal shock resistance. Such a recently proposed application is a thermochemical cycle producing elemental solid sulphur that can be combusted in air to produce high-temperature heat and hence can be used both as a renewable fuel as well as a seasonal solar energy storage means. This cycle shares the high-temperature, potentially solar thermally-driven step of the combined sulphuric acid decomposition/sulphur trioxide splitting, with the other two thermochemical cycles of the sulphur family, the Hybrid Sulphur (HyS) and the Sulphur-Iodine (SI) cycle, conceived to produce hydrogen via water-splitting. In addition to catalyst compositions and reactor designs, the particular "hostile" and corrosive characteristics of sulphuric acid chemistry impose further necessities with respect to the structural materials and components to be employed. Recent progress in materials and technical solutions' design in this field will be presented, articulated into an integrated sulphuric acid decomposer/sulphur trioxide splitting reactor from qualified materials, capable of concentrated solar irradiation-driven operation.

9:40 AM

(PACRIM-339-2021) Synthesis of MnO_2 Carbon Nanotubes catalyst with enhanced Oxygen Reduction Reaction

A. Ullah*[1]

1. University of science and technology, Energy, Republic of Korea

Polymer Electrolyte Membrane Fuel Cell (PEMFC) is an electrochemical cell which undergoes oxygen reduction reaction to produce energy. Platinum (Pt) metal has been used for catalysis since its inception but expensiveness is the major obstacle in commercialization of fuel cell. Herein a non-precious group metal (NPGM) is employed instead of Pt to reduce the cost of PEMFCs. Manganese dioxide based carbon nanotubes (MnO_2 CNTs) is a catalyst having excellent electrochemical properties and offers a better alternative to the Platinum based PEMFC. The catalyst is synthesized by impregnating the transition metal on large surface carbonaceous CNTs by hydrothermal synthesis techniques. To enhance the catalytic activity and increase the volumetric current density, the sample was pyrolyzed at 800 ^{0}C temperature under nitrogen atmosphere. Linear Sweep Voltammetry (LSV) depicts a excellent current density of -4.0 mA/cm^2 and an over potential of -0.3V vs. Standard Calomel Electrode (SCE) in 0.1M KOH electrolyte. Rotating Disk Electrode (RDE) was conducted at 400, 800, 1200, and 1600 rpm. The catalyst exhibited a higher methanol tolerance and long term durability with respect to commercial Pt/C. The results of MnO_2CNT show that the low cost catalyst will supplant the expensive Pt/C catalyst in the fuel cell.

PACRIM Symposium 27: Advanced Materials and Technologies for Electrochemical Energy Storage Systems

All Solid State Batteries/Li/Na-Sulfur Batteries

Room: Balmoral

Session Chair: Palani Balaya, National University of Singapore

8:40 AM

(PACRIM-340-2021) Development of All-Solid-State Lithium-Sulfur Batteries Using Interconnected Mesoporous Carbon (Invited)

A. Sakuda*[1]; M. Tatsumisago[1]; A. Hayashi[1]

1. Osaka Prefecture University, Department of Applied Chemistry, Graduate School of Engineering, Japan

The need for an effective design of composite electrodes in all-solid-state Li-S batteries is warranted because of their slow charge–discharge reactions. By employing a composite of interconnected mesoporous carbon and sulfur, we developed an effective all-solid-state Li-S battery that demonstrated the advantages of exhibiting a high capacity and good cyclability; the cell showed a reversible twoelectron reaction between S and Li_2S under high current densities of more than 1 mA cm^{-2} at 25°C. The recent progress on the study of lithium electrode is also discussed.

9:10 AM

(PACRIM-341-2021) Stack Pressure Effect in Sulfide Electrolyte Based Alkali Metal Solid-State Cells and the Dramatic Implication of Interlayer Growth (Invited)

C. Hänsel[2]; D. Kundu*[1]

1. UNSW Sydney, Chemical Engineering, Australia
2. ETH Zurich, Switzerland

The application of a constant stack pressure during the Li/Na all-solid-state battery (ASSB) cell operation is crucial for intimate electrode-solid electrolyte (SE) contact, which not only minimizes the interfacial resistance but also ensures a homogenous current distribution during Li/Na stripping and plating. A constant stack pressure can also prevent the formation of voids at the alkali electrode interface through the plastic deformation of the soft alkali metals. The required optimal stack pressure depends on the SE's and the electrodes' mechanical properties. Yet, the influence of the stack pressure must not be viewed in isolation, considering that electro-chemo-mechanics of interphase formation and growth also influence the same aspects of the ASSB operation. Through this presentation, we unravel the influence of stack pressure on the cell operation for the sulfide SEs in the context of alkali metal-SE interfacial chemical reactivity. While the cells are prone to shorting by a mechanically induced extrusion of the soft alkali metal even under a moderate stack pressure, chemical reaction at the interface and consequent interphase formation can dramatically influence the proceedings. Our findings can guide in the evaluation of potential approaches to mitigate the mechanically induced short-circuit of alkali metal ASSBs.

9:40 AM

(PACRIM-342-2021) Advanced sulfur cathode for high performance lithium sulfur battery (Invited)

V. Thangadurai*[1]

1. University of Calgary, Chemistry, Canada

Theoretically, lithium sulfur (Li-S) battery provides 5 times energy density (2600 Wh kg^{-1}) than that of conventional Li-ion battery chemistry. A major challenge encountered in Li-S battery is the sluggish cathode interface kinetics. To circumvent the latter, nano materials that can provide active sites for accelerating electrode interface reactions are essential. Herein, we report amorphous MoS_3 that exhibit sulfur-like electrochemical behavior and produces MoS_2 electrocatalyst for propelling redox kinetics in Li-S battery chemistry. MoS_3 cathode was obtained via thermal decomposition of ammonium tetra thiomolybdate ($(NH_4)_2MoS_4$) at 250°C. The cathode demonstrated a long term cyclability of 200 cycles at a rate of 2 C with a gravimetric capacity of ~ 800 mA.h g^{-1}. In addition to that, a homogeneous sulfur composite cathode was synthesized through an energy-efficient deposition method. A stable and highly ionic conductive composite polymer-ceramic electrolyte was employed for a solid-state lithium-sulfur battery (SSLSB). The SSLSB delivered a remarkable specific capacity of 1108 mA h g^{-1} with an areal capacity of 1.77 mA h cm^{-2} and demonstrated stable cycling for 50 cycles. Also, benefiting from fast reaction kinetics, the SSLSB was able to operate at 20 °C and established an excellent rate capability. In this talk, overview of our work on Li-S battery will be presented

PACRIM Symposium 34: Glass and Ceramics for Nuclear Waste Treatment and Sequestration

Waste Form Matrices-Synthesis and Characterization V

Room: Regency F

8:30 AM

(PACRIM-343-2021) Mesoscale model of radionuclide leaching kinetics in hierarchical nuclear waste materials (Invited)

S. Hu*[1]; Y. Li[1]; S. R. Phillpot[2]; N. Shustova[3]; H. zur Loye[3]

1. Pacific Northwest National Lab, USA
2. University of Florida, Department of Material Science & Engineering, USA
3. University of South Carolina, Department of Chemistry and Biochemistry, USA

Different nuclear waste form materials such as Zeolite, Salt-inclusion compounds(SIMS), and metal-organic framework(MOF) were developed to absorb or sequestrate radionuclides. However, their performance including radionuclide capacity, absorption efficiency, and leaching kinetics depend on crystal structure, chemistry, material microstructure and environmental conditions. In this talk, we will present a physics-based mesoscale model of radionuclide transport in hierarchical materials to describe the effect of microstructures, species' diffusivity, chemical potentials, and operational conditions on radionuclide transport kinetics. The thermodynamic and kinetic properties of diffusive species are assessed using first principle calculations. Particularly, the model is applied to study the effect of MOF structures on UO_2^{2+} leaching kinetics. The simulation results show that a simple interface reaction model or concentration gradient driving diffusion model could not describe the UO_2^{2+} leaching kinetics in MOF. To capture the observed UO_2^{2+} leaching kinetics, the effect of concentration dependent chemical potential and mobility has to be taken into account. A parametric study on the effect of thermodynamic and kinetic properties and MOF particle aggregation on leaching kinetics will be presented to demonstrate the model capability of understanding the physics behind the UO_2^{2+} leaching in MOF.

9:00 AM

(PACRIM-344-2021) Synthesis, characterisation and accelerated leaching properties of the zirconolite wasteform for Pu immobilisation

L. Blackburn*[1]; S. Sun[1]; L. J. Gardner[1]; M. C. Stennett[1]; C. L. Corkhill[1]; N. C. Hyatt[1]

1. University of Sheffield, United Kingdom

Around 340 metric tonnes of separated Pu are held internationally, stored largely as unirradiated PuO_2, presenting a need to develop advanced ceramic and glass materials capable of acting as a disposal matrix. Zirconolite, ideally $CaZrTi_2O_7$, is an accessory mineral that has demonstrated affinity for the incorporation of actinides, and is therefore a candidate ceramic matrix for the immobilisation and sequestration of Pu. The zirconolite structure can accommodate moderate Pu loadings through homovalent or heterovalent substitution with suitable charge compensation i.e. $Ca_{1-x}Pu_xZrTi_{2-2x}Fe_{2x}O_7$. The safety case for Pu disposal is underpinned through surrogate trials, developing confidence in the immobilisation approach. In the current work, we summarise an extensive wasteform development campaign, aiming to systematically characterise the incorporation of Ce/U/Th within the zirconolite system, targeting a large array of compositions using both conventional sintering and a batch hot isostatic pressing (HIP) process. In particular, we discuss the influence of surrogate choice and secondary phase fraction on wasteform durability using a novel accelerated acid-leaching approach. This method provided a framework for down selecting of candidate compositions through measuring the extraction of surrogate fissile fraction from the wasteform to the leaching media via ICP-OES.

Waste Form Matrices-Synthesis and Characterization VI

Room: Regency F

10:00 AM

(PACRIM-345-2021) Development of glass-ceramic waste forms for the immobilization of spent fuel and actinide-rich radioactive wastes (Invited)

Y. Zhang*[1]; K. T. Lu[1]; L. Kong[1]; T. Wei[1]; P. Dayal[1]; R. Farzana[1]; D. J. Gregg[1]

1. Australian Nuclear Science & Technology Organisation, Australia

Various actinide-bearing waste streams including spent nuclear fuels present great technical challenges for their immobilization and safe long-term disposal. In general, they have long half-lives and may contain some fissile materials (e.g., U-235 and Pu-239), requiring criticality and safeguards controls. Several glass-ceramic waste forms have been developed for the immobilization of such radioactive wastes. They essentially combine the advantages of conventional borosilicate glasses to maintain their chemical and processing flexibilities to accommodate processing chemicals, with the superb chemical durability of ceramic phases to host actinides. Recent advances in the development of glass-ceramic waste forms include new glass-ceramic systems, actinide validations and processing techniques. We intend to provide an overview on some fundamental research activities at ANSTO focusing on the development of various glass-ceramic systems including formation design, fabrication, processing, and chemical durability studies.

10:30 AM

(PACRIM-346-2021) Cesium adsorption into zeolite-/geopolymer-based composite and thermal effects on immobilization performance (Invited)

V. Proust*[1]; A. Gossard[1]; J. Schaeperkoetter[2]; S. T. Misture[2]; t. david[5]; J. .Amoroso[3]; A. Grandjean[1]; H. zur Loye[4]

1. CEA, France
2. Kazuo Inamori School of Engineering, Alfred University, USA
3. Savannah River National Laboratory, USA
4. Department of Chemistry and Biochemistry, University of South Carolina, USA
5. Univ. Grenoble Alpes, CEA, LITEN, DTNM, LCAE, France

The growing quantity of nuclear waste and the emergence of the serious threats pose challenges researchers to innovate and target new waste form technologies. In the past decades, considerable efforts have been devoted to developing highly selective sorbents followed by safe disposal with the assurance of chemical stability and robust retention performance. Zeolite-containing geopolymers are regarded as a possible 2 in 1 material that is able to both capture and sequester elements such as Cs. These composites show promise for combining zeolite's tunable pore sizes and compositions, high capacity and selective adsorption, with the geopolymer's chemical and thermal stability. For the development of materials devoted to Cs immobilization, we have initially focused on the design of a porous composite consisting of an amorphous geopolymer matrix containing a crystalline zeolite. In addition, we have focused on improving our ability to crystallize the Cs containing composite using hydrothermal techniques and/or calcination. The hydrothermal process provides the benefit of being a low temperature method that helps reduce shrinkage or crack by generating a higher degree of cross-linking, and prepare the matrix for higher temperature treatments. Such heat treatments can simultaneously lower the geopolymer mesoporosity as well as limit the migration of Cs from zeolite to surface.

11:00 AM

(PACRIM-347-2021) Immobilization of Cs^+ on MXene-Hydroxyapatite Composite

S. Lee*[1]; M. ul Hassan[1]; H. Ryu[1]

1. Korea Advanced Institute of Science and Engineering (KAIST), Nuclear and Quantum Engineering, Republic of Korea

There is a volatilization problem with Cs^+ on the adsorbents because of vitrification at high temperatures. Therefore low temperature densification, such as cold-sintering becomes an emerging solution for the immobilization of volatile radioactive isotopes. MXene is two-dimensional transition metal carbides and nitrides, having hydrophilic and many terminate groups (-F, -O, -OH) and is considered a promising adsorbent for the purification of water. Hydroxyapatite has already been proved as an excellent adsorbent because of bi-cationic (Ca^{2+}, P^{5+}) and anionic (OH^-) sites. Moreover, it has been proved that it has good leaching resistance after cold-sintered at a temperature as low as 200 °C. In this study, nanocomposites of MXene and hydroxyapatite are synthesized and cold-sintered for adsorption and immobilization. The goal of the study is that investigating the durability of MXene-Hydroxyapatite with volatile ions, specifically, Cs^+ through long-term leaching tests.

Acknowledgement This study was supported by the KAI-NEET, KAIST.

11:20 AM

(PACRIM-348-2021) The effects of amidophosphonate ligand immobilization method on the uranium extraction efficiency of functionalized silica

A. Dressler*[1]; T. Le Nedelec[1]; A. Leydier[1]; F. Cuer[1]; A. Grandjean[1]

1. CEA, DES, ISEC, DMRC, Univ. Montpellier, France

Various hybrid silica materials were investigated as uranium extractants from sulfuric acid solutions. Silica supports were functionalized with amidophosphonate ligands either by peptide coupling grafting or by non-covalent impregnation. Prior to impregnation, the surface of silica supports was modified with either alkyl chains or ionic liquid chains. The impact of the functionalization methods on the uranium extraction efficiency was evaluated in low sulfate ($[SO_4]^{2-}/[U] = 50$, pH = 2) and high sulfate ($[SO_4]^{2-}/[U] = 900$, pH = 1) solutions, with compositions representative of typical effluents from uranium mines and uranium leaching solutions, respectively. The selectivity of the modified silica supports for uranium was determined in the presence of iron and molybdenum as competing cations. Our results show that both incorporation methods yield materials with good extraction efficiency and selectivity in low sulfate solutions. In high sulfate solutions, the impregnated materials extract uranium more efficiently and more selectively in the presence of Fe and Mo than the grafted one does. Indeed, the non-covalent impregnation of the ligands probably increase their mobility and allows an easier coordination around uranium.

11:40 AM

(PACRIM-349-2021) Hierarchically porous geopolymer-zeolite composites for Sr removal from nuclear wastewater

A. Gossard*[1]; N. Fabrègue[1]; L. Henriet[1]; Y. Barré[1]; A. Hertz[1]; A. Grandjean[1]

1. CEA, ISEC, France

The treatment of Sr-contaminated wastewater is still a challenge for the nuclear industry. For this purpose, extraction on solid supports in fixed bed processes is a promising technology but the development of materials highly selective for Sr with an appropriate porosity is needed. Indeed, contaminated wastewater are often very saline while a hierarchical porosity will increase the sorption kinetic and improve the hydrodynamic properties of the material. We present here the synthesis of meso-macroporous geopolymer containing LTA zeolite nanoparticles (selective for Sr). The macroporosity of the materials is obtained through the stabilization of an emulsion in a solution of geopolymer precursors. After geopolymerization, a solid material is obtained and macropores are created by eliminating the internal phase of the emulsion. Zeolite nanoparticles are introduced as soon as the emulsification step, allowing their homogeneous dispersion in the structure. The mesoporosity is directly induced by the geopolymerization mechanism, which make the zeolite accessible for the wastewater to be treated. The selective removal efficiency of these materials is demonstrated both in batch and column studies as well as the interest of the hierarchically porous network. The sorption mechanisms are highlighted, demonstrating a coupling between ionic exchange and co-precipitation in the material porous network.

Thursday, December 16, 2021

PACRIM Symposium 1: Characterization and Modeling of Ceramic Interfaces: Structure, Bonding, and Grain Growth

Correlation of Interfaces with Macroscopic Properties / Grain Boundary and Interface Structures

Room: Regency A

8:30 AM

(PACRIM-350-2021) Electronic and atomic structures of crystal defect cores determining peculiar materials properties (Invited)

K. Matsunaga*[1]

1. Nagoya University, Materials Physics, Japan

It is well known that point defects, dislocations, grain boundaries and interfaces play essential roles on structural and functional properties of ceramic-related materials systems. Here I would like to share our experimental and theoretical studies on dislocations and interfaces and their related mechanical properties in inorganic compounds. As an example, it was found that zinc sulfide, which is one of II-VI compound semiconductors, shows brittle mechanical behavior in normal surrounding light conditions whereas they display significant ductility in darkness. In this case, a number of partial dislocations with stacking faults are introduced during deformation in darkness, and yet such partial dislocations are not able to move and multiply under external light. This is because external light can induce free electrons and holes via interband transition, and the excited carriers can interact with the partial dislocations. The partial dislocations interacting with carriers can undergo atomic reconstruction at the partial-dislocation cores. The reconstructed partial-dislocation cores can make the dislocations less mobile, which may result in the observed brittle behavior of zinc sulfide. In the presentation, another examples on grain boundaries and interfaces will also be shown.

9:00 AM

(PACRIM-351-2021) Correlation between grain boundary structure and thermal conductivity: A computational and machine learning approach (Invited)

S. Fujii*[3]; T. Yokoi[1]; A. Seko[2]; M. Yoshiya[3]

1. Nagoya University, Department of Materials Physics, Japan
2. Kyoto University, Department of Materials Science and Engineering, Japan
3. Osaka University, Division of Materials and Manufacturing Science, Japan

Revealing grain boundary (GB) structure-property relationships are of great importance for controlling macroscopic properties of polycrystalline materials. Recently, to obtain a comprehensive understanding of these relationships, machine learning (ML) techniques such as structure descriptors and ML potentials (MLP) have been used in conjunction with high-throughput computational simulations. Here, we applied these techniques to phonon thermal transport across ceramic interfaces which are critical for many applications such as thermoelectrics. For highly ionic MgO, we systematically calculated thermal conductivities of 92 GBs using perturbed molecular dynamics (MD) and developed a model for accurately predicting them based on a microscopic structure metric derived from SOAP descriptor. This demonstrates that small distortions of atomic environments at GBs suppress thermal conductivity significantly. For covalent Si, we employed an MLP to a few Si GBs and performed perturbed MD and phonon wave packet simulations. A comparison of these results revealed that thermal conductivities of Si GBs are also highly dependent on their atomic structures probably due to anharmonic vibrations of the GB atoms. Such quantitative and accurate thermal conduction analyses should lead to more precise design of next-generation thermally functional materials.

9:30 AM

(PACRIM-352-2021) Artificial-neural-network interatomic potentials for predicting grain boundary structure and its physical properties (Invited)

T. Yokoi*[2]; A. Nakamura[1]; K. Matsunaga[1]

1. Nagoya University, Japan
2. Nagoya University, Department of Materials Physics, Japan

In polycrystalline materials, grain boundaries (GBs) often govern macroscopic properties through changing atomic structure and chemical composition at GB cores. To reveal the connection between GB atomic structure and its physical properties, density-functional-theory (DFT) calculations have been performed extensively. However, most DFT studies are limited to only symmetric tilt GBs with simple structural units, due to huge computational cost of GB modeling. Thus it is still unclear how atomic structure of general GBs, which is most likely to be more complicated than symmetric tilt GBs, affects polycrystalline properties. With the goal of accurately predicting GB structure and its physical properties with low computational cost, artificial-neural-network (ANN) interatomic potentials were constructed by training them with a large amount of DFT data. These potentials were then integrated into molecular simulation algorithms. Here Si and Al_2O_3 were chosen to be model systems of covalent and ionic crystals. It was found that the error of the ANN potentials with respect to training data sets sufficiently decrease, so that standard molecular simulations can be performed. In addition, the ANN potentials were found to accurately predict atomic structures and energetics of GBs that were not included in the training data set, demonstrating their excellent transferability.

10:20 AM

(PACRIM-353-2021) Atomic-scale STEM-EDS studies of grain boundary segregation in oxide materials (Invited)

B. Feng*[1]; N. Shibata[1]; Y. Ikuhara[1]

1. The University of Tokyo, Japan

Grain boundary (GB) segregation is one of the most important topics in materials science, which has been widely studied so far. With the recent development of aberration-corrected scanning transmission electron microscopy (STEM) and energy-dispersive X-ray spectroscopy (EDS), direct observion of GB segregation is possible at atomic resolution. In this presentation, we will discuss GB segregation in Al_2O_3 and ZrO_2, which represent the cases of non-solute and solute GB segregations, respectively. Al_2O_3 and ZrO_2 model GBs were fabricated by bicrystal methods, and these GBs were studied using atomic-resolution STEM-EDS. In the case of doped Al_2O_3 GBs, it was found that the dopants only segregated at specific atomic sites, which is highly depend on the size mismatch between dopants and Al^{3+}. On the other hand, Y^{3+} segregates at ZrO_2 GB with a width of several nanometers, accompanied with the co-segregation of O^{2-}. These results suggest that in the case of solute segregation, the GB segregation behavior is dominated by the long-range electric interactions between charged GB core and charged point defects. Details will be given in the presentation.

10:50 AM

(PACRIM-354-2021) The effects of grain boundaries and interfaces on defect evolution in ceramics (Invited)

I. Szlufarska*[1]; J. Xi[1]; H. Zhang[1]

1. University of Wisconsin-Madison, Materials Science & Engineering, USA

Engineering of interfaces has been shown a promising way to control the response of materials exposed to harsh environments, including radiation, mechanical stresses, high temperature, and corrosion attack. For example, a large fraction of interfaces can lead to a superior radiation resistance, since interfaces can serve as sinks for radiation-induced defects. In addition, the atomic and the chemical structures of interfaces may evolve when the material is driven out of equilibrium by high temperature or irradiation, which can affect the plethora of properties that are relevant for performance of structural materials in harsh environments. In this talk, I will discuss the role that interfaces play in ceramics in response to radiation and mechanical stresses, focusing on SiC-based ceramics. I will demonstrate that in ceramics, the effects of interfaces on radiation resistance are closely coupled to the complex energy landscape for defect reactions. I will also show that even in ceramics that form line compounds, radiation can cause segregation of constituent elements to grain boundaries without precipitating new phases, which in turn impacts corrosion resistance. Finally, I will demonstrate that radiation can be used as a tool for nano-engineering of interfaces, leading to significant improvements of fracture toughness.

11:20 AM

(PACRIM-355-2021) Co-segregation mechanism of Ca/Si on Σ13 grain boundary in α-Al_2O_3

T. Futazuka*[1]; R. Ishikawa[1]; N. Shibata[1]; Y. Ikuhara[1]

1. University of Tokyo, Institute of Engineering Innovation, Japan

α-Al_2O_3 (alumina) is widely used as a structural material, and their mechanical properties are strongly affected by small amounts of additive elements. Usually, these dopants simultaneously segregate at the grain boundaries (GBs). It is therefore important to understand the atomic structures of the GBs with multiple dopant segregation. In this study, we investigate a Ca/Si co-doped alumina Σ13 <1-210>/{10-14} GB by using scanning transmission electron microscopy (STEM) combined with density functional theory (DFT) calculation. The Ca/Si co-doped alumina Σ13 bicrystal was fabricated by diffusion bonding at 1773 K in air. The observed atomic structure by STEM is significantly different from that of previously reported pristine Σ13 GB. To understand the Ca/Si co-segregation mechanism at the Σ13 GB, we performed systematic DFT calculations and evaluated their GB energies as a function of chemical potentials of Ca/Si. It is revealed that, at the high chemical potential regions of Ca/Si, the Ca/Si co-segregated Σ13 GB has much lower GB energy than the pristine Σ13 GB. Since the ionic radius of Ca^{2+} (1.00 Å) is much larger than that of Al^{3+} (0.54 Å), the Ca segregation at the GB is energetically stable, rather than dissolving Ca ions into the bulk. In contrast, the ionic radii of Si^{4+} (0.40 Å) and Al^{3+} are very close, and Si can therefore segregate to the GB to maintain the charge neutrality between Ca^{2+} and Si^{4+}.

PACRIM Symposium 3: Advanced Structure Analysis and Characterization of Ceramics

Electron Microscopy and Probe Microscopy

Room: Plaza C

8:30 AM

(PACRIM-356-2021) Structural and dynamical investigation of ceramic grain boundaries using atomic-resolution scanning transmission electron microscopy (Invited)

B. Feng*[1]; J. Wei[1]; N. Shibata[1]; Y. Ikuhara[1]

1. The University of Tokyo, Japan

Grain boundaries (GBs) are ubiquitous defects inside polycrystalline ceramic materials, which greatly affect the macroscopic properties. Understanding the GB structure-property relationship is essential for future materials designs. Here, we will show our recent achievements on electronic bandgap characterizations and dynamic observations of α-Al_2O_3 GBs using atomic-resolution scanning transmission electron microscopy (STEM). We first carried out band gap measurements of the GBs by valence electron energy loss spectroscopy (VEELS) in the STEM. We found that the band gap energies of the GBs are generally narrowed compared with the bulk, which is strongly depend on the GB characters. By combing analysis with core-loss EELS and density functional theory calculations, it is revealed that the band gaps are closely related to the coordination numbers of Al and O at the GB cores. The second part of this presentation is to show a direct imaging of atomistic GB migration in α-Al_2O_3. We used electron beam irradiation in STEM to controllably triggered the GB migration, which allows us to directly observe the atomistic GB migrations as a stop motion movie. It was revealed that the GB migrates through a cooperative shuffling of atoms in GB ledges along specific routes. As a result, the GB passed through several GB structures with similarly low GB energies.

9:00 AM

(PACRIM-357-2021) Atomic structure and physical properties of boundaries in iron oxides (Invited)

C. Chen*[1]

1. Institute of Metal Research, Chinese Academy of Sciences, China

Clarifying how the atomic and electronic structures of boundaries affects their electrical and magnetic properties represents one of the most intriguing issues in materials science, which is of significant importance not only for understanding the electrical and magnetic behaviors of polycrystalline materials but also for the development of advanced electronic and magnetic devices. In this study, by combining atomic-resolution scanning transmission electron microscopy (STEM), differential phase contrast STEM imaging and first-principles calculations, we systematically investigate the atomic and electronic structures, the electrical and magnetic properties of antiphase boundaries, grain boundaries and heterinterfaces in iron oxides. It is revealed that the electrical and magnetic properties of the boundaries in iron oxides directly depend on their atomic and electronic structures within a few atomic layers.

9:30 AM

(PACRIM-358-2021) Investigating solution-processed ceramics with cryo-electron microscopy

N. S. Dutta[1]; C. B. Arnold*[1]

1. Princeton University, USA

Solution processing is commonly used to prepare both crystalline and amorphous ceramics for device applications. One benefit of the method is its flexibility in solvent choice, precursor materials, and additives, but with this also comes complexity. Though it has been observed in several systems that solution chemistry affects the morphology and properties of the deposited material, the manner in which these effects arise is rarely well understood, due largely to the experimental difficulty of characterizing materials in solution. Here we demonstrate how cryo-electron microscopy (cryo-EM) can be used to overcome this challenge, sharing results from amorphous (chalcogenide glass) and crystalline (metal-halide perovskite) ceramic materials. We show multiple cryo-EM analyses, including single particle analysis adapted from biology, and discuss how each can reveal unique information about the material. Our results offer a new way to characterize structural evolution in solution, enabling better informed processing of ceramic materials.

10:10 AM

(PACRIM-359-2021) Cracking and doping effect in layered cathode for sodium-ion battery

K. Wang*[1]; P. Yan[1]; M. Sui[1]

1. Beijing University of Technology, China

Layer structured sodium transition metal oxides are an important group of cathode material for sodium ion battery due to their high theoretical capacities. However, the poor cycling stability at high voltage hinders their practical applications, which is due to the multiple phase transition induced structural instability. Our recent work reveals that phase transition induced cracking is a major cause of the performance decay for layered cathodes. Bulk elemental doping has been demonstrated to be an effective approach to improve the cycling performance. Our recent investigation shows that uniformly distributed Mg dopants can migrate and aggregate into nano-sized precipitates during high voltage cycling of an Mg-doped P2-type layer-structured cathode. Intriguingly, such a dopant aggregation behavior can further enhance the cycling stability. We herein propose a precipitation strengthening strategy to enhance the mechanical properties and cycling stability of a layered cathode. We validate such a new concept by observing more commonly used dopants, Cu-doped, Ti-doped, and Zn-doped samples. We further figured out some characteristics for selecting a good dopant.

10:30 AM

(PACRIM-360-2021) Microanalysis of cracking failure in layered cathodes for rechargeable battery (Invited)

P. Yan*[1]

1. Beijing University of Technology, China

Layered alkaline transition metal oxides (LATMO), with their superior electrochemical performance, have been utilized as cathode materials in rechargeable batteries. LATMO are high energy density cathode materials with large potential to be unlocked. However, it is very challenging for LATMO to maintain high cycling stability under high specific capacity. Cracking ralated degradations plague their electrochemical performance significantly. Therefore, it is highly urgent to solve the cracking issue in order to further explore their capacity potential as well as the safety issues. In this talk, we applied Cs-corrected transmission electron microscopy and other microanalysis tools to investigate the cracking failure mechanisms in LATMO materials for lithium ion battery and sodium ion battery. We compare many LATMO materials, such as LiCoO2, ternary NMC, P2-NaNiMnO and other variants, to unveil the atomic mechanism of crack nucleation and growth process, which paves the way to finally solving cracking degradations.

11:00 AM

(PACRIM-361-2021) Characterization of MAX phases under irradiation (Invited)

C. Wang*[1]

1. Peking University, School of Physics, China

Due to the unique structures and properties, $M_{n+1}AX_n$ phases have been proposed for use in a wide variety of industrial applications, including in nuclear reactors, where they will be exposed to extreme irradiation conditions. In this talk, I will present the direct imaging of ion irradiation-induced antisite defects in $M_{n+1}AX_n$ phases using double C_S-corrected scanning transmission electron microscopy (HAADF and ABF) and provide compelling evidence of order-to-disorder phase transformations, overturning the conventional view that irradiation leads to phase decomposition from ternary $M_{n+1}AX_n$ phases to binary fcc-structured $M_{n+1}X_n$ phases. With the formation of uniformly distributed cation antisite defects and the rearrangement of X anions, disordered solid solution γ-$(M_{n+1}A)X_n$ phases are formed at relatively low ion fluences, followed by gradual transitions to solid solution fcc-structured $(M_{n+1}A)X_n$ phases. Our findings are validated by elemental distributions in atom probe tomography (APT) reconstructions and the analysis of the solid solutions' unit cell parameters obtained from the HRTEM measurements, the corresponding SAED patterns, and ab initio calculations. I will also discuss the role of different structural and compositional factors on the radiation tolerance of these materials, based on experimental and simulation results. This study provides a comprehensive understanding of the radiation responses of $M_{n+1}AX_n$ phases.

11:30 AM

(PACRIM-362-2021) Effect of Zn Doping on Microstructure Evolution in Cu_2SnS_3 Ceramics (Invited)

R. Huang*[1]

1. East China Normal University, China

Ternary compound Cu_2SnS_3 has a direct band gap of 0.9~1.3 eV, existing as one of three possible polymorphs with monoclinic, cubic, or tetragonal symmetry depending on the arrangement of its corner-sharing MS_4 (M = Cu or Sn) tetrahedral. When alloyed with ZnS, the microstructures of these materials become considerably more complex. We prepared $Cu_2Zn_xSn_{1-x}S_3$ (x = 0.00, 0.05, 0.10, 0.15, 0.20) ceramics, which exhibits good thermoelectric performance. Atomic resolution Cs-corrected STEM and EDS mapping techniques have been used to characterize the micro- and nanostructures, including secondary phases, local cation ordering, and crystal domains. A distinctive mosaic-type nanostructure consisting of coherently bonded domains with fully disordered and semi-ordered cation sublattices was observed. The semi-ordered domains correspond to a previously unknown phase in the Cu-Zn-Sn-S system. Furthermore, the microstructure evolution with the increase of Zn doping level was revealed in the $Cu_2Zn_xSn_{1-x}S_3$ ceramics. The mechanisms responsible for the very low lattice thermal conductivity and good thermoelectric performance of these materials were also discussed.

PACRIM Symposium 5: Polymer Derived Ceramics (PDCs) and Composites

Precursors and Properties

Room: Plaza B

Session Chair: Peter Kroll, University of Texas, Arlington

8:30 AM

(PACRIM-363-2021) Microstructure engineering of ligand-modified ceramic membranes derived from network-forming precursors (Invited)

M. Kanezashi*[1]

1. Hiroshima University, Chemical Engineering Program, Graduate School of Advanced Science and Engineering, Japan

An easy template was devised for the fabrication and microstructural control of ligand-chelated, silica-zirconia-based composite membranes. Hybrid materials with crosslinking between the organic moieties were successfully fabricated via the sol-gel co-condensation of 3-(trimethoxysilyl)propyl methacrylate and zirconium (IV) butoxide chelated via acetylacetone and allyl acetoacetate ligands using an azobisisobutyronitrile radical initiator as a crosslinking agent. Characterization of the films and gels via FTIR, XRD and thermogravimetry indicated that both the polymerization and hybridization of the networks was achieved. Gas permeation experiments showed that the organic crosslink-derived membranes possess molecular sieving properties (H_2 permeance: ≈10^{-7} mol m^{-2} s^{-1} Pa^{-1}; H_2/N_2 selectivity: ≈20-30) that are superior to those of non-crosslink-derived membranes (H_2 permeance: ≈10^{-7} mol m^{-2} s^{-1} Pa^{-1}; H_2/N_2 selectivity: ≈7-11).

9:00 AM

(PACRIM-364-2021) Novel Polymer derived b-SiAlON:Eu^{2+} green phosphors

Y. Gao*[1]; J. Iihama[1]; D. Hamana[1]; T. Asaka[1]; Y. Daiko[1]; S. Honda[1]; S. Bernard[2]; Y. Iwamoto[1]

1. Nagoya Institute of Technology, Life Science and Applied Chemistry, Japan
2. University of Limoges, CNRS, IRCER, France

A series of b-SiAlON:Eu^{2+} green phosphors were synthesized through the Polymer-Derived Ceramics (PDCs) route. Single source precursors for the b-SiAlON:Eu^{2+} phosphors were synthesized by chemical modification of perhydropolysilane (PHPS) with Al(Oi-$C_3H_7)_3$, $AlCl_3$ and $EuCl_2$, then pyrolyzed at 1000 °C under flowing NH_3 to give X-ray amorphous Si-Al-Eu-O-N compounds. And then the polymer-derived multicomponent amorphous compounds were heat-treated at 1850 °C in N_2 (980 kPa) to give polycrystalline b-SiAlON:Eu^2. The polymer-derived b-SiAlON:Eu^{2+} phosphors exhibited a unique green luminescent property with a wider excitation wavelength ranging from 200 to 500 nm. The detailed PDCs route during the synthesis was studied by the simultaneous TG-MS analyses combined with FT-IR and NMR spectroscopic analyses. The results revealed the following two reactions during the precursor synthesis: non-hydrolytic sol-gel reaction between Al(Oi-$C_3H_7)_3$ and $AlCl_3$, and the Friedel-Crafts reaction by i-C_3H_7Cl formed in-situ and toluene used as reaction solvent, which contributed to maintaining oxygen content and reduction of residual carbon in the phosphors. At the presentation, the relations between the chemical composition of the polymer-derived X-ray amorphous Si-Al-Eu-O-N compounds, morphology and the unique photoluminescence properties of the b-SiAlON:Eu^{2+} will be shown and discussed aiming to develop novel green phosphors through the PDCs route.

Processing and Simulations

Room: Plaza B

Session Chair: Corson Cramer, Oak Ridge National Lab

10:00 AM

(PACRIM-365-2021) Insights into the in situ formed carbon in polymer-derived ceramics (Invited)

Q. Wen*[1]; Z. Yu[2]; X. Xiong[1]; R. Riedel[3]

1. Central South University, Powder Metallurgy Research Institute, State Key Laboratory of Powder Metallurgy, China
2. Xiamen University, College of Materials, Key Laboratory of High Performance Ceramic Fibers, China
3. Technische Universität Darmstadt, Institut für Materialwissenschaft, Germany

Polymer-derived ceramics (PDCs) have been intensively studied for nearly 50 years due to their unique advantages to produce ceramic fibers, coatings, foams, nanocomposites and additive manufacturing. A phenomenon associated with the polymer-to-ceramic transformation process using organo-substituted silicon polymers as the starting material has been widely reported, namely, in situ formation of carbon within the generated silicon-based ceramic matrix. Interestingly, the precipitation of carbon depends to a great extent on the molecular structure of preceramic polymer and significantly affects the composition, crystallization and decomposition behavior, microstructural evolution as well as the related structural and functional properties of PDCs. Thus, this talk will highlight the recent progress in the PDC field with the focus on the fate and role of the in situ formed carbon. Firstly, a brief summary of the synthesis and processing of PDCs will be provided, followed by the microstructural characterization of the formed ceramics. The in situ formation of carbon, precursor-carbon-morphology relation and high-temperature evolution of the carbon will be summarized. Secondly, the influence of the segregated carbon on the microstructure and its associated properties of the PDCs will be discussed. Finally, potential advanced structural and functional applications of the PDCs related to the carbon will be presented.

10:30 AM

(PACRIM-366-2021) Ab-Initio Simulations of Polymer Pyrolysis

P. Kroll*[1]

1. University of Texas, Arlington, USA

We present ab-initio Molecular Dynamics (aiMD) simulations of the thermal conversion of polysilanes, polysiloxanes, and polysilazanes into amorphous ceramics. Models comprising 200 to 1000 atoms are investigated at elevated temperatures in simulations running for 100 ps and longer. We observe atomistic details of the Kumada rearrangement in polymethylsilanes, and similar insertions of C into the backbone of siloxanes or silazanes are followed. These processes yield mixed tetrahedral environments of Si atoms, as has been observed in previous 29Si NMR investigations. We, furthermore, discern a variety of mechanisms that develop gaseous species, in particular hydrogen and methane. Through removal of gaseous specious in constant-pressure simulations we follow mass-loss and densification of each one polysiloxane and polysilazane model. Though simulation conditions like model size, temperature, and removal process need to be considered, compositions attained after 2 ns of "high-temperature annealing" closely resembles experimental data.

10:50 AM

(PACRIM-367-2021) Reactive Force-Field Simulations of Pyrolysis of Polysiloxanes

P. Kroll*[1]

1. University of Texas, Arlington, USA

We explore pyrolysis of polysiloxanes using reactive force field (ReaxFF) simulations of systems extending 5–10 nm for 5–100 ns. We observe that formation of carbon segregations in amorphous SiCO is linked to early stages of polymer degradation when organic and inorganic portions of the polymers partition and segregate. Reactions within the organic portion then yield sheet-like or tubular carbonaceous segregations. As the "free" carbon structure forms, it leaves behind a SiCOH matrix and SiC-rich regions. Processes within the residual SiCOH matrix occur at much slower timescales than the early carbon genesis, however. We find one example in which a carbon nanotube templates the formation of a SiC-corona surrounding it. The simulations provide insight into possible structure formation in SiCO ceramics, in particular the development of a free carbon phase, its interface with the surrounding SiCO matrix, and the impact of precursor chemistry on morphology of SiCO ceramics.

11:10 AM

(PACRIM-368-2021) A reactive force field (ReaxFF) for simulation of select Si-based polymer-derived ceramics

S. Haseen*[1]; I. Ponomarev[2]; P. Kroll[1]

1. University of Texas, Arlington, Chemistry and Biochemistry, USA
2. Czech Technical University in Prague, Czechia

Polymer-derived ceramics are materials that exhibit many desirable properties such as high-temperature stability, oxidation resistance, and creep resistance. We develop a reactive force field (ReaxFF) for large-scale simulation of the polymer-to-ceramic transformation for certain polymer-derived ceramics (PDCs) with density functional theory-like (DFT) accuracy. Using our extensive library of hypothetical crystalline and amorphous structures, we first aim to reproduce DFT energy differences within ReaxFF. Next, we further optimize parameters by matching ReaxFF molecular dynamics energies and forces with those of DFT ab initio molecular dynamics. In order to evaluate and improve our parameters, we generate melt-quench (MQ) ReaxFF models and optimize these models within DFT in a feedback loop that is used to augment the training set. We use our final set of ReaxFF parameters to investigate the thermal conversion of polymers to PDCs. During the conversion process we track reactions that occur and analyze the resulting structures, including the different carbon structures that may form.

11:30 AM

(PACRIM-369-2021) Silicon Oxycarbide-Based Composites for Li-Ion Batteries: Correlation of Ceramic Microstructure with Electrochemical Properties (Invited)

M. Wilamowska-Zawlocka*[1]; D. Knozowski[1]; M. Graczyk-Zajac[2]

1. Gdansk University of Technology, Department of Energy Conversion and Storage, Poland
2. Technische Universität Darmstadt, Fachbereich Material und Geowissenschaften, Germany

Silicon oxycarbide (SiOC) ceramics have been evaluated as promising anode material for Li-ion batteries as they exhibit high capacities, slight volume changes during charge/discharge processes, and stability. The main drawback, however, is a high irreversible capacity of the first cycle. One of the approaches to tackle this issue is combining the ceramic with one or more electrochemically active components. Additionally, the composition and microstructure of the final ceramic material can be controlled by the type of functional groups attached to the preceramic polymer. This, in turn, affects the electrochemical properties of SiOC based composites. Here we present electrochemical characteristics of various SiOC/graphite composites along with their thorough structural investigation. Blending of graphite with a preceramic polymer assisted by a high-energy ultrasonic homogenization leads to uniform distribution of partially exfoliated graphite within the ceramic matrix. The impact of the microstructural dissimilarities between SiOC/graphite composites based on various preceramic precursors is pointed out. Proper design of the ceramic matrix combined with an optimized amount of graphitic component leads to an improvement of electrochemical properties - an enhanced first cycle efficiency, increased capacity values, and a superior rate capability.

PACRIM Symposium 8: Porous Ceramics: Innovative Processing and Advanced Applications

Ceramic Membranes

Room: English Bay

8:30 AM

(PACRIM-370-2021) Current Trends and Future Directions of Ceramic Membrane Technology for Water Treatment (Invited)

I. Song*[1]; J. Ha[1]; J. Lee[1]

1. Korea Institute of Materials Science, Republic of Korea

Ceramic membranes can be applied under extreme operating conditions such as low pH, high pressure and high temperature. Especially, the inherent limitations of conventional polymer membranes can be overcome by adopting ceramic membranes simply to the existing water treatment systems. Moreover, there are a lot of potential applications of ceramic membranes such as distillation, adsorption and extraction in various industrial areas. Therefore, the ceramic membrane technology is not a mere ceramic processing technology, but a highly influential technology to the overall environment technology. In this study, we focused the current status and prospects of ceramic membrane for water treatment by adopting various kinds of ceramic materials such as alumina and silicon carbide. Especially, the ceramic membrane of flat-sheet and monolith type fabricated by extrusion process was introduced in various applications such as membrane bioreactor (MBR), industrial wastewater, and pretreatment in desalination. This presentation consists of three parts: (1) Tailoring pore structures of the substrate of ceramic membrane, (2) Functionalizing the coating layer of ceramic membrane, and (3) Extruding the substrate of ceramic membrane.

Properties of Porous Ceramics

Room: English Bay

10:00 AM

(PACRIM-371-2021) Experimental and numerical study on reactive melt infiltration process (Invited)

R. Inoue*[1]; Y. Arai[2]

1. Tokyo University of Science, Mechanical Engineering, Japan
2. Tokyo University of Science, Materials Science and Technology, Japan

Reactive melt infiltration (RMI) process using Si-, Zr, and various binary alloys has been widely applied for fabrication of ceramic matrix composites (CMCs), however, optimum process parameters were not identified because there are many factors affecting on mechanical properties and microstructures of CMCs. We fabricated some kinds of CMCs to understand the effect of microstructures of porous preform, interaction between liquid infiltrant and material, and physical properties of liquid infiltrant were examined. We also simulated flow of liquid metals and alloys during processing by computational fluid dynamics (CFD) considering the effect of reaction-formed matrix. Optimum process condition in some cases were presented.

10:30 AM

(PACRIM-372-2021) Analysis and evaluation on the relationship between structures and mechanical properties for a porous carbon material using image-based modeling (Invited)

Y. Arai*[1]; R. Inoue[1]; Y. Kogo[1]

1. Tokyo University of Science, Japan

Re-entry capsule is exposed to aerodynamic heating during re-entry. Thermal protection system (TPS) is applied to protect the capsule from heating. The ablator is one of typical TPS and it is composed of substrate and impregnated resin. Generally, carbon felt is used as the substrate, however, the compressive strength of ablator using carbon felt is low (~5MPa) and mechanical erosion during re-entry occurs. We have prepared a porous carbon material with three-dimensional networked structure (TNPC) for the substrate for ablator because TNPC has high strength and high Young's modulus compared to porous carbon with the same porosity as TNPC. However, the relationship between the structure and mechanical properties are not well known. In this study, three-dimensional image-based models of TNPC is constructed by images acquired by X-ray CT and FIB-SEM. Structural characterization such as measurement of strut length, thickness and diameter of strut joints is conducted using image-based models. In addition, the relationship between structure and Young's modulus of TNPC is evaluated by finite element method by using cubic periodic models that reflects the structural feature of TNPCs.

11:00 AM

(PACRIM-373-2021) Characterizations of highly porous mullite derived from gelation and freezing with fumed silica and alumina

M. Fukushima*[1]

1. National Institute of Advanced Industrial Science and Technology (AIST), Japan

Mullite thermal insulators were fabricated by freezing gels of aqueous slurry containing fumed silica and alumina to obtain the stoichiometric composition of mullite, followed by sintering. The relationship among those process factors, microstructure and properties was investigated. The insulators obtained showed very high porosity up to 93%, enhanced compressive strength around 10-20MPa and very low thermal conductivity less than 0.25W/mK, depending on the types of solid loading, sintering temperature and type of alumina.

11:20 AM

(PACRIM-374-2021) Effect of SiC whisker on mechanical strength of porous RB-SiC via direct foaming process

A. Shimamura*[1]; M. Fukushima[1]; N. Kondo[1]; M. Hotta[1]

1. National Institute of Advanced Industrial Science and Technology (AIST), Japan

We previously proposed a powder-based direct-foaming process for porous Reaction-bonded Silicon Cabide (RB-SiC). In this process, ceramic powder was mixed with resol resin which was a thermosetting resin and thermally cures with the foaming. The resol was used as the foaming agent as well as a binder for the SiC powder. The compact was prepared from the mixed powder by press molding. The compact was heated up to the foaming temperature of the resol in order to produce porous structure into the compact. Finally, a porous SiC was obtained by applying a reaction-bonding process with using molten silicon. In this study, we studied effect of SiC whisker on mechanical strength of porous RB-SiC via powder-based direct-foaming process. Content of SiC whisker was varied ranged from 5 wt.% to 20t.%. Porous structure and mechanical strength of the obtained porous RB-SiC were investigated depending on content of SiC whisker by microstructure observation and bending strength test.

PACRIM Symposium 9: Additive Manufacturing and 3D Printing Technologies

Emerging Technology

Room: Prince of Wales

8:30 AM

(PACRIM-375-2021) Thin-Wall Additive Manufacturing for Multidisciplinary Thermal Technology (Invited)

A. Muley*[1]

1. Boeing Research and Technology, USA

Emerging trends in aviation industry are driving the need for sustainable, efficient and affordable thermal management. The capacity growth in commercial and military aircraft, along with heat management in hybrid-electric, urban air mobility and high-speed air vehicles, warrants development of high performance, compact and conformal thermal devices. This keynote will delineate a holistic and sustainable AM enabled approach for addressing thermal challenges for current and future aerospace systems. Additive Manufacturing (AM) is rapidly evolving and offering design innovations for applications across several industries, which include aerospace, automotive, and energy systems. One of the prominent AM process development areas, which is attracting considerable attention, pertains to fabrication of thin-wall complex features with thickness in the range of 100 to 200 microns. Recent breakthroughs in thin-wall metallic AM process are enabling development of innovative conformal, compact and lightweight heat exchangers and cold plates for aircraft engine precooler and electronic cooling applications, respectively. This presentation will conclude with an overview of steps required for achieving technical maturity and production readiness of AM enabled heat exchanger technology.

9:00 AM

(PACRIM-376-2021) CerAMfacturing of miniaturized ceramic satellite thrusters

U. Scheithauer*[1]; E. Schwarzer-Fischer[1]; J. Abel[1]; N. Lorenz[1]; C. Bach[2]; M. Probst[2]; J. Sieder-Katzmann[2]

1. Fraunhofer IKTS, Shaping, Germany
2. TU Dresden, Institute for Aerospace Engineering, Germany

Aerospike nozzles promise performance and application advantages compared to classic bell nozzles but are also inherently more complex to manufacture. Additive manufacturing (AM) methods drastically simplify or even enable the fabrication of those complex structures while minimizing the number of individual parts. The general advantages of ceramic components lie in their low density combined with high strength as well as chemical and thermal resistance. The applicability of ceramic AM (CerAMfacturing) on rocket engines and especially nozzles is consequently investigated in the frame of the project called MAnufactured Ceramic AeROspike Nozzle In Space (MACARONIS) funded by Sächsische Aufbaubank-Förderbank (SAB, Ref.-Nr. 100376413). A significant part of the investigations is concerned with the general applicability of the CerAM VPP (Vat Photo Polymerisation) and CerAM FFF (Fused Filament Fabrication) processes for the fabrication of ceramic aerospike nozzles. In addition, a hybridisation of the two shaping processes and a joining of the individual parts during sintering to form a monolythic component could be successfully demonstrated. In this contribution the derived nozzle geometry as well as the resulting design approaches and manufacturing strategies for CerAM VPP and CerAM FFF are presented. Subsequently, the achievable tolerances and surface roughness are demonstrated by measurements on sample components.

Slurry-Based Technique

Room: Prince of Wales

10:00 AM

(PACRIM-377-2021) Lithography-based additive manufacturing of different non-oxide ceramics

M. Schwentenwein*[1]; A. Altun[1]; T. Prochaska[1]

1. Lithoz GmbH, Austria

While the majority of ceramics studied via stereolithography is light, non-oxide ceramics are more difficult to process due to the high light absorption and light scattering of dark-colored powders with often high refractive indices. In this work, we present non-oxide ceramic parts which were manufactured for the first time using LCM, achieving material characteristics comparable to materials obtained by conventional ceramic manufacturing methods; this includes silicon nitride, aluminum nitride and silicon carbide. It was possible to fabricate silicon carbide (reaction-bonded SiC) in very high resolution (features <200μm) and to print and sinter silicon nitride with the same thermal and mechanical properties as by conventional manufacturing (iso pressing). With a relative density of 99.8%, a hardness of 1500 and a biaxial bending strength of 760MPa the tested composition is exactly at eye-level with its conventionally processed analogue (hardness of 1500 and biaxial bending strength of 770MPa). Moreover, aluminum nitride components with a thermal conductivity >160W/m.K could be realized using LCM. This is also exactly on bar with the corresponding material processed by cold isostatic pressing. These material properties in combination with the high precision of the LCM process allow the production of highly complex components that have not been feasible before and that are fully functional.

10:20 AM

(PACRIM-378-2021) μPAD makes Powder Aerosol Deposition accessible: A modular and inexpensive approach to produce dense ceramic films at room temperature

J. Exner[1]; M. Linz[1]; J. Kita[1]; R. Moos*[1]

1. University of Bayreuth, Dept. of Functional Materials, Germany

Powder aerosol deposition (PAD), also referred to as aerosol deposition method (ADM) or vacuum kinetic spray (VKS), combines several advantages over conventional ceramic coating processes. Due to its unique deposition mechanism (called room temperature impact consolidation, RTIC), film formation occurs completely at room temperature, yet enables to form fully dense ceramic films without any binders nor sintering processes involved. Developments in the past decade pointed towards a better understanding of the deposition mechanism, and a scale-up of the spray process. The latter resulted in an enhanced process control and increased coating areas, however at the cost of a significantly higher complexity of the devices. High investment costs and the difficult prospect to build the required knowhow may hinder researchers in the field of chemistry and material science to start using this otherwise very promising coating technique. In a complementary effort, we present a simplified, more accessible approach to use the powder aerosol deposition. The aim is to enable film formation on smaller scale (spots with a diameter of 1 cm) yet still adequate for material research and application testing. Therefore, we propose and will demonstrate a modular, miniaturized and easily usable coating setup capable of producing ceramic films at room temperature.

10:40 AM

(PACRIM-379-2021) Stereolithographic Additive Manufacturing of Zirconia Dendritic Electrodes for Sustainable Aluminum Refining without Carbon Dioxide Emissions

M. Takahashi*[1]; S. Kirihara[1]

1. Osaka University, engineering, Japan

Zirconia porous electrodes were fabricated by stereolithographic additive manufacturing for no carbon dioxide emissions in aluminum refining. Dendritic structures were designed as connected cylinder lattices with 4,6,8 and 12-coordination numbers. Aspect ratios were systematically modulated to increase specific surface areas. Oxygen atoms should be absorbed from alumina molten salts, and oxygen ions should be conducted through zirconia solid electrolytes. Fluid behavior of refined molten aluminum can be simulated and visualized by computer fluid dynamics of finite element method. Acryl components including zirconia nanoparticles were successfully fabricated by a stereolithographic equipment, and formed precursors were dewaxed and sintered in the air atmosphere.

11:00 AM

(PACRIM-380-2021) Stereolithographic Additive Manufacturing of Thermoacoustic Converters with Ceramic Vascular Bundles

T. Ito*[1]; S. Kirihara[1]

1. Osaka University, Japan

Ceramic vascular bundles with tessellated prism cavities were fabricated by laser scanning stereolithography as effective thermoacoustic converters. Aerial vibrations with adiabatic compressions and expansions excited in the prism cavities can form acoustic standing waves in the vascular bundles. Antiphase heat conductions should be realized at the both ends of thermoacoustic converters by isothermal compressions and expansions. In ceramic additive manufacturing, nanosized alumina particles were dispersed into photo sensitive acrylic resin to form high viscosity pastes. On a laminated paste surface, an ultraviolet laser beam was scanned, and a cross sectional pattern was created. Through the layer laminations, a solid component was processed according to designed solid graphics. Obtained precursors were dewaxed and sintered in the air atmosphere. After systematic optimizations in the lithography and sintering, high part accuracies and fine microstructures were observed by digital optical and scanning electron microscopy. Through computer aided smart manufacturing, design and evaluation (Smart MADE), The section sizes of prism cavities were graded between the both orifices of ceramic vascular bundles to increase the thermoacoustic conversion efficiencies.

11:20 AM

(PACRIM-381-2021) Additive Manufacturing of Silicon Carbide Component by Thermolithography

T. Shimizu[1]; S. Kirihara[1]; Y. Uemura*[1]

1. Osaka University, Japan

Silicon carbide (SiC) components with high temperature strengths, wear resistances and chemical stabilities were fabricated by thermolithographic additive manufacturing for aerospace applications. Fine SiC particles with bimodal size distributions were systematically mixed into thermosetting acrylic resin at optimized volume fractions. Closed packing dispersions were simulated by discrete element method (DEM). A formed paste was spread on a metal substrate by using a mechanical knife edge. An ultraviolet laser beam of 355 nm in wavelength was scanned on the surface. A cross sectional pattern was formed as a solid layer through thermal polymerization. Composite precursors were processed thorough continuous laminations. Irradiation powers could be optimized to obtain enough curing depths for reliable layers bonding. Full ceramic structures were obtained by dewaxing and sintering in the nitrogen atmosphere. Transport tubes of combustion gas were successfully fabricated. Micro lattices of ordered porous patterns were introduced to distribute cooling gas between inner and outer tube walls. Gaseous streamlines and temperature distributions were visualized by computer fluid dynamics (CFD) and finite element method (FEM).

PACRIM Symposium 13: Novel Nanocrystal Technologies for Advanced Ceramic Materials & Devices

Synthesis of Nanocrystals and Nanocomposites

Room: Georgia A

8:30 AM

(PACRIM-382-2021) Hydrothermal synthesis of $BaTiO_3$-based nanocubes for dielectric applications

K. Mimura*[1]; Z. Liu[1]; H. Itasaka[1]; K. Kato[2]

1. Innovative Functional Materials Research Institute, National Institute of Advanced Industrial Science and Technology (AIST), Japan
2. National Institute of Advanced Industrial Science and Technology (AIST), Japan

Nanocubes (NCs) attract a great deal of attention due to a lot of advantages attributed to their facet structure, fabrication of orientation films or 3D architecture of NCs by self-assembly, and so on. We previously synthesized BT NCs by hydrothermal reaction. We have reported that BT NC assemblies shows high dielectric constant above 4000. In this study, BT-based NCs were synthesized by hydrothermal reaction in order to modify the electrical properties. In case of Sr substitution of BT NCs, the X-ray diffraction peaks of Sr-substituted BT NCs (BST NCs) shift to higher angles with increasing Sr ratio, which means that the substitution of Sr at the Ba site reduced the lattice constant. Sr ions were detected in NCs by scanning transmission electron microscopy with energy dispersive X-ray spectroscopy (STEM-EDX). On the other hand, when the ratio of Ba/Ca was 0.7/0.3, Ca did not substitute Ba of BT NCs but core-shell NCs of $BaTiO_3$ and $CaTiO_3$ (BT/CT NCs) was formed. These results suggest that substitution of Ba site depends on the degrees of ionization of Sr and Ca in basic solutions. As the dissociation constant of Sr is almost same as that of Ba, Sr can substitute Ba in hydrothermal process. However, Ca cannot substitute Ba site due to small dissociation constant of Ca. As a result, CT is grown on BT surface in hydrothermal process.

8:50 AM

(PACRIM-383-2021) Solvothermal Synthesis of Dispersible $KNbO_3$ Nanocubes from $K_4Nb_6O_{17}$ Precursor Particles

S. Ueno*[1]; Y. Yamada[1]; I. Fujii[1]; S. Wada[1]

1. University of Yamanashi, Graduate School Department of Interdisciplinary Research, Japan

It has been reported that the two-dimensional $BaTiO_3/SrTiO_3$ superlattice exhibited ultrahigh dielectric constant owning to the highly-strained heteroepitaxial interface[1]. We have focused on simple-cubic-superstructured assemblies consisting of perovskite-type dielectric nanocrystals with a cube shape (nanocubes) because different nanocubes (enclosed by {100} facets) connected in a face-to-face manner can form the strained {100}//{100} heteroepitaxial interface which leads to dielectric and piezoelectric enhancement. To prepare such nanocube assemblies, we have been studying on a solvothermal synthesis of perovskite-type $BaTiO_3$, $NaNbO_3$, and $KNbO_3$ nanocubes. For this purpose, uniform and dispersible nanocubes are required for the direct face-to-face attachment. In previous $KNbO_3$ studies, nanocubes could be obtained even by simple hydrothermal and solvothermal reactions between Nb_2O_5 and KOH solutions, though they were accompanied by the severe aggregation. We attempted to solvothermally synthesize the dispersible $KNbO_3$ nanocubes from $K_4Nb_6O_{17}$. When the precursor changed from Nb_2O_5 to $K_4Nb_6O_{17}$, the optimal reaction time to obtain the $KNbO_3$ nanocubes became much shorter. The resultant $KNbO_3$ nanocubes were dispersive without aggregation after classification. $K_4Nb_6O_{17}$ can be easily converted into $KNbO_3$ because the arrangement of NbO_6 units in $K_4Nb_6O_{17}$ is relatively similar to that of $KNbO_3$.

9:10 AM

(PACRIM-384-2021) Novel $MoSi^2$ catalysts featuring surface activation as highly efficient cathode materials for long-life Li–O_2 batteries

G. Zhang*[1]

1. Shandong University, China

Searching for highly efficient and low cost electrocatalysts is an urgent challenge to provide high capacity and long cycle life Li–O_2 batteries (LOBs). In this work, the highly efficient electrocatalytic properties of $MoSi_2$ were first demonstrated using particles prepared by a solid–state reaction method. The surface was activated, playing a dominant role in the electrocatalytic performance. After acid etching to remove SiO_2 impurities arising from the surface oxidation, $MoSi_2$ exhibited excellent bifunctional electrocatalytic properties for LOBs. Mo^{6+} on the surface contributed to the initial low overpotentials of 0.45 V (3.2 V for the OER and 2.75 V for the ORR) and transformed into Mo^{4+} for a long cycle life. As a consequence, outstanding rate capabilities of 12 708 mA h g^{-1} at 100 mA g^{-1} and 10 794 mA h g^{-1} at 800 mA g^{-1}, and excellent cycle durability of 215 and 90 cycles at a limited capacity of 1000 mA h g^{-1} at 200 and 500 mA g^{-1}, respectively, were obtained for the $MoSi_2$ cathode

Fabrication of 1D-, 2D-, and 3D-Assemblies, Coating Films, and Bulk Ceramics by using Nanocrystals

Room: Georgia A

10:00 AM

(PACRIM-385-2021) Oriented assemblies of 1D metal hydroxide nanomaterials toward functional coating (Invited)

K. Okada*[1]; M. Takahashi[1]

1. Osaka Prefecture University, Japan

Oriented films consisting of inorganics and organic-inorganic hybrid materials play an important role in various applications including separations, electrics, and optics due to their improved physical and chemical properties. In general, those oriented films can be obtained by physical/chemical vapor deposition methods via epitaxial growth. Although high quality films are obtained, they require high energy process and usable substrates are limited, which limits the number of materials that can be oriented. For examples, metal-organic framework (MOF) has not been oriented because suitable substrates are not available in these methods. Herein, we report on a novel approach to achieve oriented metal oxide and MOF films from aligned metal hydroxide nanobelt films. Metal hydroxide has two unique characters: 1) convertible to metal oxide by topotactic transformation maintaining its crystallographic orientation, 2) the presence of abundant hydroxyl groups with a long-range periodicity. Oriented metal oxide films (such as TiO_2) have been obtained from oriented metal hydroxide films by topotactic transformation. This approach also afforded a fabrication of oriented MOF films through an epitaxial growth. This approach based on oriented assemblies of 1D metal hydroxide nanomaterials will open new possibilities to achieve functional films consisting of organics, inorganics, and hybrid materials.

10:30 AM

(PACRIM-386-2021) Preparation of perfectly oriented free-standing nanosheet film and photocatalytic properties (Invited)

S. Ida*[1]

1. Kumamoto University, Japan

Metal oxide nanosheet has been attracted great attention due to its ultimate thickness of single crystal unit. Owing to this anisotropic structure, extraordinarily strong van der Waals interactions are largely effective for self-assembly of face-to-face stacked 3D composites. We would like to emphasize that monolayer nanosheets can form a highly oriented film by simply stacking up the sheets without any complicated machineries or processes, while it is difficult to form oriented structures from a nanoparticle (0D) because of the point-to-point interactions along random directions. In the present paper, we shall try to create a perfectly oriented titania nanosheet stacking on Si wafer for a free-standing film via spin-coating process. Out-of-plane and in-plane XRD patterns proved the high degree of orientation along [010] plane of the titania nanosheet. The spin-coated titania nanosheet was separated as a free-standing film by means of simple cation-exchange process for further electrochemical investigation. The band structure was similar to the monolayer titania nanosheet reported previously, which indicated that the quantum effect of the monolayer was maintained even though its thickness was over 200 nm. This film showed higher photocatalytic activity compared with the parent layered oxide.

Applications and Functional Devices using Nanocrystals and Characterizations

Room: Georgia A

11:00 AM

(PACRIM-387-2021) Metal/graphene electrodes for parallel-plate capacitor structures containing $BaTiO_3$ nanocube ordered assemblies

H. Itasaka*[1]; Z. Liu[1]; K. Mimura[1]; K. Kato[1]

1. National Institute of Advanced Industrial Science and Technology (AIST), Japan

Dielectric nanocrystals are promising materials to meet increasing demands for miniaturization of dielectric devices. Although there has been dramatic progress in synthesis and assembly techniques of nanocrystals, evaluating dielectric properties of nanocrystal assembly structures is still a challenging issue. One of the factors that make the evaluation difficult is cracks forming in nanocrystal assemblies during assembly and heating processes. When metal top electrodes are deposited on such nanocrystal assemblies by an evaporation method for preparing a parallel-plate capacitor structure, metal vapor invades the cracks and results in formation of leak passes. In this study, we introduce a metal/graphene electrode as a top electrode to fabricate parallel-plate capacitor structures containing $BaTiO_3$ nanocube ordered assemblies. Cross-sectional observation by scanning transmission electron microscopy reveals that graphene layers laid on the nanocube assemblies successfully avoid formation of metal leak passes in the cracks of the nanocube assemblies.

11:20 AM

(PACRIM-388-2021) Nanoscale Electronic Properties of Ultrathin Functional Films via Tomographic AFM

T. Moran[1]; K. Suzuki[2]; T. Hosokura[2]; J. Kaszas[1]; W. Huey[1]; W. Linthicum[1]; J. Song[1]; B. Huey*[1]

1. University of Connecticut, Materials Science and Engineering, USA
2. Murata Manufacturing Co., Ltd., Japan

Nano- and meso- scale materials properties are crucial to the macroscopic performance of a wide range of functional devices. In two dimensions, ferroelectricity and even domain dynamics have thus been investigated for decades using variations of Atomic Force Microscopy. Our work and others reveals how these properties are frequently mediated by strain, orientation, grain boundaries, and other microstructural defects or heterogeneities. However, practical devices are often sensitive to, or even controlled by, sub-surface effects or thickness dependencies related to microstructure and concentration, polarization, and/or field gradients. Therefore, we are advancing Tomographic AFM for volumetric materials property mapping, with voxels of properties on the order of $\sim$10 nm^3. Charging and discharging dynamics, as well as piezoelectricity, are especially investigated for nanocrystalline $BaTiO_3$ thin films and $BaTiO_3/SrTiO_3$ superlattices. Such volumetric insight is increasingly important for engineering optimal performance and reliability of real-world, 3-Dimensional materials devices.

PACRIM Symposium 14: Functional Nanomaterials for Energy Harvesting and Solar Fuels

Innovative Processing of Functional Nanomaterials for Optoelectronic Devices / Advanced Materials for Next Generation Photovoltaic Devices

Room: Plaza A

8:30 AM

(PACRIM-389-2021) Interfacial properties in composite nano-systems for energy harvesting (Invited)

A. Vomiero*[1]

1. Lulea University of Technology, Engineering Sciences & Mathematics, Sweden

Composite nanostructures can be efficiently applied for Sunlight detection and conversion and, more in general, for energy harvesting and generation of solar fuels. In most of the applied systems (photodetectors, excitonic solar cells and (photo)-electrochemical cells), nanomaterials can play a critical role in boosting photoconversion efficiency by ameliorating the processes of charge photogeneration, exciton dissociation and charge transport. Critical role in such processes is played by the structure and quality of the interface, which needs to be properly assembled to obtain the desired functionality. Several strategies can be pursued to maximize energy harvesting and storage, including broadening of light absorbance to reduce solar light losses, fastening exciton dissociation and charge injection from the photoactive medium to the charge transporting materials, reducing charge recombination during charge transport and collection at the electrodes. In this lecture, a few examples of application of nanocomposites will be discussed, including all-oxide coaxial p-n junction nanowire photodetectors and solar cells, core-shell quantum dot fluorophores for high-efficiency luminescent solar concentrators, composite sulfides for hydrogen generation, and oriented carbon nanotube forest dispersed in polymer matrix as efficient low-temperature thermoelectric composite.

9:00 AM

(PACRIM-390-2021) Single-source-precursor synthesis and processing of silicide-containing polymer derived ceramic nanocomposites

E. Ionescu*[1]

1. Technical University Darmstadt, Materials Science, Germany

Polymer-derived ceramics (PDC-NCs) have been addressed in the last decades and were shown to possess intriguing properties which make them excellent candidates as structural and (multi)functional materials. PDC-NCs can be synthesized via thermal conversion of suitable single-source precursors, leading in a first step to amorphous single-phase ceramics, which subsequently undergo phase separation processes to furnish bi- or multiphase ceramic nanocomposites. In the present talk, recent work related to the development of silicide-containing PDC-NCs, their (micro)structural features and properties will be introduced and critically assessed. Here, mainly nanocomposites consisting of transition metal silicides (e.g., V_5Si_3, Fe_3Si, etc.) dispersed in an amorphous ceramic matrix, i.e., silicon (oxy)carbide or silicon nitride, will be presented. Special emphasis will be set on describing the intimate relationship between the molecular architecture of the single-source precursors and the phase composition / microstructural features of the resulting PDC-NCs. Preparative concepts for the knowledge-based design of silicide-containing PDC-NCs with tailored phase compositions and property profiles as well as selected prospective applications will be highlighted and discussed.

9:20 AM

(PACRIM-391-2021) Copper-based Dichalcogenide Materials for electronics and clean environmental applications (Invited)

D. Chua*[1]

1. National University of Singapore, Materials Science & Engineering, Singapore

Transition metal dichalcogenides (TMD) materials have attracted much attention due to their unique properties, ranging from low dimensional effects, good structural integrity, high electrical and thermal conductivity, and chemical stability. It is also important to synergize rational design with the synthesis process of material so as to achieve the desired results. With a good coupling between design and engineering the synthesis of TMD materials, useful material, electronic and optical properties could be obtained which may not have been expected in the first place. Focussing on a single covelite or CuS, controlling the morphology and stoichiometry, CuS was found to have good electrical switching properties for low voltage applications in resistive random access memories and even as an optical attenuation material. By redesigning the morphology, we are further able to show that fabricating the vertical pillars were found elicit localized surface plasmon resonance allowing it to be used as passive photothermal vaporizers and even as excellent thermal reflectors for clean environmental applications..

10:10 AM

(PACRIM-392-2021) Morphology Control of Functional Metal Oxide Nanomaterials in Aqueous Solutions

Y. Masuda*[1]; P. Choi[1]; A. Tsuruta[1]; T. Akamatsu[1]; T. Itoh[1]

1. National Institute of Advanced Industrial Science and Technology (AIST), Japan

Morphology control of functional metal oxide nanomaterials have been realized in aqueous solutions. For instance, a sensor has been developed for detecting 1-nonanal gas present the breath of lung cancer patients by combining SnO_2 nanosheets with SnO_2 nanoparticles and noble metal catalysts. A significant change in the electrical resistance of this sensor was observed with increasing 1-nonanal gas concentration; the resistance decreased by a factor of 1.12 within the range of 1 to 10 ppm at 300 degrees C. The recovery of the sensor's resistance after detecting 1-nonanal gas concentrations of 0.055, 0.18, 1, and 9.5 ppm was determined to be 86.1, 84.2, 80.4 and 69.2%, respectively. This high sensitivity is attributed to the accelerated oxidation of 1-nonanal molecules caused by the (101) crystal faces of the SnO_2 nanosheets and should provide a simple and effective approach to the early detection of lung cancer.

10:30 AM

(PACRIM-393-2021) Plasmonic Enhancement of Luminescence from Er^{3+}/Yb^{3+}/Nd^{3+} Co-doped Upconversion Nanoparticles by Aluminum Lattice Mode

Y. Gao*[1]; S. Murai[1]; K. Shinozaki[2]; K. Tanaka[1]

1. Kyoto University, Japan
2. National Institute of Advanced Industrial Science and Technology (AIST), Japan

Rare-earth-doped up-converters with Yb^{3+} as a sensitizer show a high internal conversion efficiency under the excitation at wavelength λ = 980 nm. For the solar cell applications of up-converters, however, there is a problem that the solar light at λ = 980 nm is absorbed by the water molecule in the atmosphere. While Nd^{3+} co-doping allows the up-conversion at the transparent window of water around λ = 800 nm, the absorption coefficient of Nd^{3+} is still small due to the forbidden nature of 4f–4f transition. Here, we achieve amplified up-conversion enhancement of 44-fold (8.9-fold) compared to the same layer on the flat glass (on the flat aluminum film) by coupling the Nd^{3+}-sensitized up-conversion nanoparticles with the periodic lattice of aluminum nanocylinders under 15 W/cm^2 of λ = 808 nm excitation. The size of nanocylinders and the period of the lattice are designed to realize the lattice resonance around λ = 800 nm, facilitating the Nd^{3+} $^4I_{9/2}\rightarrow{}^4F_{5/2}$ transition to yield amplified up-conversion. Numerical simulation indicates that the up-conversion enhancement is induced by an intensified near field accompanied with the excitation of lattice modes, where the in-plane light diffraction facilitates the radiative coupling of localized surface plasmons in each nanocylinder.

PACRIM Symposium 16: Advanced Structural Ceramics for Extreme Environments

Advanced Structural Ceramics for Extreme Environments

Room: Regency E

8:30 AM

(PACRIM-394-2021) High Entropy Ultra-High Temperature Ceramics: Synthesis, Processing, and Properties (Invited)

W. Fahrenholtz*[1]; L. Feng[1]; G. Hilmas[1]

1. Missouri University of Science & Technology, Dept. of Materials Science and Engineering, USA

High entropy ultra-high temperature ceramics are the subject of increasing research activity. Synthesis methods affect the powder particle size and purity, which also impact the ability to densify the resulting powder. Densification is typically conducted with applied pressure by hot pressing or spark plasma sintering. To date, the most commonly reported properties are hardness and modulus measured by nanoindentation. In our laboratory, high entropy carbides were synthesized by carbothermal reduction of oxides and high entropy borides were synthesized by boro-carbothermal reduction. Both processes produce sub-micron powders with oxygen contents below 1 wt%. Densification was accomplished by hot pressing of carbides and spark plasma sintering of borides. Finally, the presentation will summarize systematic studies of the mechanical and thermal properties of the high entropy ultra-high temperature ceramics that have been produced.

9:00 AM

(PACRIM-395-2021) Irradiation Damage of High-Entropy Carbide Ceramics in Extreme Environments (Invited)

F. Wang[1]; T. Wang[2]; L. Shao[2]; M. Nastasi[2]; Y. Wu[3]; Y. Lu[1]; B. Cui*[1]

1. University of Nebraska-Lincoln, USA
2. Texas A&M University, USA
3. Boise State University, USA

The concept of entropy stabilization has created promising opportunities for the design of new ceramic materials for advanced nuclear reactors. This research investigates the phase stability, formation of irradiation-induced defect clusters, and irradiation hardening in the novel high-entropy carbide ceramics (HECC) to provide fundamental mechanisms governing their irradiation damage behavior. The model material $(Zr_{0.25}Ta_{0.25}Nb_{0.25}Ti_{0.25})C$ was synthesized using spark plasma sintering and irradiated by 3 MeV Zr ions to 20 dpa at 25, 300, and 500 degree C, respectively. Grazing incidence X-ray diffraction analysis showed that $(Zr_{0.25}Ta_{0.25}Nb_{0.25}Ti_{0.25})C$ maintained a high phase stability without phase transformation after irradiation, while about 0.2% lattice parameter expansion was observed. The irradiation-induced microstructures were comprised of defect clusters with diameters of several nanometers, without void formation or radiation-induced segregation. The defect clusters were characterized by transmission electron microscopy as two types of dislocation loops, including perfect loops and faulted Frank loops. The growth of dislocation loops may be suppressed by the strong local lattice distortion. Nanoindentation tests showed irradiation-induced hardness increase, which was possibly caused by dislocation loops and lattice strain.

9:30 AM

(PACRIM-396-2021) Solvothermal Synthesis of High Entropy Metal Carbides: A New Class of Ultrahigh Temperature, Irradiation Resistant Ceramics

V. Vakharia*[1]; L. Zhang[1]; S. Sufy[1]; O. A. Graeve[1]

1. University of California, San Diego, Mechanical and Aerospace Engineeering, USA

High-temperature ceramics are of great interest for hypersonic leading edges as well as aeronatuical structural materials. We have implemented a solvothermal synthesis process, a procedure novel for these materials, to synthesize high entropy metal carbides (HEMCs). Specifically, (MoNbTaVW)C was produced because it has been predicted to have a high "entropy forming ability." The solvothermal process can obtain ceramic nanopowders and can be easily modified to introduce dopants. It is also helpful for producing nanopowders of cubic morphology from which ultra-dense samples can be produced with minimal porosity. The technique consists of mixing metal chlorides and carbon powder in the presence of molten lithium. En route to this HEMC, we have produced binary, ternary, quaternary, quinary, and senary metal carbides through systematic manipulation of elemental composition. Solid solutions of (NbTa)C, (NbTaW)C, and (MoNbTaW)C have been confirmed by a combination of X-ray diffraction (XRD) and energy dispersive spectroscopy (EDS). Rietveld refinement was used to deconvolute the XRD data and obtain approximate phase compositions of the different carbides to reestablish precursor amounts and obtain a solid solution HEMC. Through a spark plasma sintering (SPS) process, dense samples of compositionally complex ceramics have been produced.

10:10 AM

(PACRIM-397-2021) Thermochemistry of Reactions between Ceramic Coating Materials and Silicate Melts at High Temperature

G. Costa*[1]; B. Kowalski[1]; B. J. Harder[1]; N. P. Bansal[1]; J. L. Stokes[1]; K. Lilova[2]; T. Subranami[2]; S. Ushakov[2]; A. Navrotsky[2]

1. NASA Glenn Research Center, USA
2. Arizona State University, USA

The thermochemistry involved in the corrosion of ceramic coating materials at high temperature is crucial for thermodynamic modeling of phase equilibria to improve their durability in aerospace applications. One such emerging area is the corrosion of gas turbine engine coatings by molten silicate. Here, we summarize and discuss our previous and current studies using high temperature calorimetry to evaluate the energetics of reactions of coating materials in silicate melts with varying network former SiO_2 mol% in the $CaO-MgO-Al_2O_3-SiO_2$ (CMAS) system. 7YSZ and $Gd_2Zr_2O_7$ (GZ) exhibited exothermic enthalpies of solution in molten silicate, indicating higher reactivity and lower energetic stability against molten CMAS corrosion. The enthalpy of solution of GZ becomes slightly less exothermic with increasing SiO_2 mol%, indicating lower reactivity towards CMAS corrosion in more acidic or more polymerized melts. Enthalpies of solution become less endothermic and enthalpies of mixing more exothermic for $Y_2Si_2O_7$ and $Yb_2Si_2O_7$ with increasing SiO_2 mol%, indicating that these disilicates are increasingly reactive with more polymerized CMAS melts. We also report and discuss the calorimetric measurements of the enthalpies of solution of the constituent binary oxides from the ceramic coating materials and their corrosion products.

10:30 AM

(PACRIM-398-2021) The effect of oxygen on the vacancy ordering and stability of UHTC transition metal carbides

T. Davey*[1]; Y. Chen[1]

1. Tohoku University, School of Engineering, Japan

Ultra-high temperature ceramic transition metal carbides, such as zirconium and hafnium carbide, find use in aerospace and nuclear industries thanks to their extremely high melting temperatures and tuneable properties with varying carbon vacancy concentration. At low temperatures, structural ordering of the carbon vacancies has been predicted theoretically using methods such as the Cluster Expansion Method and the USPEX evolutionary algorithm, and has been sporadically reported experimentally. However, it remains poorly understood why experimental observations are so limited. In our previous work, first-principles calculations were used to show that inevitable oxygen contamination is a factor affecting the stability of vacancy-ordered zirconium carbides, destabilising the ordered arrangement of vacancies relative to a random distribution of vacancies. In this work, the effect of dissolved oxygen impurities on selected group IV and group V transition metal carbides is explored using first-principles calculations.

10:50 AM

(PACRIM-399-2021) The role of fluorination during the physicochemical erosion of yttria in fluorine-based etching plasmas

M. Kindelmann*[1]; M. L. Weber[2]; M. Stamminger[3]; R. Buschhaus[3]; M. Bram[1]; O. Guillon[1]

1. Forschungszentrum Juelich, IEK-1: Materials Synthesis and Processing, Germany
2. Forschungszentrum Juelich, PGI-7: Electronic materials, Germany
3. Ruhr Universität Bochum, Institute for Experimental Physics II, Germany

Even though advanced ceramics are widely applied as consumables in semiconductor etching processes, the erosion mechanisms and connected surface phenomena are not fully understood. A physicochemical reaction between the plasma and the surface controls the erosion of polycrystalline ceramics in fluorine containing etching plasmas. However, the influence of microstructural properties like crystallographic orientation on the erosion behavior has not been investigated and understood so far. Therefore, in this study, pure Y_2O_3 and Y_2O_3/YOF composites were exposed to a fluorine etching plasma under fixed plasma conditions to investigate the erosion response at the microscopic scale. A straight-forward relocalization technique enables the correlation of crystalline orientation with the material response. Our study reveals an orientation dependent surface fluorination of Y_2O_3, which controls the etching resistance and morphology formation, leading to a plateau-like etch topography. Furthermore, a direct comparison with the behavior of the oxyfluoride YOF highlights the importance of the fluorocarbon layer formation and its subsequent surface reaction on the overall mechanism. The presented results help to extend the mechanistic understanding of the plasma-material interaction of polycrystalline Y_2O_3 ceramics.

PACRIM Symposium 18: Advanced Wear Resistant Materials: Tribology and Reliability

Mechanical and Tribological Behavior of Advanced Materials

Room: Balmoral
Session Chair: Junichi Tatami, Yokohama National University

8:30 AM

(PACRIM-400-2021) Novel sustainable composites for tribological applications

S. Javaid*[1]; M. Dey[1]; C. Matzke[1]; S. Gupta[1]

1. University of North Dakota, Mechanical Engineering, USA

In this presentation, we will present some of the recent work on the design of polylasctic acid (PLA) based composites by using biomass from corn-feedstock as precursors. The manufacturing method for fabricating these composites will be presented. In addition, detailed microstructure, mechanical, and tribological behavior of these composites will be presented.

8:50 AM

(PACRIM-401-2021) Multi-scale Tribological Behaviors of Cold Sprayed Ti_2AlC coatings (Invited)

A. Agarwal*[1]; D. John[1]; C. Zhang[1]; T. Paul[1]

1. Florida International University, Mechanical and Materials Engineering, USA

Ti_2AlC MAX phase attracts great research interests due to its superior properties. Due to its unique layered structure, Ti_2AlC combines both high hardness of a ceramic as well as high ductility of a metal and can be considered as ductile ceramic. In the present work, thin layers of Ti_2AlC coatings were deposited on Al6061 and HDPE substrates using cold spray technique. Unlike plasma spray and HVOF, cold spray utilizes much lower temperature during the deposition, so it prevents oxidation and decomposition of Ti_2AlC. Both nitrogen and helium were used as carrier gases. Tribological and scratch-induced deformation behaviors of cold sprayed Ti_2AlC coatings as well as single splats are investigated at multiple load scales. The effects of carrier gas and substrate material on the deformation and bonding mechanisms of Ti_2AlC during cold spray are carried out on the single splats using in-situ nano-scratch technique. Micro-scratch testing is performed, and real time coefficient of friction (COF) is obtained to elucidate the tribological behaviors such as deformation, inter- and intra-splat sliding.

9:20 AM

(PACRIM-402-2021) Wear of Thermally-Sprayed Ceramic Coatings in Fuel Lubrication and the Effect of Ultrasonic Nanocrystal Surface Modification (Invited)

S. Berkebile*[1]; A. Amanov[2]; M. L. Ferrera[1]; R. Karimbaev[3]

1. US Army Research Laboratory, DEVCOM, USA
2. Sun Moon University, Department of Mechanical Engineering, Republic of Korea
3. Sun Moon University, Department of Fusion Science and Technology, Republic of Korea

Mechanical components operating in a fuel environment are subjected to harsh conditions that can lead to early failure. Due to the low viscosity of fuel (as compared to oil), mechanical interfaces experience heavy mixed and boundary lubrication at most sliding speeds, with only a thin layer of molecules between the two opposing materials or direct material contact between the materials. Especially as the fuels become less viscous and have fewer lubricious molecular species, longer-term wear increases and rapid scuffing becomes more likely. Hard coatings have the potential to mitigate failure due to wear and scuffing in fuel-lubricated mechanical interfaces. We report on the wear and scuffing behavior of ceramic thermal spray coatings (WC-Co, Cr_3C_2, Cr_2O_3, Y_2O_3, YSZ, yttrium aluminum garnet, Al_2O_3/TiO_2) in reciprocating mechanical contacts lubricated by F-24 jet fuel, dodecane, and ethanol. The thermal spray coatings were measured both as deposited and after treatment with ultrasonic nanocrystalline surface modification (UNSM), which is a peening process that reduces surface roughness and coating porosity. The various coatings showed widely varying amounts of wear under identical conditions. UNSM treatment reduced the wear in all cases, with UNSM-treated WC-Co and Al_2O_3/TiO_2 showing the least amount of wear.

10:10 AM

(PACRIM-403-2021) Mechanical properties of Si_3N_4 ceramics in meso-scale measured by bending test using microcantilever beam specimens

J. Tatami*[1]; M. Uda[1]; M. Iijima[1]; T. Takahashi[2]

1. Yokohama National University, Japan
2. Kanagawa Institute of Industrial Science and Technology, Japan

Since wear of ceramic materials is the fracture of a small area of the surface, it is important to obtain knowledge about the mechanical properties near the surface of the material. In particular, it is useful to know the strength and fracture toughness at mesoscale, which is the same size as the crystal grains and grain boundaries of ceramics. In this study, the mechanical properties of Si_3N_4 ceramic surfaces were measured by bending test using microcantilever beam specimens fabricated by focused ion beam technique. As a result, it was found that the intergranular fracture toughness of Si_3N_4 ceramics depended on the added rare earths. The strength at mesoscale was extremely high compared to that of the bulk Si_3N_4 ceramics. The strength was found to be decrease significantly when the surface was in contact with acid for several hours. These findings are useful to understand the wear of ceramics and to develop superior wear resistant materials.

10:30 AM

(PACRIM-404-2021) Design of ternary ceramics from biomass based precursors

M. Dey*[1]; S. Javaid[1]; C. Matzke[1]; S. Gupta[1]

1. University of North Dakota, Mechanical Engineering, USA

Novel sustainable methods of manufacturing are needed for pushing the world towards circular economy (CE). Greening of ceramics is a major goal for making the manufacturing environmentally friendly. Different groups have designed binary carbides by using biomass as precursors. In this present, we will present design of novel ternary ceramics by using different types of biomass as precursors. Detailed microstructural and phase characterization of powders produced by this method will be reported.

PACRIM Symposium 23: Transparent Ceramic Materials and Devices

Transparent Ceramic Materials I

Room: Oxford
Session Chair: Yiquan Wu, Alfred University

10:00 AM

(PACRIM-405-2021) Modeling the effects of crystal structure, dopants and microstructure on the scattering and absorption of light in polycrystalline materials (Invited)

J. E. Garay*[1]

1. University of California, San Diego, Dept. of Mechanical and Aerospace Engrg., USA

M. H. Shachar, G. Uahengo, E. H. Penilla, Y. Kodera and J. E. Garay As transparent polycrystalline materials become more important in optical applications, evaluation of their optical properties across a wide range of wavelengths (or frequencies) is crucial for device design. Transmission measurements are often used to assess the suitability of materials for a given optical application. We have developed analytical models that describes reflection, scattering, and absorption losses that commonly affect transmission. Modelling allows us to demonstrate the effects that parameters such as absorption type (Lorentzian or Gaussian), scattering regime (Raleigh–Gans–Debye or Rayleigh), and optical path length have on transmission spectra. In addition, the models capture effects of anisotropic crystal structures and grain size effects. We use the models to describe and interpret transmission data from real ceramics systems that we and others have fabricated in order to highlight key differences in material type and suggest design strategies. The models presented are analytical making them relatively easy to implement.

10:30 AM

(PACRIM-406-2021) Single crystal fiber lasers (Invited)

S. Bayya*[1]; W. Kim[1]; D. Gibson[1]; D. Rhonehouse[1]; R. Nicol[2]; C. Askins[2]; T. Zhou[3]; B. Shaw[1]; J. Myers[1]; F. Kung[3]; G. Villalobos[1]; J. Kolis[4]; B. Stadleman[4]; J. Sanghera[1]

1. Naval Research Laboratory, USA
2. Jacobs, USA
3. University Research Foundation, USA
4. Clemson University, USA

Single crystal fiber composed of rare earth doped YAG offers the potential for high power scaling of fiber lasers due to lower intrinsic stimulated Brillouin cross-sections and higher thermal conductivity. The use of rare earth doped YAG fibers also mitigates issues of photodarkening, multiphonon edge absorption, and OH-quenching observed in Ho doped silica fiber lasers operating at eye safe wavelengths. A model based on Yb and Ho single frequency fiber laser showed that YAG crystal fiber could carry as much as 10 times higher power than silica fiber. Rare earth doped single crystal YAG and sesquioxide fibers of various diameter and length have been drawn using a state-of-the-art Laser Heated Pedestal Growth system. We have developed claddings on these fibers using various crystal growth and deposition methods. We report on the various fabrication methods, morphology, and optical characterization of these single and double clad all-crystal fibers.

11:00 AM

(PACRIM-407-2021) Crystal structure and densification of transparent magnesium aluminate spinel doped with LiOH

A. Talimian*[1]; V. Pouchlý[3]; A. Najafzadehkhoee[2]; K. Maca[3]; D. Galusek[1]

1. Centre for Functional and Surface Functionalized Glass (FunGlass), VILA, Slovakia
2. Institute of Inorganic Chemistry, Slovak Academy of Sciences, Joint Glass Centre of the IIC SAS, TnUAD and FChPT STU, Slovakia
3. CEITEC BUT, Brno University of Technology, Czechia

Sintering additives are often used to promote the densification of transparent magnesium aluminate spinel and prevent carbon contamination during spark plasma sintering. However, the impact of the incorporation of additives' ions into the spinel structure has remained unclear. In the present work, we performed a systematic study of lithium addition to magnesium aluminate spinel on its crystal structure and densification. Commercial powders were doped with limited amounts of lithium hydroxide and densified by spark plasma sintering. The effects of lithium were measured in terms of densification upon sintering, structural inversion and strain, and optical properties. The inversion of the spinel structure, produced by lithium incorporation, may play a prominent role in the diffusion of ions and affect the densification of samples. The correlations between the inversion of spinel structure and the properties of prepared spinel ceramics are critically discussed.

11:20 AM

(PACRIM-408-2021) Multifunctional material borosilicate glass - applications based on chemical, mechanical and thermal properties and enabled by its special structure

J. Brandt-Slowik*[1]

1. SCHOTT Technical Glass Solutions GmbH, Germany

Borosilicate glass is a multifunctional material, which is used in various applications. In contrast to standard soda lime glass, in classical borosilicate glass the amount of network former is with ~ 94 wt% very high and the amount of network modifier with ~ 4 wt% very low. That composition makes the large-scale production challenging, but it is the basis for very interesting material properties. The use of the glass goes far beyond its popular use as laboratory glass. The chemical resistance of borosilicate glass is close to that of pure quartz glass. When preserving cultural assets for long time, fluctuations in temperature and in humidity require stable materials. The preservation of 3500-year-old Papyrus Ebers as well as the restauration of portraits from the 18th century using borosilicate glass will be shown. Usually due to breakage, truck windshields need to be exchanged 1-2 times per year, which causes costly downtimes. The sharp impact and abrasion resistance of borosilicate glass enables the usage in this new innovative application. Such windshields have superior stone impact resistance, therefore increased durability. The low coefficient of linear thermal expansion of borosilicate glass leads to a high resistance to thermal shock and gradients. This enables for example the usage as inner ply in the pyrolytic oven door.

11:40 AM

(PACRIM-409-2021) Beta-SiC for Rugged Window Material

W. Kim*[1]; S. Bayya[1]; G. Villalobos[1]; B. Sadowski[2]; T. Zhou[3]; J. Sanghera[1]

1. Naval Research Laboratory, USA
2. Jacobs, USA
3. URF, USA

Cubic β-SiC, due to its high strength, low coefficient of thermal expansion, very high thermal conductivity, high thermal shock resistance, and good vis-IR transparency, is of great interest for window materials for extreme environments. Chemical and phase pure powders are required to fabricate high quality ceramic windows. However, the synthesis of pure β-SiC powder is very difficult and commercial β-SiC powders typically contain small amounts of secondary phase impurities and chemical impurities that cause scattering and absorption in the sintered ceramics, respectively. This paper discusses the synthesis and optical/physical properties of phase pure β-SiC powder using a unique two step sol-gel method. Different powder post-processing techniques were also applied to purify the commercial powder.

PACRIM Symposium 24: Solid Oxide Fuel Cells and Hydrogen Technologies

High Temperature Electrolysis and Electrodes

Room: Georgia B
Session Chair: Tae Ho Shin

10:00 AM

(PACRIM-410-2021) Current Progress in LSGM based SOC as the Advanced Electrochemical Devices at KICET (Invited)

T. Shin*[1]

1. Korea Institute of Ceramic Engineering & Technology, Energy Materials Center, Republic of Korea

Unlike past decades, the durability and reliability of solid electrochemical devices operated at high temperatures become more important issues for their commercialization in the energy market. Consequently, several efforts to reduce the operating temperature have been made through alternative fast ionic conductors, which have secured the life cycle of all surrounding components by lowering the temperature. Particularly, the discovery of LSGM as a perovskite oxide electrolyte has provided a stepping stone for next-generation SOFC/SOECs. Perovskite oxides have significant promise as alternatives to oxide ion conductors, and the above-discussed future approaches are expected to contribute to developing a better perovskite system. It remains a major challenge to understand the defect chemistry in the tuned lattice and design the electrochemical properties, especially regarding incompatibility with NiO and cost-effective manufacturing designs for the commercial industry. Superior performance has to be achieved with the commercial devices, which have succeeded in thin perovskite electrolyte by applying an inexpensive large-area coating process enable of suppressing reaction with other electrodes through outstanding robust tuned composition from the previous one including LSGM and should be studied. In this study, we will discuss with our improved the achievements of the LSGM's devices at KICET.

10:30 AM

(PACRIM-411-2021) High temperature electrolysis and reversible solid oxide cells: Novel manufacturing of electrodes/electrolytes (Invited)

N. Kostretsova[2]; M. Machado[2]; L. Bernadet[2]; A. Pesce[2]; M. Lira[2]; M. Núñez[2]; F. Baiutti[2]; A. Morata[2]; M. Torrell[2]; A. Tarancón*[1]

1. IREC-ICREA, Spain
2. IREC, Spain

Conventional manufacturing technologies such as tape casting or screen printing enabled the existing generation of high temperature solid oxide cells. This successful story ended up in a feasible technology currently reaching the MW scale. Complementary, novel fabrication approaches are recently presented pushing the limits of some classical constraints such as simple shape or slow startup opening new application scenarios. Among others, in this talk we will present recent advances in 3D printing of solid oxide cells that increase the performance of SOCs made of state of the art materials as well as the interest of incorporating thin film layers for improving the performance and durability of currently existing designs.

11:00 AM

(PACRIM-412-2021) Characteristics of oxide-based electrodes for CO_2 reduction for high temperature solid oxide electrolysis cells

S. Lee*[1]; M. Kim[1]; T. Shin[1]

1. Korea Institute of Ceramic Engineering and Technology (KICET), Energy & Environmental Division, Republic of Korea

Solid oxide electrolysis cells (SOECs) have received extensive attention in recent years because of the depletion of fossil fuels. In order to generate H_2 as green energy, many efforts have been made to develop high-temperature steam electrolysis. Besides steam electrolysis, CO_2 electrolysis by using SOECs are currently gaining increasing research efforts worldwide. High-temperature SOECs works very similarly to solid oxide fuel cell works, which has attained significant technological and materials development. Similar to the fundamentals of high-temperature steam electrolysis, CO_2 electrolysis involves the dissociation of CO_2 to carbon monoxide and oxygen ions, which will diffuse through the ion-conduction electrolyte to the anode side, where the oxygen anions proceed to combine with each other to form O_2 and then the dissociation of CO_2 occurs on the cathode of SOEC. Therefore, the studies of electrode materials, the kinetics of reaction, and electrochemical reaction mechanisms of CO_2 electrolysis are important. In this study, we consider oxide-based electrode materials for CO_2 electrolysis. The oxide electrodes have been characterized by X-ray diffraction (XRD), scanning electron microscopy (SEM), and electrical conductivity measurements. Finally, in order to analyze their CO2 reduction kinetics, the electrochemical performance is discussed in detail.

11:20 AM

(PACRIM-413-2021) Investigation of $Sm_{1-x}Ba_{0.5}Sr_{0.5}Co_2O_{5+d}$ Layered Perovskites for IT-SOFCs cathode

S. Woo*[1]; J. Shim[2]; J. Kim[3]; T. Shin[1]

1. Korea Institute of Ceramic Engineering and Technology, Republic of Korea
2. Korea University, Republic of Korea
3. Hanbat National University, Republic of Korea

Solid oxide fuel cells (SOFCs) are energy converters that directly convert from the chemical energy of oxygen and hydrogen to electrical energy at high temperatures. In the case of SOFCs, high efficiency and long-term stability have been reported. Recently, the perovskite and layered perovskite which are the crystal structure of ABO_{3-d} and $AA'B_2O_{5+d}$ (A: lanthanide, A': Ba, Sr, and Ca B: Co, Fe, Mn, Ni, and Cu) are introduced as a functional electrode for use in SOFCs. These materials reported that possess excellent electrical conductivity and ionic conductivity in high and low temperature range. In this study, non-stoichiometric layered perovskite oxide of $Sm_{(1-x)}Ba_{0.5}Sr_{0.5}Co_2O_{5+d}$ (x = 0 - 0.25) was synthesized by partially removing the A-site (= Sm). And the synthesized were analyzed the phase characteristics, electrical conductivity, cell performance, electrochemical analysis and X-ray photoelectron spectroscopy (XPS). For example, composition cathode materials which mixed with SBSCO-1(=excellent electrical properties) and SBSCO-0.95(=excellent electrochemical properties) at the same weight ratios were measured in SOFC sing cell test. The power density of single cell showed 2.84, 2.11, 1.38, and 0.69 W/ cm2 at 850, 800, 750 and 700 °C. It had higher power performance than other single cells which was applying a cathode of a single composition.

11:40 AM

(PACRIM-414-2021) Realization of highly durable O_2-electrode for solid oxide fuel cells by isovalent doping

B. Koo*[1]; J. Seo[1]; J. Kim[1]; W. Jung[1]

1. Korea Advanced Institute of Science and Technology, Department of Materials Science and Engineering, Republic of Korea

Surface Sr segregation and phase separation are the key reasons behind the chemical instability of Sr-containing perovskite oxide surfaces and the corresponding performance degradation of solid oxide fuel cell O_2-electrodes, but to date, practical solutions to prevent this phenomenon are limited. Here, we investigate how isovalent doping (in this case, Zr substitution of Ti) changes the surface morphology, chemical composition, and thus the O_2 activation kinetics under actual operating condition. Thin films of $SrTi_{0.5}Fe_{0.5}O_{3-\delta}$ as a representative model perovskite O_2-electrode, with Zr doping, are fabricated via pulsed laser deposition and their surface oxygen exchange rates are then characterized via electrical conductivity relaxation assessments. Zr dopants strengthen the Sr-O bonds in the oxide lattice, inhibiting the formation of surface SrO_x clusters and significantly reducing the deterioration of the oxygen exchange rates compared to the results from undoped film at 650°C for 30 h. These observations suggest a new strategy for ensuring the surface stability of Sr-containing perovskite oxides for fuel cell O_2-electrodes.

PACRIM Symposium 28: Atomic Structure and Electrochemical Property Diagnosis Toward Full Crystal Rechargeable Batteries

All Solid-state Battery System

Room: Regency B

8:30 AM

(PACRIM-415-2021) Developing oxide-based solid electrolytes with superior electrochemical/chemical properties for operating all solid-state battery (Invited)

B. Kang*[1]

1. Pohang University of Science and Technology(POSTECH), Republic of Korea

The demand for rechargeable Li batteries having more safety and higher energy density has been increased to meet the strong demand of novel applications such as electric vehicles and energy storage system. In this aspect, lithium ion batteries containing typical liquid electrolytes have fundamental limitations because liquid electrolytes can act as fuels in thermal runaway behavior leading to a fire or an explosion of battery and can be decomposed at high potential (> 4.5V) leading to the restricted use of high potential cathodes. To address these problems, there are several approaches. One of promising approaches is to apply proper oxide-based solid electrolytes (SEs) instead of liquid electrolytes. Among oxide-based SEs, garnet-type SEs have been a lot of attraction because they can have several advantages over liquid electrolytes in terms of electrochemical window, chemical stability with Li metal, and safety and over other SEs with respect to ionic conductivity and Li metal reaction. In this talk, I will discuss about the newly developed garnet-type SE that has superior electrochemical/chemical properties with respect to the wettability with Li metal, ionic conductivity, and chemical stability with air. Also, I will talk about the efforts in our group to build up all solid-state battery by using the developed garnet-type SE.

9:00 AM

(PACRIM-416-2021) Mixed-Fluorine Anion Effects on Electrochemical Energy Storage Characteristics of High-Nickel Layered Cathode Materials

N. Zettsu*[1]; T. Kondo[1]; K. Hara[1]; T. Sudare[1]; H. Shiiba[1]; K. Teshima[1]

1. Shinshu University, Japan

Designing and synthesizing metastable surfaces is a major challenge and core of interests in modern material science. Exploration of mixed anion compounds has been attracted many interests. It becomes currently one of most promising candidate approachs for materials designing due to offer many fascinating physical properties. It thanks to significant changes in their electronic structures in association with huge crystal field splitting of transition metal-octahedral complexes, comparing to conventional cation engineering. Here we applied non-equilibrium surface fluorination reaction with XeF_2 at room temperature for making metastable oxyfluoride surface to $LiNi_{0.8}Co_{0.1}Mn_{0.1}O_2$ system. Small amounts of fluorine substituting enhance the C-rate capability, cyclability as well as increase of specific capacity at higher cut-off voltage ranges (2.5-4.8 V, vs Li/Li^+). We demonstrate that the incorporated F- highly mitigate the side reaction with electrolytes through the stabilization of the $LiNi_{0.8}Co_{0.1}Mn_{0.1}O_2$ surface in association with the modified density of states near Fermi-levels of transition metals. Futhermore, the incorporated F- offers different deintercalatation mechanism composed of both minor two-phase coexsisting reaction and major solid-solution reaction via the stabilization of intermediate compotion phase in delithiated $Li_{1-y}Ni_{0.8}Co_{0.1}Mn_{0.1}O_2$.

9:20 AM

(PACRIM-417-2021) Distribution of Fe/Co in the oxygen evolution reaction catalyst Ca_2FeCoO_5

K. Nakayama*[1]; R. Ishikawa[1]; A. Kuwabara[2]; S. Kobayashi[2]; T. Motohashi[3]; N. Shibata[1]; Y. Ikuhara[1]

1. The University of Tokyo, Japan
2. Japan Fine Ceramics Center, Japan
3. Kanagawa University, Japan

Metal-air batteries have higher theoretical energy densities (> 1000 Wh/kg) than current lithium-ion batteries (380 Wh/kg). However, to realize such high energy densities, it is necessary to improve the activity of oxygen-evolution-reaction (OER) catalysts currently used at the air electrode. $Ca_2Fe_{2-x}Co_xO_5$ ($0 \le x \le 1$) with higher Co content is currently one of the best OER catalysts. Identifying the Fe/Co occupancies at the octahedral (O_h) and tetrahedral (T_d) sites in the crystal structure is important for the understanding of the role of Co and further improvement of OER catalysts. In this study, we investigate the Fe/Co distribution in Ca_2FeCoO_5 by atomic-resolution energy dispersive X-ray spectroscopy in scanning transmission electron microscopy. Co is found to be slightly enriched at the T_d and O_h sites in the samples of as-synthesized (1100 °C) and 800 °C annealed, respectively. The observed Co site preferences are considered to be the Fe/Co ionic size effect and ligand field effect, which are competitive around a crossover point at a certain temperature between 800 and 1100 °C. We also elucidate that the as-synthesized (1100 °C) sample with Co enrichment at the T_d sites shows a better OER activity, and the optimum annealing temperature for more OER active Ca_2FeCoO_5 should be higher than the crossover temperature. This study was supported by NEDO-RISING2 project (JPNP16001).

Solid-state Electrolytes

Room: Regency B

10:00 AM

(PACRIM-418-2021) Fluoride Ion Conducting Ceramics for Battery Application (Invited)

A. Mineshige*[1]

1. University of Hyogo, Japan

Recently, fluoride ion (F^-) conducting ceramics attract attention as solid electrolytes to be used in fluoride ion batteries. In the present study, fluoride ion conducting behavior in ceramics with fluorite-type and its related structure was reinvestigated in order to think of a strategy for the development of fast ion conductors. Especially, enhancing the ionic conductivity by using a mechanochemical synthesis was examined. First, fluorite-type BaF_2, $(Ba_{0.5}Ca_{0.5})F_2$ and $(Ba_{1-x}Sn_x)F_2$ fluorides were prepared via a mechanical milling process by changing chemical composition, milling media, weight ratios of ball to powder (B/P), and milling time. It was found that the size of grains, the degree of solid solution of Ca or La into the BaF_2 lattice, and higher pressure for pelletizing are the three main factors to determine the conductivity. Furthermore, to obtain higher F^- conductivity, the $(Ba_{1-x}Sn_x)F_2$ system with the fluorite-related structure was also investigated. By employing the mechanical milling method, we succeeded in obtaining the highest conductivive $(Ba_{1-x}Sn_x)F_2$. These BaF_2-based electrolytes are successful candidates for the electrolyte materials in high performance fluoride ion batteries. This work was supported by the projects, RISING2 (JPNP16001) and RISING3 (JPNP21006) commissioned by the New Energy and Industrial Technology Development Organization (NEDO), Japan.

10:30 AM

(PACRIM-419-2021) Sinterability and Grain Boundary resistivity of Garnet-type Li ionic conductor (Invited)

D. Mori*[1]; K. Ohmori[1]; K. Sato[1]; R. Katsu[1]; R. Ito[1]; Y. Matsuda[2]; S. Taminato[1]; N. Imanishi[1]

1. Mie University, Department of Chemistry for Materials, Japan
2. Osaka Institute of Technology, Department of Applied Chemistry, Japan

Fast ionic conductors attract much attention as an indispensable material for the development of next generation energy systems such as all solid-state lithium batteries and rechargeable aqueous lithium-air batteries. Garnet-like $Li_7La_3Zr_2O_{12}$(LLZ) and its derivatives are promising candidates because their electrochemical stability against lithium metals. The high ionic conductive phase of LLZ is stabilized by the substitution of Al^{3+} and Ga^{3+} for Li^+ sites and/or Nb^{5+} and Ta^{5+} for Zr^{4+} sites. For the development of all solid-state batteries with high energy density, the resistivity of grain boundary, interfacial resistance between electrodes and electrolytes, and Li dendrite formation are residual important issues. In previous study, we demonstrated that Ga and Sr co-substitution for LLZ improved sinterability and ionic conductivity in grain boundary. Moreover, the interfacial resistance between the electrolytes and Li metal electrode clearly decreased. In this study, we investigated the doping effect of alkali earth elements in sinterability and interfacial resistance against Li metal electrode of LLZ. In addition, to suppress the short circuit due to Li dendrite formation along grain boundaries, we investigated the relationship between composition and sinterability of grain boundary in LLZ. Acknowledgement: this work was partially supported by JSPS KAKENHI Grant Number 19H05793.

11:00 AM

(PACRIM-420-2021) Li-ion conductivities, atomic and electronic structures of $(Li_{3x}La_{2/3-x})TiO_3$ tilt grain boundaries

S. Sasano*[1]; R. Ishikawa[1]; H. Ohta[2]; N. Shibata[1]; Y. Ikuhara[1]

1. The University of Tokyo, Japan
2. Hokkaido University, Japan

$(Li_{3x}La_{2/3-x})TiO_3$ (LLTO; $0 < x < 0.16$) is one of promissing candidates for the solid-state electolytes owing to the high Li-ion conductivity in the bulk (1.0×10^{-3} S cm^{-1}). However, the Li-ion conductivity is largely reduced at the grain boundaries (GBs) (7.5×10^{-5} S cm^{-1}), which prevents practical applications. Here, we investigate the origin of low Li-ion conductivity at the LLTO GBs by using electrochemical strain microscopy (ESM) combined with scanning transmission electron microscopy (STEM). The LLTO thin films were heteroepitaxially grown on $SrTiO_3$ (STO) bicrystal substrates ($\Sigma5$, $\Sigma13$) by PLD method. We have qualitatively measured local Li-ion conductivities at the GBs by ESM, and investigated the GB atomic and electronic structures by STEM. It was found that the Li-ion conductivity was largely reduced at the $\Sigma13$ GB compared with the $\Sigma5$ GB. Moreover, a large amount of oxygen vacancies is introduced at the $\Sigma13$ GB compared with the $\Sigma5$ GB. These results suggest that local Li-ion concentration could be reduced at the $\Sigma13$ GB, inducing the reduction of the observed Li-ion conductivity at the $\Sigma13$ GB. Detailed discussion would be given in this presentation.

11:20 AM

(PACRIM-421-2021) Anisotropic ionic conductivity and its lithium concentration dependence in $Li_{3x}La_{2/3-x}TiO_3$ single crystal grown by the TSFZ method

Y. Maruyama*[1]; M. Ali[1]; M. Nagao[1]; S. Watauchi[1]; I. Tanaka[1]

1. University of Yamanashi, Japan

Double-perovskite-type $La_{2/3-x}Li_{3x}TiO_3$ (LLT) is one of candidates of the solid-state electrolyte for all-solid-state lithium ion batteries since LLT has a high ionic conductivity ($\sim10^{-3}$ S/cm) at room temperature and high stability at high temperature. Especially, LLT single crystals are required for experiment to clarify their basic physical properties such as anisotropic ion conductivity and to apply a substrate for application in high-performance all-solid-stare batteries. In this study, the single crystal of LLT with different nominal Li composition ($x = 0.083 \sim 0.15$) were grown by the TSFZ method. Crack- and inclusion-free single crystals with $x = 0.042\sim0.12$ were obtained. The ionic conductivity first increased with increasing Li composition to $x = 0.059$ and then decreased with increasing Li composition. The ionic conductivity along [100] direction, $\sigma_{[100]} = 1.75 \times 10^{-3}$ S cm^{-1} is higher than the [001] directions $\sigma_{[001]} = 7 \times 10^{-4}$ S cm^{-1}. The anisotropy of ionic conductivity was about 2.5 which is independent with lithium concentration. The activation energy slightly increased from 0.33 eV to 0.41 eV with increasing the Li composition.

11:40 AM

(PACRIM-422-2021) Fluoride ion conductivity of potassium tetrafluoroantimonate ($KSbF_4$)

K. Kawahara*[1]; R. Ishikawa[1]; K. Nakayama[1]; N. Shibata[1]; Y. Ikuhara[1]

1. University of Tokyo, Institute of Engineering Innovation, Japan

All solid-state fluoride ion batteries (FIBs) are considered to be higher energy density and safer energy storages than current lithium ion batteries. There is a serious issue for practical use of FIBs that the room temperature F^- ion conductivities of solid electrolytes and active materials are quite low ($< 10^{-6}$ S cm^{-1}), and operating temperature of FIB is limited to 420 K or higher. Therefore, the faster F^- ion conductors is strongly desired. Potassium tetrafluoroantimonate ($KSbF_4$) is an air stable F^- ion conductor that can be synthesized from aqueous solution. In this study, we synthesized $KSbF_4$ polycrystals under various sintering temperatures from 490 K to 550 K, and evaluated F^- ion conductivity by using electrochemical impedance spectroscopy. The F^- ion conductivity is increased as a function of annealing temperature, and $KSbF_4$ annealed at 550 K shows the highest F^- ion conductivity of 1.0×10^{-4} S cm^{-1} at room temperature that is 60 times larger than that of $KSbF_4$ annealed at 490 K. Mass spectrum during the heat treatment indicates that SbF_3 evaporates from $KSbF_4$ during the heat treatment with retaining the crystal structure. This result suggests that F site vacancies are introduced in the $KSbF_4$ lattice by the heat treatment and promote F^- ion diffusivity in $KSbF_4$. This study was supported by NEDO-RISING2 project (JPNP16001).

PACRIM Symposium 32: Ceramics for Enabling Environmental Protection: Clean Air and Water

Ceramics for Enabling Environmental Protection: Clean Air and Water

Room: Stanley

8:30 AM

(PACRIM-423-2021) Wet mechanochemical synthesis of morphology-controlled functional particles (Invited)

T. Kozawa*[1]; A. Kondo[1]; M. Naito[1]

1. Osaka University, Joining and Welding Research Institute, Japan

A ball milling process not only affords particle size reduction by grinding but also particle synthesis induced by mechanical actions at ambient temperature. Wet mechanochemical reactions is a simple and scalable method to produce thermodynamically unstable hydrous materials including hydrates and hydroxides. We report a wet mechanochemical process in water to synthesize functional hydrous materials with various shapes (nanosheets, nanotubes, platelets, flakes, rods, and nanoparticles). Morphology-controlled particles can be obtained by only adjusting the milling conditions such as milling time, ball size, and centrifugal acceleration. The formation mechanism is based on a dissolution-precipitation reaction. Furthermore, the hydrous products can be converted into electrode materials for Li-ion batteries, retaining the precursor morphologies. The study results will promote wet mechanochemical processes to be widely used as classic but advanced particle synthesis method.

9:00 AM

(PACRIM-424-2021) Turning CO_2 Into a Cementitious Binder by Mineralization

M. Bauchy*[1]; G. Sant[1]

1. University of California, Los Angeles, Civil and Environmental Engineering Department, USA

Concrete, which is by far the most manufactured material in the world, faces many challenges, including a substantial embodied carbon intensity. To mitigate the carbon footprint of concrete, we present a new carbon immobilization process based on a 3D-printable portlandite-based cementitious binder that permits CO_2 uptake by mineralization. This process greatly reduces the carbon footprint associated with traditional concrete production. In turn, by using CO_2 as a resource for producing concrete, this approach could change our perception of CO_2—wherein CO_2 is reimagined as a resource for the production of value-added products rather than as a waste.

PACRIM Symposium 1: Characterization and Modeling of Ceramic Interfaces: Structure, Bonding, and Grain Growth

Microstructure Evolution and Grain Growth / Advances in Interface Characterization and Modeling

Room: Regency A

1:30 PM

(PACRIM-425-2021) Non-Arrhenius grain growth in Strontium Titanate: The impact of space charge on grain boundary motion (Invited)

W. Rheinheimer*[1]

1. Jülich Research Ceter, Institute of Energy and Climate Research - Materials Synthesis and Processing, Germany

Perovskite materials are widely used in a variety of electronic devices, e.g. capacitors, oxygen conductors, PTC (positive temperature coefficient) heaters and piezoelectric actuators. Many macroscopic properties of polycrystalline perovskites are governed by their grain boundaries and, depend strongly on microstructure evolution during processing. Several perovskite ceramics (ABO_3) are known to show non-Arrhenius grain growth. Particularly strontium titanate has a grain growth transition, with decreasing growth rates between 1350°C and 1425°C. Other materials such as barium titanate, lithium lanthanum titanate and barium strontium titanate show similar grain growth transitions. The growth transitions are coupled to the formation of bimodal microstructures and are postulated to be caused by grain boundary structural transitions. This talk argues that the grain growth transition of strontium titanate is a space charge transition, i.e., a solute drag effect. Evidence for this was found in the morphology of the microstructure, the occurrence of grain growth stagnation under certain defect-chemical conditions, the impact of dopants and the thermodynamics of space charge. Since space charge involves segregation of metal vacancies and impurities, a diffusional drag arises for those grain boundaries with space charge resulting in bimodal microstructures and the grain growth transition.

2:00 PM

(PACRIM-426-2021) Pressureless sintering of nano Y_2O_3: Grain growth and densification

A. Najafzadehkhoee*[1]; A. Talimian[4]; J. Sedláček[2]; P. Hvizdoš[3]; D. Galusek[4]

1. Joint Glass Centre of the IIC SAS, TNUAD, and FChPT STU, Slovakia
2. Institute of Inorganic Chemistry, Slovak Academy of Sciences, Slovakia
3. Institute of Materials Research, Slovak Academy of Sciences, Slovakia
4. Centre for Functional and Surface Functionalized Glass, Alexander Dubček University of Trenčín, Slovakia

The fabrication of highly dense fine-grained bodies is a prerequisite for obtaining desired thermo-mechanical properties in transparent polycrystalline ceramics. In the present work, the effect of the sintering regime on the densification and grain growth of nano yttria were studied. A commercial nano Y_2O_3 was shaped by dry pressing and subjected to pressure-less sintering with various combinations of sintering temperatures (1350-1700°C) and soaking times (0-24 h). The sintering trajectory of the powder was constructed according to the relative density of sintered samples as a function of grain size distribution. The grain growth was suppressed by performing two-step sintering consisting of the primary step at 1700°C followed by secondary dwelling at 1400 °C. The influence of subsequent hot isostatic pressing (HIP) on grain growth and the physical properties of pre-sintered samples is critically discussed.

2:20 PM

(PACRIM-427-2021) Segregation-controlled densification and grain growth in rare earth doped Y_2O_3

M. Kindelmann*[1]; K. Ran[2]; W. Rheinheimer[1]; K. Morita[3]; J. Mayer[2]; M. Bram[1]; O. Guillon[1]

1. Forschungszentrum Juelich, IEK-1: Materials Synthesis and Processing, Germany
2. RWTH Aachen University, Central Facility for Electron Microscopy (GFE), Germany
3. National Institute for Materials Science (NIMS), Japan

Cation doping of Y_2O_3 is an established approach for tailoring densification and grain growth during sintering. Here, we utilize two isovalent dopants (La^{3+} and Gd^{3+}) and focus on the impact of their segregation behaviour on the processing properties. We investigate the densification as well as the grain growth kinetics of both doped and undoped Y_2O_3 during field assisted sintering / spark plasma sintering (FAST/SPS). While Gd^{3+} is showing no significant effect on densification, La^{3+} resulted in a strongly reduced sintering activity. The analysis of the grain growth during sintering and on pre-densified samples revealed a decrease of the grain growth coefficient, with La^{3+} having the strongest impact. The grain boundaries were investigated by aberration corrected STEM. While no structural change was caused by doping, the chemical analysis showed a strong segregation of La^{3+} to the grain boundary, which could not be observed for Gd^{3+}. The results indicate that segregated La^{3+} causes a drastic decrease of grain boundary migration rates through solute drag as well as much slower sintering kinetics, likely caused by a decrease of the grain boundary self-diffusion due to segregation. This study further underlines the importance of the elastic contribution to cation segregation and establishes a clear relationship to grain growth and sintering kinetics, which are both decreased by segregation.

2:40 PM

(PACRIM-428-2021) Design of high-performance Al_2O_3–Ce:YAG ceramic converters for white LEDs by optimization of cerium doping

A. A. Vornovskikh*[1]; D. Y. Kosyanov[1]; A. P. Zavjalov[1]; A. A. Leonov[2]; W. Li[3]; X. Liu[3]; J. Li[3]

1. Far Eastern Federal University, Russian Federation
2. Far Eastern Branch, Russian Academy of Sciences, Institute of Automation and Control Processes, Russian Federation
3. Shanghai Institute of Ceramics, Chinese Academy of Sciences, Key Laboratory of Transparent Opto-functional Inorganic Materials, China

High-performance Al_2O_3–Ce:YAG ceramic converters were obtained via reactive vacuum sintering using α-Al_2O_3, Y_2O_3, and CeO_2 powders as starting materials, and SiO_2, MgO as a complex sintering additive. The effect of cerium content (0.05-0.3 at%) on the structural-phase state, optical and photoluminescent properties of ceramic phosphors were investigated comprehensively. It is shown that an increase in the cerium doping level of the composites initiates their partial recharging into the non-luminescent Ce^{4+} state and the parasitic introduction of magnesium ions into the aluminum oxide phase due to the stabilization of the crystal structure of Mg,Si,Ce:YAG solid solutions. The highest value of the PL intensity is observed for the 0.1 at% Ce^{3+} sample, which is caused by the optimal ratio of its absorbance and radiative recombination rate compared to other samples. The obtained phosphors show excellent thermal quenching behavior – reduction in the PL-emission intensities did not exceed 5% at heating temperatures to 150°C. Al_2O_3–Ce:YAG samples with 0.05-0.1 at% Ce^{3+} doping have an optimal balance between high CRI, LE, and appropriate CCT values and may find application in high-power and high-brightness natural-WLEDs excited by blue-LED chips. Acknowledgments: This work was supported by the Russian Science Foundation (Project No. 20-73-10242; for D Yu K, A A V and A P Z).

3:20 PM

(PACRIM-429-2021) Charged Interfaces: Equilibrium, Phase Transitions, and Microstructural Evolution (Invited)

R. Garcia*[1]

1. Purdue University, Materials Engineering, USA

Surfaces and interfaces in ionic solids play a pivotal role in defining the transport properties and microstructural evolution in many of the existing and emerging material applications, including energy-related systems such as fuel cells, rechargeable batteries, as well as advanced electronics such as those found in semiconducting, and ferroelectric applications. Here, a variational framework is presented to understand the effects of ionic species and point defects on the structural, electrochemical and chemomechanical stability of grain boundaries in polycrystalline ceramics. The proposed theory predicts the equilibrium and phase transition conditions of charged interfaces, and quantifies the properties induced by the broad region of electrochemical influence in front of a grain boundary, and includes the effect of other driving forces, including stresses, misorientations, and structural disorder. These effects are capable to span anywhere from a few angstroms to entire grains. The effects of charge on grain boundary drag by the interfacially accumulated ionic species and defects during grain growth under electrical, chemical, mechanical, or structural driving forces are quantified. The microstructural mechanisms leading to the onset of the flash during electric field assisted sintering and the resultant grain growth are predicted and directly compared against the experiments.

3:50 PM

(PACRIM-430-2021) Oriented Surface Nucleation, Volume Nucleation Ability and Interfacial Energy

C. Tielemann*[1]; R. Müller[1]; S. Reinsch[1]

1. BAM Federal Institute for Materials Research and Testing, 5.6 Glass, Germany

We found a good correlation between the calculated surface energy of crystal faces, oriented surface nucleation and the mode of crystal (surface vs. volume) crystallization. Surface energies were estimated assuming that crystal surfaces resemble minimum energy crack paths along the given crystal plane. This concept was successfully applied by Rouxel in calculating fracture surface energies of glasses. Several oriented nucleation phenomena can be herby explained assuming that high energy crystal surfaces tend to be wetted by the melt. This would minimize the total interfacial energy of the nucleus. Likewise, it was shown that volume nucleation tendency increases with decreasing average crystal surface energy.

PACRIM Symposium 3: Advanced Structure Analysis and Characterization of Ceramics

Spectroscopic and Scattering Methods

Room: Plaza C
Session Chairs: Nishant Garg, University of Illinois Urbana-Champaign; Scott Misture, Alfred University

1:30 PM

(PACRIM-431-2021) Understanding the irradiation damage tolerance of zirconium carbide ceramics by means of multiscale characterizations (Invited)

J. Zhang*[1]; L. Chen[1]; C. Wang[2]; J. Wang[1]

1. Institute of Metal Research, CAS, Advanced Ceramics and Composites, China
2. Center for Applied Physics and Technology, State Key Laboratory of Nuclear Physics and Technology, China

Zirconium carbide is considered as an attractive structural material for new generation reactors operating at high temperature, due to its promising properties, including high melting point, high hardness, good wear and corrosion resistance, high thermal conductivity at high temperature, high thermal-chemical stability, as well as low cross section capture for neutrons. Besides, zirconium carbide is a typical non-stoichiometric compound with a high carbon vacancy concentration. Accordingly, a series of non-stoichiometric ZrC_x were synthesized by PVD method. By means of synchrotron X-ray diffraction as well as first-principle calculations, the nearest neighboring carbon vacancy pairs was traced in the sub-stoichiometric zirconium carbides. The microstructure evolution of zirconium carbides with different stoichiometry induced by 3 MeV Au ion irradiation were investigated by multiscale characterizations. Understanding the irradiation damage tolerance of zirconium carbide against heavy ions, it is expected to provide informatics for ceramic material screening in the application of Generation IV nuclear power systems.

2:00 PM

(PACRIM-432-2021) Effect of silicon dopant on the hexagonal boron nitride interphase for SiC_f/SiC composites

S. Zhang*[1]; J. Wang[1]; J. Wang[1]

1. Insitute of Metal Research, Chinese Academy of Sciences, Advanced Ceramics and Composites Division, China

In urgent need of new generation high-temperature materials, SiC_f/SiC composites for thermal-structural and security-critical components in the aerospace field that feature an excellent level in their thermophysical properties and high chemical stability under extreme conditions have gained great attentions. With the prime object of reaching high damage tolerance when a crack propagation based on energy dissipation effects, h-BN interphase has proven to be effective by achieving a weak fiber-matrix interfacing. However, sensitiveness to moisture and moderate-temperature degradation has brought great challenges for h-BN interphase and an incorporation of Si is the best solution regarding a suitable compromise between mechanical properties and chemical stability by optimizing silicon content. In the present work, rewarding theoretical simulation of $(BN)_{64}Si_x$ (x=4, 8, 16, 32) models has been conducted to uncover and allow a visualization of the atomistic origins of structural evolution and intrinsic physical properties. The analysis, including RDF, coordination environment and elastic properties have provided pictorial description for Si-doped BN models. Results have verified that, $(BN)_{64}Si_{16}$ model with 22.1 wt% silicon dopant satisfies the high requirement of oxidation resistance and also realize an optimal debonding as a mechanical fuse.

2:20 PM

(PACRIM-433-2021) The effect of structural phase transition on phase separation in glass

K. Nakazawa*[1]; Y. Tsukada[2]; K. Mitsuishi[1]; S. Amma[3]; K. Shibata[4]; T. Mizoguchi[4]

1. National Institute for Materials Science (NIMS), Japan
2. Nagoya University, Japan
3. AGC Inc., Japan
4. University of Tokyo, Japan

Phase separation phenomena are used to modify the physical or chemical properties of glasses. The effect of structural phase transition such as crystallization or amorphous-amorphous phase transitions on phase separation in glass is not fully understood. Especially, the effects of the structural phase transitions in the initial or intermediate stages of phase separation are not understood well. The reason for the lack of understanding is that it was difficult to observe the phase-separated structures of the initial and intermediate stages of phase separation whose phases and the compositional difference between the phases are small. Here, we used scanning transmission electron microscopy(STEM) which has high spatial resolution and chemical sensitivity. STEM can observe the phase-separated structures of the initial and intermediate stages. We selected the BaO-SiO_2 glass system and conducted in situ heating experiment. As a result, the formation and growth processes of a phase-separated structure are observed. We found a rapid and local change in phase-separated structures. We compared the result of in situ observation with phase-field simulation and revealed that the rapid and local change is caused by the structural phase transitions. Details are discussed in the presentation.

2:40 PM

(PACRIM-434-2021) Investigation of the Defect-Chemistry of Ceria-Zirconia Mixed Oxides (CZO) Using Microwaves

C. Steiner*[1]; I. Kogut[2]; G. Hagen[1]; H. Fritze[2]; R. Moos[1]

1. University of Bayreuth, Department of Functional Materials, Germany
2. Clausthal University of Technology, Institute of Energy Research and Physical Technologies, Germany

Ceria-zirconia mixed oxides (CZO) are oxygen storage materials. They are an elementary part of three-way catalysts (TWC) in modern gasoline vehicles. Previous studies showed that the oxygen storage level of a TWC can be operando determined using microwave technology. To investigate the dielectric behavior of CZO under microwave conditions in more detail, $Ce_{1-x}Zr_xO_{2-\delta}$ powders were studied in a microwave resonator specifically designed for high temperatures in a wide oxygen partial pressure (p_{O2}) range. Dielectric properties of the CZO powders were analyzed between $p_{O2} \approx 10^{-25}$ bar and $p_{O2} \approx 0.2$ bar at 600 °C. Under oxidizing conditions, a lower p_{O2} led to an increase in conductivity, which is likely related to an activated small-polaron hopping mechanism due to oxygen vacancy formation. Under reducing conditions, only small changes in conductivity were observed, which is probably attributed to a suppression of the hopping mechanism by higher non-stoichiometries δ ($[Ce^{3+}] \approx [Ce^{4+}]$). Additionally, comparison with data of pure ceria suggests that the formation of oxygen vacancies in CZO is hardly dependent on the microstructure. The results of the study provide further insight into CZO and may contribute to better understand the defect chemistry of CZO and to improve the control of three-way catalysts using microwave technology.

3:20 PM

(PACRIM-435-2021) Novel Raman imaging protocol for high-resolution phase mapping of granite

K. Polavaram[1]; N. Garg*[1]

1. University of Illinois Urbana-Champaign, Civil and Environmental Engineering, USA

Accurate characterization of a material's chemical and phase composition is central to our fundamental understanding of how ceramics perform and respond to external environment. Specifically, for phase mapping silicates, there has been a strong interest in adopting automated, objective mapping methods based on SEM-EDXS (Scanning Electron Microscopy – Energy Dispersive X-ray Spectroscopy) as well as Electron Probe Micro Analysis (EPMA). However, mineral phase maps obtained from these newer approaches rely on conversion of elemental compositions to mineralogicalcompositions and are hence not fully effective in distinguishing mineral polymorphs of the same chemical composition but different atomic structure. Secondly, these techniques often require laborious sectioning and polishing sample preparations which are time-consuming. Here, we develop a new Raman imaging protocol that is capable of mapping unpolished samples with an auto-focusing Z-mapping feature that allows direct fingerprinting of different polymorphs. Specifically, we report a novel methodology for generating high fidelity phase maps by exploiting characteristic peak intensity ratios which can be extended to any multi-phase, heterogenous system. Collectively, these enhancements allow us to rapidly map an unpolished granite specimen (~ 2 x 2 mm) with an exceptionally high accuracy (>95%) and an extremely fine spatial resolution (< 0.3 - 2 μm).

3:40 PM

(PACRIM-436-2021) Correlative Raman spectroscopy and EBSD for orientation and strain state analysis in MAX phases

J. Lyons*[1]; F. Giuliani[1]; F. Bouville[1]

1. Imperial College London, Materials, United Kingdom

MAX phases are a highly anisotropic material, giving rise to a large variation in their properties depending on orientation of the crystal lattice. Typically, orientation can be determined through Electron Backscatter Diffraction (EBSD), is a powerful technique with very high spatial resolution and the ability to determine strain in a crystal structure. However, it requires a high level of sample finish, long scan times and can be limited but sample size dependant on the machine used. Raman spectroscopy is highly sensitive to the anisotropic nature of the MAX phases, leading to significant changes to full width half maximums and intensities of peaks. Raman has a lower spatial resolution but requires essentially no sample preparation and allows for rapid acquisition of data compared to EBSD. Therefore, we have been able to correlate orientations provided by EBSD to Raman spectra taken in those grains, allowing for rapid orientation analysis of samples. As Raman is sensitive to changes in bond length, it can analyse strain at the bonding level, as any change in bond length will lead to a shift in the Raman peak. This makes it an incredibly powerful technique for MAX phases as peaks relating to the M-A and M-X are segregated into two different regions of the Raman spectra. This means it is possible to determine how strain is being accumulated in the different layers of the MAX phase.

4:00 PM

(PACRIM-437-2021) On the tunable coefficient of thermal expansion of rare earth di-silicates: A combination of experimental and theoretical investigations (Invited)

L. Sun*[1]; Y. Luo[1]; J. Wang[1]

1. Institute of Metal Research, Chinese Academy of Sciences, Advanced Ceramics and Composites Division, China

Silicon carbide fiber reinforced SiC composites (SiC_f/SiC_m CMCs) are leading candidates to replace supperalloy hot-section components in new generations of gas turbine engines. It is believed that environmental barrier coating (EBC) systems are necessary in order to prevent SiC-based CMCs suffering from severe recession due to water-vapor reaction and calcium-magnesium-aluminosilicates (CMAS) corrosion. Rare-Earth (RE) mono- and di-silicates (RE_2SiO_5 and $RE_2Si_2O_7$) are most concerned candidates due to their excellent capability in harsh combustion environments. However, there would be huge gaps in coefficient of thermal expansion (CTE) from inner to outer coating layers and it would be significant to tailor the CTE of rare earth silicates for their practical applications. In this work, lattice dynamics calculations and experimental investigation, including HT-XRD and TEM, are performed on $RE_2Si_2O_7$ compounds and the relationships among the structure, phonon behavior and thermal expansion are discussed. Hereby a possible route for control the CTE of $RE_2Si_2O_7$ is proposed. This study provides new insights into the mechanisms driving diverging CTE among $RE_2Si_2O_7$ polymorphs and facilitate the realization of tunable CTE for rare earth di-silicates.

4:30 PM

(PACRIM-438-2021) X-ray Total Scattering Study of Cation Clustering in Hollandite Radionuclide Hosts

R. Koch[1]; J. Schaeperkoetter*[2]; J. Amoroso[3]; K. Brinkman[4]; T. M. Besmann[5]; M. Zhao[4]; S. T. Misture[2]

1. Brookhaven National Laboratory, USA
2. Alfred University, USA
3. Savannah River National Laboratory, USA
4. Clemson University, USA
5. University of South Carolina, USA

The hollandite phase is a potential nuclear waste form that can accommodate alkali and alkaline-earth metals such as Cs and Ba by chemical incorporation into the tunnels available in the crystalline lattice. Recent investigations reveal multi-scale ordering of these materials that may be important in the structure-function relationship, with local ordering around the radionuclide playing a key role in effective sequestration. Here we present results from X-ray and neutron total scattering studies on Cs entrapped in hollandite-type materials, $A_xB_8O_{16}$. The A-site is occupied variously by Ba, Cs, Na and the B-site by Ti, Al, Mg. Pair distribution function (PDF) analysis reveals clear differences in the long- and short-range order, signaling that local structure is present that is not reflected in the overall average crystal structure. Sliding boxcar fitting followed by reverse Monte Carlo simulations were employed, and some new software developed using ASE and python to interrogate the large box models generated using RMC. We assess our structural models in light of the measured thermodynamic stability and atomistic modeling efforts used to describe the overall average bulk solids.

PACRIM Symposium 5: Polymer Derived Ceramics (PDCs) and Composites

Processing and Applications I

Room: Plaza B

1:30 PM

(PACRIM-439-2021) Processing of polymer-derived ceramics (PDCs) for fibers, monoliths, and ceramics matrix composites (CMCs) (Invited)

C. L. Cramer*[1]; T. G. Aguirre[1]

1. Oak Ridge National Laboratory, Manufacturing Sciences, USA

Polymer-derived ceramics (PDCs) are used to make fibers, monolithic ceramics, and ceramic matrix composites (CMCs). They are used in pre-impregnation (prepreg) and infiltration of fiber preforms, infiltration of powder preforms, and they are also amenable to additive manufacturing (AM) since they can be processed similar to thermosetting polymers. The challenges for polymer precursor processing of fibers, monoliths, and CMCs are infiltration times, shrinkage, cracking, obtaining final stoichiometry and microstructure, and achieving high-temperature stability. PDCs cannot retain their shape in neat form, and once fully cross-linked, they will become solidified. As such 3D printing of PDCs has been mainly limited to stereolithography with the addition of acrylates and photoinitiator or direct ink writing with particles and small amount of cross-linker. Here, it will be shown how catalysts and cross-linkers can help achieve proper rheology and increase the polymer to ceramic yield, how infiltration can be improved, how fibers can be processed, as well as show how polymer-based slurries can be used for printing of CMCs and high temperature composites. From precursor selection, mixing, and modifying to curing, pyrolyzing, and crystallizing; the processing of PDCs for many applications will be addressed.

2:00 PM

(PACRIM-440-2021) Freeze Casting of porous Polymer-derived Ceramics (Invited)

D. Schumacher[1]; P. H. da Rosa Braun[1]; H. Zhang[1]; M. Dreyer[1]; K. Rezwan[1]; M. Wilhelm*[1]

1. University of Bremen, Production Engineering, Germany

Freeze casting has proven to offer a wide variety of possibilities to largely modify the resulting pore structure and associated properties such as permeability or mechanical strength. In addition, polymer derived ceramics (PDC) are reported to offer a unique combination of properties, such as low thermal conductivity, tailored surface properties and the ability to easily modify the chemical structure. Consequently, the combination of flexible freeze-casting with the tailored polymer-derived ceramics is very promising to address the challenges of various applications in the field of porous monoliths. Preceramic polymers offer the interesting property that they can be processed both in suspension-based freeze casting with preceramic particles dissolved in a solvent and in solution-based freeze casting dissolved in organic solvents. In addition, a hybrid process in which polysiloxane-derived fillers are suspended in a dissolved preceramic polymer allows further structural as well as functional properties such as pore wall roughness, specific surface area, strength and hydrophilicity to be modified and added. Further adjustments of the pore structure can be achieved by changing the freezing conditions. These processes thus offer a wide range of possibilities, which will be explained in this presentation, especially with regard to the development of monolithic structures used in capillary transport applications.

2:30 PM

(PACRIM-441-2021) Hydrogen transport properties of polymer-derived cobalt cation-doped amorphous silica

S. Tada*[1]; Y. Daiko[1]; S. Honda[1]; S. Bernard[2]; Y. Iwamoto[1]

1. Nagoya Institute of Technology, Japan
2. CNRS, IRCER, France

Hydrogen transport properties induced by divalent cobalt cation (Co(II)) in amorphous silica network were intensively studied with the aim to improve high temperature hydrogen-permselectivity of microporous amorphous silica-based membranes. Co(II)-doped amorphous silica materials were synthesized through polymer-derived ceramics (PDCs) route: (i) Perhydropolysilazane (PHPS)—as a non-oxidic preceramic polymer—was reacted with Co(III)-acetylacetonate(Co(acac)$_3$) at various Co/Si atomic ratios to investigate the atomic and molecular structure of the ternary Si–Co–O system, and (ii) the synthesized Co-modified PHPS sample was converted to Co(II)-doped amorphous silica by oxidization in aqueous ammonia vapor at room temperature, followed by pyrolysis in air at 600 °C. Hydrogen (H)/deuterium (D) isotope exchange behavior at the surface silanol groups (Si–OH/OD conversion) of the polymer-derived non-doped and Co(II)-doped amorphous silica was in situ monitored by measuring diffuse reflectance infrared Fourier transform (DRIFT) spectra at 500 °C. At the presentation, enhanced hydrogen transport properties of the polymer-derived ternary Si–Co–O system will be shown and discussed based on the local structure of the polymer-derived Co(II)-doped amorphous silica intensively characterized by several spectroscopic analyses and high-resolution transmission electron microscopy observation.

Processing and Applications II

Room: Plaza B

Session Chair: Corson Cramer, Oak Ridge National Lab

3:10 PM

(PACRIM-442-2021) Additive manufacturing of preceramic polymers by vat photopolymerization

H. Elsayed[1]; Y. Feng[2]; K. Huang[1]; G. Franchin[1]; P. Colombo*[1]

1. University of Padova, Industrial engineering, Italy
2. Shandong University of Technology, China

The use of different approaches for the fabrication of SiOC and SiCN complex structures using vat photopolymerization technologies will be described. Ceramic parts were manufactured using either a chemically modified precursor or the blending of a preceramic polymer with a photocurable polymer, employing Digital Light Processing (DLP). For the first time, we tested the use of a photocurable preceramic polymer solution with a novel linear volumetric 3D printing method denominated Xolography, adapting the feedstock to suit the specific technology's requirements in terms of viscosity, transparency and reactivity. The promising results of the initial experiments will be reported.

3:30 PM

(PACRIM-443-2021) SiCN Ceramics via Reactive Atmosphere Pyrolysis under Hydrogen

A. B. Hande*[1]; P. Kroll[1]

1. University of Texas, Arlington, Chemistry and Biochemistry, USA

Starting with a polysilazane polymer (Ceraset PSZ20) we synthesize SiCN ceramics via pyrolysis under different atmospheres. Visual differences are observed depending on the pyrolysis conditions. Samples pyrolyzed under nitrogen turn black already at 800°C. In contrast, pyrolysis under pure hydrogen maintains transparent products at 600°C and yields semi-opaque samples at 800°C. Vibrational spectroscopy indicates the persistence of aliphatic carbon in samples pyrolyzed under hydrogen, with significant amounts of carbidic carbon integrated into a SiCN host structure. Samples pyrolyzed under nitrogen, on the other hand, exhibit mostly aromatic carbon. Therefore, high activity of hydrogen during pyrolysis of polysilazanes promotes carbon towards sp^3-like configurations and prevents early formation of extended aromatic segregations in SiCN ceramics. The results further indicate it is possible to synthesize transparent amorphous SiCN ceramics using reactive atmosphere pyrolysis under hydrogen and to produce ceramics with high content of SiC.

PACRIM Symposium 8: Porous Ceramics: Innovative Processing and Advanced Applications

Innovative Processing Route for Porous Ceramics

Room: English Bay

Session Chair: Mary Anne White, Dalhousie University

1:30 PM

(PACRIM-444-2021) Highly Porous Polymer-derived Biosilicate-Carbon Composites: Phase Design and Thermal Characterization (Invited)

F. Dogrul[2]; H. Elsayed[1]; M. Michalek[2]; D. Galusek[2]; D. Del Col[1]; S. Bortolin[1]; E. Bernardo*[1]

1. University of Padova, Department of Industrial Engineering, Italy
2. University of Trencin, FunGlass - Centre for Functional and Surface Functionalized Glass, Slovakia

Silicone polymers have been recently established as precursors for several bioceramics, especially when comprising oxide fillers, introduced as carbonates, hydroxides, phosphate and borate salts. The firing at 900-1200 °C may yield a glass-ceramic material by direct reaction between the silica-rich residue - from the decomposition of silicones - and fillers. The synthesis of can be coupled with the shaping - before firing - of highly porous foams and reticulated scaffolds, exploiting the water vapour release from hydrated fillers or the direct ink writing of silicone-based pastes. The present paper is dedicated to porous composites (porosity >75%), featuring a Biosilicate glass-ceramic matrix (a highly biocompatible and bioactive silico-phosphate system) embedding pyrolytic carbon, from the treatment in N2, instead of air, of silicone mixtures. The mixtures were adjusted according to the silica fraction in the SiOC residue of two commercial silicones (MK and H44), in order to achieve the desired matrix. Despite the different nominal content of pyrolitic C of the two polymers, the obtained composites exhibited a similar, remarkable IR absorption (800-2000 nm), later exploited in photothermal experiments. IR radiation, in fact, was found to lead to a rapid heating of porous bodies (up to 70-75 °C), which could be exploited for sterilization before use as bone implants.

2:00 PM

(PACRIM-445-2021) Sol–Gel Synthesis of Preceramic Polymer Gels and Their Conversion to Porous Ceramics (Invited)

G. Hasegawa*[1]

1. Nagoya University, Japan

Porous ceramics based on reduced crystal phases attract attention due to their unique physical properties. In general, it is more difficult to synthesize porous reduced ceramics compared to their oxide counterparts in terms of the morphological control. The preceramic polymer route is a powerful tool to manufacture various reduced ceramics, which are called polymer-derived ceramics (PDCs). By combining the preceramic polymer route with the other techniques such as hard-templating and blowing, a number of porous reduced ceramics have been developed to date. However, it is still challenging to tailor small pores (less than 10 micrometers) in PDCs. We have proposed a synthetic approach to obtain reduced ceramics in monolithic shape with three-dimensionally interconnected macropores ranging from submicrons to several microns: macroporous inorganic-organic hybrid gels are prepared via the sol–gel process accompanied by phase separation, which are subsequently subjected to calcination under inert atmosphere to allow for the polymer-to-ceramic conversion. The crystal transition behaviors and the change of pore properties on calcination is to be discussed.

Porous Ceramics for Thermal Management

Room: English Bay

3:20 PM

(PACRIM-446-2021) Form-Stable Phase Change Materials Using Porous Ceramics (Invited)

M. White*[1]; J. Noel[1]

1. Dalhousie University, Canada

Form-stable phase change materials (PCMs) can provide containerless thermal energy storage with enhanced thermal conductivity relative to the pure PCMs. We show that many different freeze-cast matrices (alumina, titania, carbon black, alumina/carbon black, chitosan, carbonized chitosan) can be used as matrices for PCMs (fatty acids, PEG, paraffin, esters, sugar alcohols, salt hydrates), to prepare a wide variety of form-stable phase change materials. The porous ceramic matrix did not substantially influence the PCM melting points, but did control the fill, thermal conductivity and mechanical properties of the form-stable PCM. Alumina at a 38 mass% loading of PCM gave the best thermal conductivity, up to 3.2 $W\ m^{-1}\ K^{-1}$, significantly enhanced compared with 0.15 W m-1 K-1 for the pure PCM, and with excellent hardness and mechanical stability after 1000 cycles. Chitosan and carbonized chitosan scaffolds gave form-stable PCMs with the highest PCM loading, up to 94 mass%.

3:50 PM

(PACRIM-447-2021) Preparation of porous diatomite ceramics and hollow silica nanofiber for heat insulating materials (Invited)

Y. Nakashima*[1]; M. Fukushima[1]

1. National Institute of Advanced Industrial Science and Technology (AIST), Japan

In this presentation, new methods have been proposed for producing porous diatomite ceramics and hollow silica fibers for improving heat insulating materials. Porous diatomite ceramics were prepared by alkali treatment of the raw powder without any sintering procedure, in which the alkali treatment was carried out by dissolution of the surface by KOH solution to form solid crosslinking among raw particles near room temperature. Variation of the treatment factors was found to be effective for controlling properties such as porosity, strength and thermal conductivity. The resultant porous diatomite ceramics showed comparable compressive strength and thermal conductivity to those of the sintered materials. The proposed method is a eco-friendly, useful and versatile tool for high-performance porous ceramics prepared around room temperature. Hollow silica nanofibers were prepared by a template method in which chitosan nanofibers were used as a templating material for silica coating. Varying the processing factors was found to be effective in controlling the morphology of the silica nanofibers. The resultant samples had a rough silica walls composed of densely assembled silica nanoparticles, with a high specific surface area (338 m^2/g). Therefore, this prepared fiber can be applied as a high-performance heat insulating material.

PACRIM Symposium 9: Additive Manufacturing and 3D Printing Technologies

Multi-Material Process

Room: Prince of Wales
Session Chair: Bai Cui, University of Nebraska–Lincoln

1:30 PM

(PACRIM-448-2021) A Novel Solidification Microstructure Simulation Method for Metal Additive Manufacturing (Invited)

Y. Zhao*[1]

1. McGill University, Mechanical Engineering, Canada

Metal additive manufacturing (MAM) technology is widely applied in the manufacturing industry where microstructure simulation shows its importance. Traditional solidification microstructure simulation methods have their merits as well as drawbacks when applied in MAM. In this work, a deterministic solidification microstructure model, named "invasion model", is established to avoid the essential defects of traditional methods. This model focuses on the interaction between the neighboring bi-crystals instead of simulating the growth kinetics of each columnar grain or deriving the field form of variables. Within a bi-crystal system, the grain boundary tilt from thermal gradient direction is understood as a transient invasion behavior of one grain to another, and the competitive grain growth behavior along the buildup process of MAM is a summary of all the invasions in bi-crystal systems. To fill the gaps in the rapid solidification theory, a database of the anisotropic growth effect under a rapid directional solidification condition was established with the help of an artificial neural network. Wire direct energy deposition fabricated Ti6Al4V thin-wall samples with full dendritic columnar grains are used as a benchmark to test the validation of the novel simulation model which demonstrate good simulation accuracy. The model can also be applied to other MAM processes.

2:00 PM

(PACRIM-449-2021) A novel fully LTCC-based differential scanning calorimeter with high resolution and high heating rates

R. Moos*[1]; J. Kita[1]; R. Werner[1]; M. Gerlach[2]; M. Gollner[2]; F. Linseis[2]

1. University of Bayreuth, Dept. of Functional Materials, Germany
2. Linseis Thermal Analysis, Germany

Differential scanning calorimetry (DSC) is widely-used to analyze thermodynamic material properties. However, materials that form aggressive or corrosive reaction products cannot be easily analyzed with conventional DSC devices, since they contaminate or damage these highly complex and cost-intensive apparatuses. Moreover, classical DSC devices are designed as stationary table-top units and therefore are not suitable for mobile usage. To allow for mobile DSC applications and to detect aggressive materials, the objective of our work was to develop a low-cost miniaturized all-ceramic differential scanning calorimeter device. The need of an external furnaces as it is today's standard in table-top apparatuses should be avoided. The new device is fully manufactured using LTCC (Low Temperature Co Fired Ceramics) and thick film technology. The chip arrangement consists of heater and temperature sensors in a chemically inert ceramic arrangement with metallic heater and temperature sensor. This arrangement allows high reproducibility and, due to its low mass, excellent temperature control and heating rates of over 500K/min. The integrated sensor is easily replaceable and available at low cost. In the talk, the steps of the ceramic development (including finite-element modelling), the final setup, and some typical DSC-curves will be shown.

2:20 PM

(PACRIM-450-2021) CerAMfacturing of single- and multi-material ceramic components

U. Scheithauer*[1]; S. Weingarten[1]; E. Schwarzer-Fischer[1]; J. Abel[1]; C. Berger[1]; W. Kunz[2]; A. Fussel[2]

1. Fraunhofer IKTS, Shaping, Germany
2. Fraunhofer IKTS, Materials, Germany

Additive manufacturing (AM) technologies allow the preparation of tailor-made structures for manifold applications. Becoming state of the art for polymers and metals, ceramic components are manufactured additively, as well. But the ceramic process chain is much more complex because after the AM process debinding and sintering steps are needed to realize the unique properties of ceramic components. At Fraunhofer IKTS we use different direct (CerAM FFF – Fused Filament Fabrication, CerAM MMJ – Multi Material Jetting) and indirect (CerAM VPP – Vat photo-polymerization, CerAM BJT – Binder Jetting, CerAM Replica) AM technologies to realize single- and multi-material ceramic and metal-ceramic components. Beside material and process development, machine and component development are very important topics as well. The presentation will give an overview about the current developments in AM of ceramic-based components at Fraunhofer IKTS Dresden. Starting with the design and modelling of ceramic components for various applications (static mixers, micro reactors, …), the opportunities of the different AM technologies will be discussed as well as available materials (e.g. silicon nitride, silicon carbide, aluminum nitride). Furthermore, latest results of a specific characterization method for AM components will be presented.

Laser Processing

Room: Prince of Wales
Session Chair: Yaoyao Fiona Zhao, McGill University

3:20 PM

(PACRIM-451-2021) Selective Laser Sintering of Hexagonal Barium Titanate Ceramics

X. Zhang[1]; F. Wang[1]; Z. Wu[1]; M. Nastasi[2]; Y. Chen[3]; Y. Lu[1]; B. Cui*[1]

1. University of Nebraska-Lincoln, USA
2. Texas A&M University, USA
3. Oak Ridge National Lab, USA

A direct selective laser sintering (SLS) process was combined with a laser preheating procedure to decrease the temperature gradient and thermal stress, which was demonstrated as a promising approach for additive manufacturing of $BaTiO_3$ ceramics. The phase compositions in $BaTiO_3$ ceramics fabricated by SLS were investigated by X-ray and neutron diffractions. The surface morphologies and cross-section microstructures were characterized by scanning electron microscopy (SEM) and transmission electron microscopy (TEM). A dense hexagonal h-$BaTiO_3$ layer was formed on the surface and extended to a depth of 500 μm, with a relative density of higher than 97% and absence of pores or microcracks. SLS resulted in the formation of the high-temperature phase, h-$BaTiO_3$, which was retained at room temperature possibly due to the high cooling rate. The grain boundaries of SLSed h-$BaTiO_3$ ceramics consist of a Ti-rich secondary phase. Compared with that of the pressureless sintered t-$BaTiO_3$ ceramics, the Vickers hardness of SLSed h-$BaTiO_3$ was 70% higher.

3:40 PM

(PACRIM-452-2021) Enhanced near-infrared absorption for direct laser additive manufacturing using reduced graphene oxide

I. Elizarova*[1]; C. Leung[2]; M. Isaacs[3]; S. Marathe[4]; E. Saiz[1]; P. Lee[2]

1. Imperial College, Materials, United Kingdom
2. University College London, Mechanical Engineering, United Kingdom
3. Research Complex at Harwell, Rutherford Appleton Laboratory, United Kingdom
4. Diamond Light Source Ltd, United Kingdom

Direct laser additive manufacturing, or DLAM, is a sustainable 3D printing technology that builds objects through the layer-by-layer laser fusion of powder particles. This process binds particles directly without using binders, which facilitates accuracy of the build, reduces waste and saves energy on post-processing treatments. It is, however, restricted in the range of suitable feedstocks due to variation in the laser absorption among compounds. Here we explore the possible route for expanding the library of processable materials by employing an absorbance enhancing additive – chemically modified graphene (CMG). We utilise in-situ and in-operando X-ray characterisation and chemical analysis to study the effects of CMG on the near-infrared laser processability of doped SiO_2 glass, which in its pure form does not absorb the laser efficiently. We monitor the sintering/melting process in real-time to study the melt pool behaviour (consistency of the formed tracks, formation of defects). In this work, we observe that CMG increases near-infrared laser absorbance of silica three times greater than commonly used nano-carbon additive and enables direct sintering of otherwise unprocessable glass. This finding paves the way to widen the DLAM feedstock palette to potentially include all classes of materials, transforming the technology into a truly versatile production method.

4:00 PM

(PACRIM-453-2021) Spatially Filled Polyhedrons with Glass Blocks and Air Cavities for Virus Sterilization Filters Using Deep Ultraviolet Rays

H. Tsuduki[1]; S. Kirihara*[1]

1. Osaka University, Japan

Deep Ultraviolet rays (DUV) of 200 nm in wavelength can be used for effective virus sterilizations in air cleaners, and Light Emitting Diodes (LEDs) with continuous spectrums and arbitrary peak lengths are developed instead of mercury lamps. Three-dimensional air filters connected with DUV-LEDs were fabricated by using stereo lithographic additive manufacturing. Spatially filled polyhedrons with tetrahedron and octahedron composed of glass blocks and air cavities were systematically designed to increase emission surfaces and optimize air filtrations. Optical expansion paths and air flow streamlines can be simulated and visualized by ray tracing method (RTM) and computer fluid dynamics (CFD). Formed acryl precursors including glass microparticles were dewaxed and sintered in the air atmosphere. Practical compatibilities of sterilization performances and filtration effects will be discussed while considering multiple resonations in air flows.

PACRIM Symposium 13: Novel Nanocrystal Technologies for Advanced Ceramic Materials & Devices

Applications and Functional Devices using Nanocrystals

Room: Georgia A

1:30 PM

(PACRIM-454-2021) Tin oxide thin film consisting of nanosheet-type structure for gas sensing (Invited)

P. Choi*[1]

1. National Institute of Advanced Industrial Science and Technology (AIST), Japan

Detection of chemical substances has become a significant technology to protect the environment and for human healthcare. Tin oxide (SnO_2) is one of important materials used in chemical substances sensing devices. Since the fundamental operational mechanism of the devices is adsorption and desorption on the surface, tuning of the surface structure and the morphology can modify and improve the sensing ability. Tin oxide thin films consisting of nanosheet-type structure were prepared by a simple aqueous process and were adopted to a gas sensing material. Since the tin oxide thin film consisting of nanosheet-type structure was mainly exposed (101) crystal facet, it exhibited a specific selectivity and a superior gas response as compared to normally used tin oxide, that is mainly exposed the most stable (110) crystal facet. It expects that this specific and superior sensing properties of the tin oxide thin film consisting of nanosheet-type structure will contribute to detection of hazardous substances, biomarkers, and so on.

2:00 PM

(PACRIM-455-2021) Photonic Curing to Fabricate Oxide Thin Films on Flexible Substrates (Invited)

J. W. Hsu*[1]

1. UT Dallas, USA

To achieve high-throughput, low-temperature processing of thin film oxide electronics, we introduce photonic curing, using short but high intensity pulses of broadband light, to convert metal oxide films from sol gel precursors. In the first example, we show the results of using photonic curing to fabricate ZrO_2 dielectric films on flexible PEN substrates. Photonic curing has not previously been used to fabricate oxide films because the precursor is not light-absorbing. We demonstrate that ZrO_2 can be converted over the entire PEN substrate without cracks. Moreover, by reducing the radiant energy, the oxide film can be selectively converted on top of Al electrodes, which act as selective area heat absorbers, but not on the bare PEN substrate, resulting in a self-aligned oxide pattern. We achieve a ZrO_2 dielectric with breakdown field of 8 MV/cm and areal capacitance of 200 nF/cm^2 with little frequency dispersion. In the second example, we synthesize NiO hole transport layer from sol-gel precursor on flexible Willow glass using photonic curing to enable high-speed fabrication of halide perovskite solar cells. We also apply "stitching" to demonstrate how to achieve uniformity when the sample is conveyed under the lamp at high speed while the lamp pulses are discrete. In both examples, the conversion time is reduced by several orders of magnitude with a pathway to achieve a roll-to-roll web speed of > 10 m/min.

PACRIM Symposium 18: Advanced Wear Resistant Materials: Tribology and Reliability

Design of Novel Functional Materials

Room: Balmoral

1:30 PM

(PACRIM-456-2021) Design and Application of Nano-Layered Functional Coatings for Macro-Scale Tribology (Invited)

O. V. Penkov*[1]

1. Zhejiang University, ZJU-UIUC Institute, China

Among the numerous strategies developed for reducing friction and wear, coatings have been successfully utilized in various engineering applications to mitigate tribological problems. One of the benefits of coatings is that they may be fabricated using multiple materials in different forms and structures to satisfy the requirements of the operating conditions. Periodical multilayer and multi-material coatings have gained considerable interest due to the added degree of freedom in tailoring the coating property. In this presentation, design concepts of tribological multilayer coatings are firstly discussed. Then, relationships between structure, mechanical and tribological properties of nanolayer periodical coatings were reported. It is shown that even structures with thicknesses of individual layers of several nanometers provide high macro scale wear resistance. High tribological performance is accompanied by unique functional properties, including biocompatibility, transparency, and flexibility.

2:00 PM

(PACRIM-457-2021) Novel Functional Glass-ceramic Coatings on Titanium Substrates from Glass Powders and Reactive Silicone Binders

E. Bernardo*[1]; H. Elsayed[1]; J. Kraxner[2]; D. Galusek[2]

1. University of Padova, Department of Industrial Engineering, Italy
2. FunGlass - Centre for Functional and Surface Functionalized Glass, Alexander Dubček University of Trenčín, Trenčín, Slovakia, Slovakia

'Silica-defective glasses', combined with a silicone binder, have been already shown as a promising solution for the manufacturing of glass-ceramics with complex geometries. A fundamental advantage is the fact that the binder connects the glass powders from room temperature up to the firing temperature, at which, operating in air, it converts into silica. A specified 'target' glass-ceramic formulation is achieved through the interaction between glass powders and the binder-derived silica. The present paper is dedicated to the extension of the approach to the coating of titanium substrates (to be used for dental and orthopaedic applications), with a bioactive wollastonite-diopside glass-ceramic layer, by simple airbrush of suspensions of glass powders in alcoholic silicone solutions. The interaction between glass and silica, from the decomposition of the binder, led to crack-free coatings, upon firing in air; in argon, glass/silicone mixtures yielded novel glass-ceramic composite coatings, embedding pyrolytic carbon. The latter phase is intended to enable absorption of infrared radiation from the coating, to be exploited for disinfection purposes.

2:20 PM

(PACRIM-458-2021) On the design of novel composites for tribological applications

S. Gupta*[1]

1. University of North Dakota, Mechanical Engineering, USA

There is a need for triboactive materials which can withstand adverse environment. MAX and MAB phases have emerged as novel solids which can be used in both bulk and particulate form. In this presentation, I will review different types of novel microstructures which can be designed by using these phases. Systematic evaluation of these microstructure for tribological properties will be presented. Some examples of microstructures are where MAX or MAB phases are embedded in metal or polymeric composites with 0-3 connectivity or MAX or MAB phases connected with potential metal or polymeric composites in 3D and interpenetrating connectivity.

PACRIM Symposium 23: Transparent Ceramic Materials and Devices

Transparent Ceramic Materials II

Room: Oxford

Session Chair: Shyam Bayya, Naval Research Laboratory

1:30 PM

(PACRIM-459-2021) Laser-driven color converters for high-power and high brightness solid state lighting (Invited)

R. Xie*[1]

1. Xiamen University, China

Laser lighting and laser displays are recognized as a new generation of lighiting and emissive dsiplays, in which laser diodes (LDs) are used instead of light-emitting diodes (LEDs) as a primary source to pump luminescent materials (or color converters). As LDs have a much higher power density than LEDs, the color converters under laser irradiations are easily quenched or damaged. Therefore, re-deisgn of color converters for laser lighting and laser displays is essential. In this presentation, we will firstly clarify some key issues that affect the luminance saturation of color converters, then propose matreials descriptors for designing laser phospors, and finally report some nitride phosphors that could be used as laser-driven color converters.

2:00 PM

(PACRIM-460-2021) Long persistent luminescence and blue photochromism in Eu^{2+}-Dy^{3+} co-doped barium silicate glass ceramic phosphor (Invited)

S. Tanabe*[1]

1. Kyoto University, Japan

Eu^{2+}-Dy^{3+} co-doped barium silicate glass was prepared by a melt quenching method and cerammed at various temperatures between 800 and 1250 °C to obtain the glass ceramic (GC) persistent phosphors. The crystal phase of $Ba_5S_{18}O_{21}$ was primarily precipitated in the GC samples. The GC sample cerammed at 1250 °C (GC1250) exhibited the longest persistent luminescence over 16 h until the luminance becomes 2 mcd/m^2. The color of GC1250 sample was changed from white to blue after UV light illumination, and then bleached efficiently by red light (660 nm) illumination or heat treatment at 200 °C. From thermoluminescence (TL) glow curve measurements, four different TL glow peaks were observed at 191 K, 278 K, 446 K, and 494 K. On the other hand, after red-illumination following UV-illumination, the TL intensity of the only 446 K-peak was selectively decreased in the TL glow curve. It was shown that the blue photochromism of GC 1250 sample is due to an electron trapping defect strongly related with this glow peak.

2:50 PM

(PACRIM-461-2021) Pressure sintering processes adated to the control of the microstructure and optical properties of transparent ceramics (Invited)

R. Boulesteix*[1]; L. Viers[1]; M. Vandenhende[2]; A. Maitre[1]; R. Belon[3]; P. Munsch[4]; S. Le Floch[5]; T. Gaudisson[5]; Y. Le Godec[6]

1. University of Limoges, IRCER, UMR CNRS 7315, France
2. CNRS, IRCER UMR CNRS 7315, France
3. CILAS, France
4. University de Toulouse, IRAP UMR CNRS 5277, France
5. University Lyon1, ILM, UMR CNRS 5306, France
6. Sorbonne University, IMPMC, UMR CNRS 75090, France

This work aims to show that non-conventional sintering processes such as HIP (Hot Isostatic Pressing) or SPS (Spark Plasma Sintering) are well suited to the production of transparent ceramics. In fact, better control of the microstructure (grain size, porosity) is possible by the application of pressure during sintering. The thermomechanical cycle must however be adapted to the densification and grain growth kinetics of the material. With this aim, the study and modelling of sintering kinetics, the determination of rate-limiting mechanisms, and the construction of sintering maps is very useful. HIP and SPS sintering processes have been used successfully to manufacture transparent ceramics of the rare-earth oxides garnets family like YAG ($Y_3Al_5O_{12}$) or rare-earth sesquioxides (Lu_2O_3), and spinel ($MgAl_2O_4$). Today, our research is focused on the study of new compositions and / or new processes. A new very high pressure sintering process, SPS-VHP (Very High Pressure Spark Plasma Sintering) is described as well as the optical and mechanical properties of transparent alumina ceramics obtained by this method at a temperature of 600°C for 10 min under a pressure of 1 GPa. The mean grains diameter is remarkably small (i.e. 220 nm) for a dense material. The corresponding optical transmittance to a wavelength of 640 nm is 65% for a thickness of 1 mm.

3:20 PM

(PACRIM-462-2021) Ion mobility in silicates and their effect on electrical performance (Invited)

M. Yuan*[1]; C. Nieves[1]; E. Furman[1]; M. Lanagan[1]; P. Clem[2]; E. Schrock[2]

1. Penn State University, Materials Science and Engineering, USA
2. Sandia National Laboratories, USA

Silicate glass has outstanding capabilities to operate at high temperatures and large electric fields, and is therefore widely used in energy storage and power conversion under harsh environments. It is known that the high field dielectric and electrical properties are largely affected by the trace amounts of ions and their movement within the dielectrics. In this study, we will present a systematic study on a series of silicate glasses containing different types of ions (alkali, proton, non-bridging oxygen (NBO), etc.). The particular focus is to understand the effects of ion migration (activated by high voltages and/or temperatures) on the dielectric/electrical properties of the silicates. The experimental results obtained from absorption currents, thermally stimulated depolarization currents (TSDC), and impedance spectroscopy are well correlated, which are further explained and elaborated by simulation. This study provides fundamental insights and understandings on how the presence of ions and ion movements can affect the local e-field distribution and local conductivity of inorganic dielectrics, and ultimately determine their high field dielectric performance and reliability.

PACRIM Symposium 24: Solid Oxide Fuel Cells and Hydrogen Technologies

Reliability and Degradation

Room: Georgia B

1:30 PM

(PACRIM-463-2021) First-principles investigation on the BaM_2NiO_5 precipitates in Ba(Zr, M)O_3 solid electrolyte in Proton-Conducting Solid Oxide Fuel Cell (Invited)

K. Nakamura*[1]; M. MORI[1]; Y. Okuyama[2]

1. Central Research Institute of Electric Power Industry, Japan
2. University of Miyazaki, Japan

Ba(Zr, M)O_3 perovskites, in which B-sites are substituted with rare earth elements M, has been investigated for use as a solid electrolyte in Protonic Ceramic Fuel Cells (PCFC). However, co-sintered electrodes-derived Ni solution causes second phase formation of BaM_2NiO_5 and decrease of rare earth element solution level in $BaZrO_3$. This problem limits the operating temperature of the PCFCs and the selection of additive elements. Thus, in this study, we investigated the formation tendency of BaM_2NiO_5 in Ba(Zr,M)O_3 by using first-principles calculations. Calculated formation energies of BaM_2NiO_5 (M = B, Al, Ga, In, Tl, Sc, Y and 15 types rare-earth element) could well explain experimental precipitation tendency. Rare-earth element, except for Lu and Yb, were predicted to form precipitates of BaM_2NiO_5. In the case of M=Sc and In, imaginary vibrational modes were observed in the phonon dispersion curves of BaM_2NiO_5, but not in the case of M=rare-earths. Detailed precipitation mechanism of BaM_2NiO_5 in Ba(Zr, M)O_3 will be discussed at conference.

2:00 PM

(PACRIM-464-2021) Industrial scale SOC stacks manufacturing and lifetime improvement: Achievements and challenges (Invited)

D. Montinaro*[1]

1. SOLIDpower SpA, Italy

Solid Oxide Cells (SOC) are one of the most effective technologies for energy conversion, recognised for high efficiency and fuel flexibility (H2, NG, LPG, biogas, ammonia) in SOFC-mode as well the capability to generate hydrogen by steam electrolysis or syngas by co-electrolysis of steam and CO_2. SOLIDpower SpA is an Italy-based industry that has acquired HTceramix SA in Yverdon (Switzerland) in 2007 and the assets and team of Ceramic Fuel Cells Germany GmbH in Heinsberg in 2015, and operates also an office in Melbourne, Australia. During more than 20 years experience, SOLIDpower validated competitive manufacturing technologies by extensive lab testing and demonstration in a pilot line, scaling the production up to an industrial automated plant of 25MW per year. This work presents the experience of SOLIDpower on the implementation of conventional ceramic manufacturing technologies on the large scale production of SOC and stack components and gives an overview of the results related to stack power scale up to 8 and 25 kW in SOFC and SOE mode, respectively, in a single tower. Data related to lifetime improvement in SOFC and SOE modes will be presented, providing an overview of some degradation phenomena, including degradation of the barrier layer at the electrolyte/oxygen-electrode interface and manganese accumulation at the fuel/steam electrode.

SOFC Prototypes, Electrodes and Sealants

Room: Georgia B

Session Chair: Federico Smeacetto, Politecnico di Torino

3:20 PM

(PACRIM-465-2021) Glass-ceramic sealants for solid oxide cells: Integration and compatibility with metallic interconnects

F. Smeacetto*[1]

1. Politecnico di Torino, Applied Science and Technology, Italy

The role of the sealants is essential in order to assure high efficiency and durability when joining both similar and dissimilar materials in solid oxide cells, since they must have thermo-chemical and thermo-mechanical compatibility with the adjacent materials, as well as stability at the relevant operating conditions. This contribution provides an overview on different glass-based sealants, reviewing different aspects in terms of compatibility and integration of metallic interconnects and ceramics. Electrical characterization of glass-ceramic sealant-metallic interconnect joined samples under solid oxide electrolysis cell conditions, with the influence on the microstructure and composition at the different polarized interfaces, is also reviewed and discussed.

3:50 PM

(PACRIM-466-2021) Discussion about the operation of the first industrial size biogas-fed SOFC plant in Europe (Invited)

M. Santarelli*[1]

1. Politecnico di Torino, Energy, Italy

The EU-funded DEMOSOFC project has spent 4 years in demonstrating the technical and economic feasibility of operating an industrial-size ceramic electrochemical system (SOFC, 60+50 kWe modules) in a wastewater treatment plant (WWTP). The fuel for the SOFC modules is biogas from the collected sludge. A heat-recovery loop from the hot SOFC exhaust gases (50 kW) is used for the sludge pre-heating. The plant includes 2 SOFC modules started respectively in October 2017 and October 2018. The first SOFC module has reached more than 5500 hours of operation on site and the second SOFC module more than 9800 hours. Results from the operation of these modules confirmed the expected high-level performance of the fuel cell system: average measured SOFC efficiency from compressed biogas to AC power is 50%, with peaks at 56% at partial power (40 kWe). Some degradation on the stacks has been also experienced. An emissions measurements campaign has been performed. Results show no NOx (detection limit of the method 20 mg/m3) or SO2 (d.l. 8 mg/m3) and particulate lower than ambient air values (< 0.01 mg/m3). An economic analysis has been also performed to understand the current site preparation and the potential reduction of a second replication, thanks to the lesson-learned from the demonstration site. A total potential reduction of costs of 56% could be achieved according to the analysis performed.

4:20 PM

(PACRIM-467-2021) The invesitation of the new materials, $Gd_{0.135}Yb_{0.015}Bi_{0.02}Ce_{0.83}O_{1.915}$ (GYBC), as a buffer layer

H. Kim*[1]; J. Bae[1]; Y. Lee[2]; S. Lee[1]; T. Shin[1]

1. Korea Institute of Ceramic Engineering and Technology (KICET), Republic of Korea
2. Gyeongsang national university, Republic of Korea

Solid oxide fuel cells (SOFCs) have high attention in the hydrogen energy market because of their high energy conversion efficiency, fuel flexibility and low emission. SOFCs consists of two porous electrodes of anode and cathode, separated by dense yttria-stabilized zirconia (YSZ) electrolyte. Because of its high electronic and oxygen ion conductivities at low temperature, the strontium and cobalt doped lanthanum ferrites, $La_xSr_{1-x}Co_yFe_{1-y}O_{3-\delta}$ (LSCF), have been commonly used as cathode materials. However, the LSCF cathode could be reacted with YSZ electrolyte forming secondary phase such as LZO and SZO. Therefore, the Ce doped fluorite materials were used to barrier layer to prevent reaction. In this study, we investigated the possibility as the barrier layer for the new materials, $Gd_{0.135}Yb_{0.015}Bi_{0.02}Ce_{0.83}O_{1.915}$ (GYBC) which was Ce doped material to improve the cell performance. Moreover, the effects of buffer layers were observed by means of microstructure through scanning electron microscopy (SEM) and reported using electrochemical analysis.

4:40 PM

(PACRIM-468-2021) Active Ceramic Anode for Direct Propane Fueled Solid Oxide Fuel Cells

M. Kim*[1]; T. Shin[1]

1. Korea Institute of Ceramic Engineering and Technology (KICET), Energy Efficiency Materials Center, Republic of Korea

Propane, which is the main component of Liquified Petroleum Gas (LPG), is now commonly used in various industrial fields because of its high volumetric energy density and simplicity for storage and transportation in a liquid state. Thus, direct utilization of this propane is one of the attractive projects in Solid Oxide Fuel Cells (SOFCs) due to its high energy conversion efficiencies and possibility for simplification of the systems. However, direct use of propane as a fuel has been limited due to several challenging issues such as low anodic activity and severe carbon deposition. We previously reported $Ce(Mn,Fe)O_2$ (CMF) and $La(Sr)Fe(Mn)O_3$ (LSFM), oxidation-tolerant oxide anode for SOFC with reasonably high performance and good tolerance to coking, even when using dry propane as a fuel. In this study, to enhance catalytic performance and electronic conductivity of previous oxide anodes, core-shell structured composite oxide of LSFM@CMF was prepared. Shell-structured CMF on nano-sized core LSFM surfaces remarkably enhanced the catalytic properties of LSFM even when using dry propane as a fuel, showing excellent power density of 1.25W cm^{-2} at 1073K.

PACRIM Symposium 28: Atomic Structure and Electrochemical Property Diagnosis Toward Full Crystal Rechargeable Batteries

Characterization of Battery Materials and Interfaces I

Room: Regency B

1:30 PM

(PACRIM-469-2021) Understanding degradation mechanism via diverse analyses at the interphase of solid-electrolytes (Invited)

H. Lee*[1]

1. UNIST, School of Energy and Chemical Engineering, Republic of Korea

Research on the interface between solid electrolytes and electrode materials is of vital importance for the development of various battery systems. The complex electro-chemo-mechanical evolution of the electrode/solid-electrolyte interface can diminish performance. In the first part, I would like to provide insights into the interfacial dynamics during the operation of a hybrid Na-seawater battery under harsh operating conditions. Furthermore, chemical stability variations at a steady state can severely affect battery performance. Seawater batteries fabricated with NASICON in immersed DI water for 1 year exhibit a large resistance region from the first cycle; this system breaks down before 200 h, unlike a cell fabricated using NASICON immersed for 1 year in a marine environment. In the second part, I would like to introduce stack pressure measurements for all-solid-state batteries to provide new insight into the effects of applied stack pressure and electrolyte processing on the interfacial behavior of two representative solid-state electrolytes, $Li_{10}SnP_2S_{12}$ and Li_6PS_5Cl. The findings advance our understanding of the interfacial evolution of two important classes of solid-state electrolytes, and they demonstrate the utility of electro-chemo-mechanical measurements to understand solid-state battery behavior.

2:00 PM

(PACRIM-470-2021) Cryo-TEM study of sensitive battery materials (Invited)

M. Gu*[1]

1. southern University of Science and Technology, Materials Science and Engineering, China

Li-ion batteries have revolutionized people's life through its wide applications in daily electronics. Nowadays the electric transportations call for safer and higher capacity batteries. The formation of Li dendrites and unstable solid electrolyte interphase (SEI) layer seriously threatens the safety of batteries. Researchers propose that the usage of Li metal anode in a solid-state battery enables superior safety and high-energy density. However, the ionic transport and interfacial structure between the electrode and solid electrolyte is largely unclear, impeding the fast development of better and safer batteries. Here, we developed low-dose cryo-transmission electron microscopy (cryo-TEM) imaging to extract the detailed cation mixing, lattice distortions, defect and new phase formation at the interface with atomic scale resolution. In my talk, I am going to explore the effects of different additives in reshaping the morphology and stability of Li dendrite and SEI. We found that the mechanical strength and flexibility to fold and unfold during Li stripping and re-plating are important to maintain long life of batteries. We also used 3D STEM tomography to examine the fast Li stripping directions during Li stripping and origin of dead Li formation.

Characterization of Battery Materials and Interfaces II

Room: Regency B

3:20 PM

(PACRIM-471-2021) Characterization of Interfaces in Li-ion Secondary Battery Electrodes Using Scanning Transmission Electron Microscopy (Invited)

S. Kobayashi*[1]

1. Japan Fine Ceramics Center, Nanostructures Research Laboratory, Japan

Improving the electrochemical performance of Li-ion batteries for a wide range of applications from mobile devices to electric vehicles is a major goal of materials research. In order to optimize battery performance, it is necessary to understand the interface morphologies, e.g., delithiation fronts, surfaces, and interfaces between electrolyte and active materials. $LiFePO_4$ is one of the most intensively studied cathode materials for lithium ion batteries, and thus improving its performance further requires a greater understanding of lithium intercalation in the vicinity of interfaces between lithiated and delithiated phases, viz., $Li_{1-\alpha}FePO_4$ and $FePO_4$, during charge/discharge. We formed such a biphase interface by chemically delithiating $LiFePO_4$ from a single crystal surface and analyzed Li concentrations in its vicinity using electron energy loss spectroscopy and high-resolution scanning transmission electron microscopy techniques. The results illustrate how nano-scale and atomic-resolution microscopy analysis provides greater insights into the nature of the intermediate phase, phase transformation processes, and relaxation mechanisms in $LiFePO_4$ during delithiation.

3:40 PM

(PACRIM-472-2021) Structural Disorder in (Li,La)NbO_3 Solid Electrolytes: A Molecular Dynamics Study

C. Fisher*[1]; X. Hu[1]; Y. H. Ikuhara[1]; H. Moriwake[1]; Y. Ikuhara[2]

1. Japan Fine Ceramics Center, Japan
2. The University of Tokyo, Institute of Innovative Engineering, Japan

(Li,La)NbO_3 solid-electrolytes are layered cation-deficient perovskites with a complex hierarchy of structural features. Quenching of (Li,La)NbO_3 single crystals has been shown to improve their Li-ion conductivity, with some La atoms displaced into A2 layers [4]. To gain insights into the effect of this disorder, we performed classical molecular dynamics simulations of 18x18x9 supercells of $La_{(1-x)/3}Li_xNbO_3$ for 0.0 <= x <= 0.20 (~25,000 atoms) using the Moldy program. Pseudo-ordering of vacancies in A1 layers, similar to that observed using electron microscopy, was confirmed to be energetically more favorable than randomly distributed or clustered vacancies. Although slightly less stable, structures with 5% or 10% of La atoms moved randomly onto A2 sites were found to induce some of the Li ions to move from A1 layers into A2 layers, resulting in a modest increase in 2D Li-ion diffusion in both. The ion conductivity was found to peak at around x ~ 0.10, consistent with experiment. The results provide a plausible mechanism for explaining the experimental observations.

4:00 PM

(PACRIM-473-2021) Cathodic thin films on solid electrolyte (Li,La)NbO_3 prepared by chemical solution deposition

Y. H. Ikuhara*[1]; S. Kobayashi[1]; X. Hu[1]; C. Fisher[1]; A. Kuwabara[1]; H. Moriwake[1]; Y. Ikuhara[2]

1. Japan Fine Ceramics Center, Japan
2. The University of Tokyo, Japan

All-solid-state lithium ion batteries are being actively researched to provide high power density and long cycle life. For this, multilayer structure analysis of cathode, electrolyte, and anode components is needed to optimize the processing-microstructure-performance relationships. In particular, heterointerfaces between components are known to influence charge/discharge rates, cyclability, and battery lifetime. Thin-film quality and its performance are also known to be affected by the processing method and conditions. Chemical processing using metalorganic precursors is advantageous for preparing thin films with desired stoichiometry, controlled purity, and compositional homogeneity. In this study we deposited spinel-type cathode thin films on solid electrolyte (Li,La)NbO_3 substrates using a chemical solution deposition method. We have characterized the interfacial atomic structure and the orientation relationship between film and substrate by scanning transmission electron microscopy and show that the present method can be used to produce highly oriented thin-film electrodes.

4:20 PM

(PACRIM-474-2021) In-situ transmission electron microscopy of battery electrode materials during their charging (Invited)

J. Yuk*[1]

1. KAIST, Republic of Korea

Rechargeable batteries receive intensive research effort, with a wide range of application space including mobile electronic devices, electric vehicles and grid-scale energy storage. The limit in battery capacity partly arises from our lack of understanding on the crystal structure and morphology evolution of the electrode materials during operation. In this talk, we show in-situ transmission electron microscopy studies that investigate this missing link with unprecedented temporal and spatial resolution. The first example studies the sodiation pathways of copper sulfide particles. In particular, we reveal not only sequential phase transition pathways that deviate from the thermodynamic equilibrium, but also the semi-coherent phase boundaries during the conversion reaction. We rationalize that this unique interface behavior alleviates the particle pulverization during cycling and thus explains the high reversible capacity without excessive morphology optimization. In the second example, we show stress-driven composition change in core-shell SnO_2-Sn nanoparticles. Especially, the generation of non-uniform composition field is directly visualized with in-situ graphene liquid cell electron microscopy.

4:40 PM

(PACRIM-475-2021) Direct uncovering of multiphase evolution during hydrated Zn-ion insertion in vanadium oxide

P. Byeon*[1]; S. Chung[1]

1. Korea Advanced Institute of Science and Engineering (KAIST), Republic of Korea

An initial crystalline phase can transform into another phases as cations are electrochemically inserted into its lattice. Precise identification of phase evolution at an atomic level during transformation is thus the very first step to comprehensively understand the cation insertion behavior and subsequently achieve much higher storage capacity in rechargeable cells, although it is sometimes challenging. By intensively using atomic-column-resolved scanning transmission electron microscopy, we directly visualize the simultaneous intercalation of both H_2O and Zn during discharge of Zn ions into a V_2O_5 cathode with an aqueous electrolyte. In particular, when further Zn insertion proceeds, multiple intermediate phases, which are not identified by a macroscopic powder diffraction method, are clearly imaged at an atomic scale, showing structurally topotactic correlation between the phases. The findings in this work suggest that smooth multiphase evolution with a low transition barrier is significantly related to the high capacity of oxide cathodes for aqueous rechargeable cells, where the crystal structure of cathode materials after discharge differs from the initial crystalline state in general.

5:00 PM

(PACRIM-476-2021) Investigating the benefits of fluorine anion exchange in O3-type $NaFe_{0.4}Ni_{0.3}Mn_{0.3}O_2$ cathode

Y. Charles-Blin[1]; N. Zettsu*[1]

1. Shinshu University, Japan

The consolidation of Na-ion batteries in the world-wide market mainly depends on increasing their energy density and stability. Among the promising sodiated cathode materials, O3-type layered oxides, like $NaFeO_2$ provide moderate specific capacity in comparison with the P2-type layered oxides, around 100 $mAh.g^{-1}$ for 0.5 moles of Na. The $NaFe_{0.4}Ni_{0.3}Mn_{0.3}O_2$ (NaFNMO) phase has for advantage to get rid of cobalt. However this phase present an inferior rate capability than with cobalt-substituted system. The capacity decay of such compound is mainly due to irreversible structural changes accompanied by migration of transition metal ions, from octahedral site in the slab to tetrahedral site and then to face-shared octahedral site in interslab space. Thus, our strategy consists of strengthening the chemical bonds within the slab structure by anion exchange, controlled amount of oxygen anions is replaced by fluorine anions, in order to mitigate transition metal migration. In addition, the fluorine exchange could increase Na insertion potential and reduce the SEI resistance and thus improve the kinetic properties of NaFNMO, has we already shown for Li-ion systems. Both theoretical and experimental approach for a thorough study of the influence of the fluorine anion exchange has been conducted.

PACRIM Symposium 33: Photocatalysts for Energy and Environmental Applications

Photocatalysts for Energy and Environmental Applications

Room: Stanley

1:30 PM

(PACRIM-477-2021) Metal NPs loaded $SrTiO_3$-Al supported with Rh/Cr_2O_3 and CoOOH cocatalysts for overall water splitting

M. H. Abd Elkodous*[1]; G. Kawamura[1]; W. Tan[1]; A. Matsuda[1]

1. Toyohashi University of Technology, Electrical and Electronic Information Engineering, Japan

Generation of fuels like hydrogen from water splitting process using particulate nanomaterials is of significant importance. In this context, Al-doped $SrTiO_3$ microspheres were prepared by molten flux method, followed by the photodeposition of Rh/Cr_2O_3 and CoOOH cocatalysts by UV irradiation, for hydrogen and oxygen evolution, respectively. In addition, three metal NPs (Au, Cu, and Pt) were further photodeposited onto the reductive active sites of the main photocatalyst ($SrTiO_3$-Al). The effect of the order of deposition of the cocatalysts and metal NPs, and metal NPs' concentration on the overall water splitting efficiency were studied. Our results revealed that, $SrTiO_3$ microspheres possessed a mean diameter of about 150 nm as confirmed by TEM analysis Moreover, SEM analysis presented the uniform distribution of the cocatalysts and metal NPs onto the surface of the main photocatalyst. Overall water splitting evaluation showed that, Au NPs-supported samples exhibited the highest H_2 evolution efficiency with high stability. Unlike other samples, changing the cocatalyst deposition order of Cu supported samples, significantly affected the H_2 evolution efficiency. Moreover, O_2 with a (1:2) stochiometric ratio with respect to H_2 was generated over optimized Au supported samples after nearly 45 min.

1:50 PM

(PACRIM-478-2021) Influences of applied magnetic field during deposition process on photoelectrochemical water splitting

H. Lee*[1]; V. Nahrstedt[1]; S. Mathur[1]

1. University of Cologne, Institute of Inorganic Chemistry, Germany

To improve photoelectrochemical (PEC) efficiency has been an attractive issue since this technology developed. There are several strategies to improve the efficiency such as surface modification, impurity doping and band gap modulation by contacting other semiconductor. The surface modification is considered as one of the simplest and most significant strategies. Therefore, it is easily and frequently adopted to a research to enhance the efficiency because surface characteristics is varied by synthetic conditions. However, although a lot of researches have been proceeded to modify surface, fundamental effects caused by electronic behavior have been rarely studied. In this report, we applied external magnetic field during film deposition process to modify surface and control magnetic moment which results from electronic spin. It is expected that the magnetic moment is aligned along the external magnetic field direction and this aligned magnetic moment might affect electron kinetics on the film surface. As a result, the film grown under magnetic field is going to applied to PEC system.

Tuesday, December 14, 2021

PACRIM Posters

Room: Regency A/B

5:30 PM

(PACRIM-P001-2021) New bioactive coatings based on glasses from Si(P)-O-C system on metallic substrates

M. T. Sitarz*[1]; M. Gaweda[1]; P. Jelen[1]; J. Marchewka[1]

1. AGH University of Science and Technology, Faculty of Materials Science and Ceramics, Poland

The main problems associated with the application of metallic implants are often their unsatisfactory bioactivity and biocompatibility as well as their insufficient resistance to physiological fluids. We propose to solve the mentioned problems by the usage of coatings of modified by phosphorus ions amorphous Si-O-C materials (black glasses). The main problem with the introduction of phosphorus ions into the black glass structure is due to the high volatility of organic phosphorus-containing monomers. Therefore, we proposed co-doping with aluminum and boron ions to enable the formation of very stable $AlPO_4$ and BPO_4 units. Prepared by sol-gel method precursors and ceramized materials in the bulk form as well as layers on asthenic steel prepared via dip-coating were examined using numerous of spectroscopic and microscopic methods. Using X-ray diffraction, FTIR, Raman and EDS it was possible to specify how the dopants changed the structure of the preceramic precursors as well as the materials in the bulk or final layered form. The confrontation of results obtained for the multi-ion modified SiOC materials with black glasses doped only with phosphate ions gave information about the efficiency of the proposed methodology. This work has been supported by the NCN project No. 2019/35/B/ST5/00338.

(PACRIM-P002-2021) Porous alumina for 3D structural catalysts

J. Marchewka*[1]; I. Rutkowska[1]; P. Bezkosty[1]; P. Jelen[1]; A. Gancarczyk[2]; J. Paczkowska[3]; M. T. Sitarz[1]

1. AGH University of Science and Technology, Poland
2. The Institute of Chemical Engineering, Polish Academy of Sciences, Poland
3. Jagiellonian University, Poland

Aluminum oxide due to its physicochemical properties is an extremely perspective material. It is characterized by a very high surface area, strong adhesion to metal substrate and the presence of acid-base active centers. These parameters are desired in terms of the application as a catalytic material. This work focuses on the preparation and characterization of alumina coatings on the 3D metal foams. Organic and aqueous sols are prepared using organoaluminium compounds and deposited using two different methods, namely dip-coating and electrophoretic deposition. After subsequent solvent evaporation associated with polycondensation they are transformed into gels. Then heating treatment results in their conversion into aluminum oxide coatings on the metal substrates. By modifying the parameters of both methods the 3D structural catalysts with alumina coatings of various thicknesses, up to several dozen micrometers, could be obtained. SEM analyses indicated the preparation of tight coatings on the steel substrates. Their specific surface area was determined using the BET method. Spectroscopic investigations showed the presence of bands derived from Al-O bonds for aluminum in octahedral and tetrahedral configurations. Catalytic tests confirmed the desired activity of the obtained materials. The work has been supported by the National Science Center (Poland) grant no. 2020/37/B/ST8/02859.

(PACRIM-P003-2021) Transition Metal Dichalcogenides Reinforced SiOC Fibermats for Rechargable Battery Electrodes

S. Dey*[1]; G. Singh[1]

1. Kansas State University, Department of Mechanical and Nuclear Engineering, USA

Transition metal dichalcogenides (TMDs) are potential candidates for Alkali metal-ion battery electrodes as they possess properties such as weak force between layers - of interest for faster ion-intercalation, and conversion chemistry with lithium-ions. However, shortcomings such as capacity decay, and degradation reaction with electrolytes at low discharge potential continues to hinder their use in rechargeable batteries. Polymer derived ceramics (PDCs) such as silicon oxycarbide (SiOC) exhibit stable longer term cyclability towards Li^+ due to PDCs' robust nanodomain structure. SiOC functionalized TMDs via polymer pyrolysis route can be beneficial in overcoming the shortcomings of neat TMDs; thus, leading to a composite electrode material possessing long term stable cycling in Li and other alkali metal-ion batteries. Herein, we report fabrication of SiOC functionalized TMDs (e.g., MoS_2, $MoTe_2$, $MoSe_2$) in fibermat form which can be used as a freestanding electrode in Li-ion cells. Characterization of the structure of SiOC functionalized TMD fiber mats was carried out via electron microscopy and spectroscopic techniques (such as SEM, TEM, XPS, Raman Spectroscopy and FTIR). The electrochemical analysis of the electrodes showed improved cycle stability which are the signs of a promising Li-ion battery electrode.

(PACRIM-P004-2021) High-pressure synthesis of new quadruple perovskite oxides $LaMn_3Ni_2Ru_2O_{12}$ and $LaMn_3Co_2Ru_2O_{12}$

A. Morimura*[1]; I. Yamada[1]

1. Osaka Prefecture University, Japan

Quadruple perovskite oxides, $AA'_3B_4O_{12}$, exhibit functional properties such as giant dielectric constants and highly active electrochemical catalysis, most of which are related to the structural feature of the quadruple perovskite. Conventional A-site ions (e.g., Ca^{2+}, La^{3+}) and transition metal ions (e.g., Cu^{2+}, Mn^{3+}) respectively occupy icosahedral A- and pseudosquare planar A'-sites in $AA'_3B_4O_{12}$, thus electronic interactions are derived from different transition metal ions at distinct crystallographic sites of A'- and B-sites. Further cationic ordering at B-sites may be achieved in selected couples of B-site ions, as represented in the formula of $AA'_3B_2B'_2O_{12}$, which leads to intriguing electronic properties. $LaMn_3Mn_2Ni_2O_{12}$ has a rocksalt-type ordered structure at B-site (Mn^{4+}, Ni^{2+}), whereas $LaMn_3Mn_2Co_2O_{12}$ has a disordered structure (Mn^{4+}, Co^{2+}). In this study, we successfully synthesized new quadruple perovskite oxides containing Ru and 3d metal ions at B-sites, $LaMn_3Ru_2Ni_2O_{12}$ and $LaMn_3Ru_2Co_2O_{12}$, by using high-pressure and high-temperature conditions. Structural analysis using X-ray diffraction data indicates that both compounds exhibited random distributions of B-site ions, in contrast to the order/disorder in $LaMn_3Mn_2B'O_{12}$ family. We discuss the probable factors dominating the ordering of B-site ions in $LaMn_3Ru_2B'_2O_{12}$ system, together with the electronic properties.

(PACRIM-P005-2021) High-pressure synthesis of novel quadruple perovskites oxides containing group 9 elements

Y. Kato*[1]; A. Tanaka[1]; T. Odake[1]; M. Oshita[1]; S. Kawaguchi[2]; I. Yamada[1]

1. Osaka Prefecture University, Japan
2. JASRI, Japan

Quadruple perovskite oxides, $AA'_3B_4O_{12}$, are complex oxides consisting of multiple transition metal ions at different crystallographic sites of A'- and B-sites, in which valence-variable transition metal ions may be included. $CaCu_3B_4O_{12}$ (B = group 9 element; Co, Rh, Ir) possess different valence states: $Ca^{2+}Cu^{3+}{}_3Co^{3.25+}{}_4O_{12}$, $Ca^{2+}Cu^{\sim2.8+}{}_3Rh^{\sim3.4+}{}_4O_{12}$, and $Ca^{2+}Cu^{2+}{}_3Ir^{4+}{}_4O_{12}$, indicating that the valence states in $AA'_3B_4O_{12}$ are not uniquely determined because of severe competitions of d band energy. To examine the valence states and electronic properties in various $AA'_3B_4O_{12}$ oxides (B = Co, Rh), we synthesized novel quadruple perovskite oxides using high-pressure synthesis method. $ACu_3Co_4O_{12}$ (A = La, Bi), $YMn_3Co_4O_{12}$, $CeMn_3Rh_4O_{12}$, and $YCu_3Rh_4O_{12}$ were synthesized under high-pressures of 12-15 GPa. All the samples crystallized in the quadruple perovskite structure. $ACu_3Co_4O_{12}$ (A = La, Bi) were paramagnetic insulators as well as $YCu_3Co_4O_{12}$. $YMn_3Co_4O_{12}$ was a ferrimagnet in which A'-Mn^{3+} and B-Co^{3+} spins are probably aligned antiparallelly. $CeMn_3Rh_4O_{12}$ exhibited a temperature-induced structural transition at ~300K. $YCu_3Rh_4O_{12}$ was a diamagnet as well as $LaRhO_3$, proposing the lack of localized spins in low-spin state of Cu^{3+} ions.

(PACRIM-P006-2021) Enhancement of mechanoluminescence characteristic by tuning $SrAl_2O_4$ crystallization process assisted with organic acid

Y. Fujio*[1]; C. Xu[1]; N. Terasaki[1]

1. National Institute of Advanced Industrial Science and Technology (AIST), Sensing System Research Center, Japan

A mechanoluminescence (ML) is the phenomenon of luminescence induced by mechanical actions, such as compression, tension, friction, or torsion. Such mechano-optical conversion characteristics in the elastic deformation range have been attracting attention as a sensing signal capable of two-dimensionally visualizing physical phenomena. Among various reports regarding ML materials, an enhancement of the ML intensity has become an important topic in ML material research. The key to material design for ML intensity enhancement is known to control a trapped energy state attributable to crystal defects. However, the ML material is commonly synthesized by a solid-state reaction method at a rather high temperature using carbonates and oxides so that the detailed control of the crystallization process seems to be difficult. In the present study, therefore, we focused on the organic acid-assisted method that can be applied to multi-element composition materials among the liquid phase methods and investigated the effect of the crystallization process on the ML characteristics. As a result, the crystal structure, particle size, and shape of the $SrAl_2O_4$:Eu precursor can be tuned by the synthesis conditions such as a kind of organic acid and pH of the reaction solvent, and the ML characteristics of $SrAl_2O_4$:Eu were improved by using such a specific crystallization process.

(PACRIM-P007-2021) Routes of CaO-MgO-SiO_2 Bioactive Glass Ceramic Synthesis

C. Yamagata*[1]; A. D. Rodas[2]; S. R. Mello-Castanho[1]; O. Z. Higa[1]; S. T. Reis[3]

1. Energy and Nuclear Research Institute, Material Science and Technology, Brazil
2. Federal University of ABC, Brazil
3. Missouri University of Science & Technology, USA

Three routes were studied for the synthesis of CaO-MgO-SiO_2 powder: sol gel process associated with coprecipitation from Na_2SiO_3 in acid catalyzed reaction or alkali catalyzed reaction and conventional sol-gel method from TEOS. Bioactive glass ceramic powder composed by 49,13% SiO_2, 43,19% CaO, and 7,68% MgO in wt % was synthesized. The samples synthesized in acid and alkaline medium were denoted as SAS and SBS respectively. The sample designated as SAT was achieved by conventional sol-gel method from TEOS. The SAS, SBS and SAT samples were sintered at 1200 °C for 2 h to attain the correspondingly glass ceramic. Diopsite and wollastonite were observed as the main crystalline phases for all. The end resulted glass ceramics by the three synthesis process were also compared. In vitro bioactivity of glass ceramic was assessed by soaking them in SBF for different time periods. Analysis of FTIR spectroscopy and observation by SEM of the surface of the glass ceramics samples revealed it was completely covered with hydroxyapatite after 21 days of immersion. Cytotoxicity test of the glass ceramics showed they are non-cytotoxic, according to ISO 10993-5:2009. In summary, the three studied routes of synthesis produced bioactive and non-cytotoxic ceramic powders. Therefore, the obtained products have potential application in the biomedical area.

(PACRIM-P008-2021) Tough Double Interpenetrating Network Hydrogel with Dual Reinforcement Mechanism

W. Fan*[1]; L. R. Jensen[2]; D. Yu[1]; M. M. Smedskjaer[1]

1. Aalborg University, Department of Chemistry and Bioscience, Denmark
2. Aalborg University, Department of Materials and Production, Denmark

Tough hydrogels, containing 3D cross-linked polymer networks and large amount of water, are highly rated for their wide applications in bioengineering, soft robotics and agriculture. Double interpenetrating networks show extraordinary toughness compared with single ones. Alginate, as a natural anionic polymer with advantages such as biocompatibility, low toxicity and low cost, could potentially be applied in biomedical applications. Ionic crosslinking methods with divalent cations (e.g., Ca^{2+}) is the most common route to prepare alginate hydrogels under mild conditions. In this work, we mixed hydrophilic polyacrylic acid (PAA) and alginate hydrogel to form a tough double interpenetrating network (DIPN) structure. In order to improve the toughness of the hydrogel, SiO_2 was added and in this way, we successfully achieved toughen DIPN hydrogel with ionic crosslinking and nanocomposites toughening mechanism simultaneously. We have systematically the impact of the amount of Ca^{2+} and SiO_2 in the hydrogel on the mechanical performances. Compared with PAA hydrogel with elongation of 240% and stress to failure at 0.36 MPa, we obtained a alginate (Ca^{2+})-PAA (SiO_2) hydrogel with elongation of 1100% and stress to failure at 0.80 MPa. Our work paves the way for the construction of tough hydrogels with multiple reinforcement methods, with potential applications of biocompatible hydrogels into drug delivery and tissue engineering.

(PACRIM-P009-2021) Valence Changes of Tin Telluride for Efficient Electrocatalysis in Lithium Oxygen Batteries

A. Xu*[1]

1. Shan Dong University, China

Valence changes is one of the most effective way to accelerate the transfer of electron for improving the electrocatalytic activities of tin telluride. Recently, efficacious research has been conducted to prove that the tellurides have well catalytic performances. However, there is few research on tellurides in the direction of electrocatalysis. Herein, this is the appearance of SnTe accompanied by good electrocatalytic performances. In this study, the Octahedral structure tin telluride with a smooth surface was synthesized by rotating hydrothermal method. The presence of the valence changes of the tellurium and tin elements enable the clipping electron transfer and charge-transport properties of tin telluride on the basis of the experimental results and theoretical calculations, thus improving electrocatalytic activities. Therefore, the Li-O2 batteries delivers electrochemical performance with large specific capacities, good cycling stability and low overpotentials for the oxygen evolution reaction/oxygen reduction reaction,demonstrating excellent electrocatalytic performance.

(PACRIM-P010-2021) High efficient Nb_2C MXene cathode catalyst with uniform O-terminated surface for Lithium-Oxygen Batteries

G. Li*[1]

1. Shandong University, China

High-efficient cathode catalyst is the key to improve the high rate cycle stability, avoid the side reaction and lower the overpotential of lithium-oxygen batteries. MXenes are predicted as one of the most impressive materials for energy applications. In this work, we demonstrated the high efficient catalytic capability of Nb_2C MXene with uniform O-terminated surface as cathode materials for LOBs. Easily fabricated uniform O-terminated surface, high catalytic activity of Nb_2CO_2 sites, and unique reaction kinetics contributed to the excellent electrocatalytic performance of Nb_2C MXene. The uniform O-terminated surface on Nb_2C MXene was obtained after heat treatment. DFT calculations revealed the superior catalytic activity of Nb_2CO_2 compared to other anchored groups and bare surface, and the multi-nucleation and growth/decomposition mechanism for discharge products on Nb_2CO_2 surface corresponding to the results characterized by ex-situ and in-situ measurements. The spatial-direction accumulated porous discharge products at high current density contributed to the excellent high-rate cycle stability. For example, the cathodes exhibited stable cycle stability of 130 cycles at ultrahigh current density of 3 A g^{-1}. The present work provides intrinsic insights toward the modulation of catalytic capability and rational design of high-performance MXenes based electrocatalysts.

(PACRIM-P011-2021) High Energy Density Ti_2C MXene as efficient cathode catalysts for Lithium-Oxygen Batteries

J. Li*[1]

1. Shandong University, China

Ti_2C MXene with only few layers was fabricated through a etching-exfoliating method. A mixed surface condition containing decorated -O and -F groups and bare surface was obtained for the Ti_2C MXene after heat treatment. The Ti_2C MXene electrode exhibited excellent electrocatalytic performance with a long cycle stability over 250 cycles at 200 mA g^{-1}. The 2D layer structure and low inside resistance of MXene electrode provided high electrical conductivity for efficient transmission of O_2 and electrons. DFT calculation reveals that the catalytic capability of Ti_2C MXene is mainly derived from the decorated O group, which enables a stable and controllable adsorption-nucleation-decomposition process for Li_2O_2 on Ti_2C MXene surface. Inversely, decorated F group and bare surface lead to the degradation of catalytic capability due to insufficient/excessive binding with Li_2O_2. Due to the inhomogeneous surface condition, a polarized nucleation and growth of discharge products are supposed for the formation of nanoflakes at the initial discharge stage and the porous structure of discharge products, which can provide an efficient pathway for mass transfer and promote the cycle stability of Ti_2C MXene electrode. This work provides an insight to understanding the catalytic capability of Ti2C MXene and can be extended to the rational design of high performance MXene based materials in Li-O_2 batteries.

(PACRIM-P012-2021) Multiple thermal resistance induced extremely low thermal conductivity in porous SiC-SiO_2 ceramics

E. Kang*[1]; Y. Kim[1]

1. The University of Seoul, Department of Materials Science and Engineering, Republic of Korea

Maximizing interfaces for blocking heat conduction in porous ceramic structures is an efficient approach for achieving maximum thermal resistance. We adopted three strategies simultaneously for maximizing thermal resistance in porous SiC-SiO_2 ceramics: (1) addition of nano-sized silica into nano-sized SiC powder for generating SiC/SiO_2 interfaces, (2) addition of nano-sized carbon into nano-sized SiC powder as the template, and (3) sintering in air, which leads to the partial oxidation of the nano-sized SiC particles and burn-out of nano-sized carbon, giving rise to the generation of the SiC-core/SiO_2-shell structure and formation of micro-, meso-, and macro-sized pores. The addition of fumed silica was beneficial for increasing the specific compressive strength of porous SiC-SiO_2 ceramics. The specific compressive strength of the SiC-SiO_2 ceramics processed at 1000 °C with 40 wt% silica with low thermal conductivity (~0.121 Wm^{-1} K^{-1}) was 11.1 MPa cm^3/g, which is 5.3 times higher than that of the SiC foam (2.1 MPa cm^3/g) with the equivalent thermal conductivity. Typical thermal conductivity, compressive strength, porosity, and specific compressive strength values of the porous SiC-SiO_2 ceramics processed at 800 °C with 50 wt% silica were 0.081 Wm^{-1} K^{-1}, 5.5 MPa, ~71 %, and 7.6 MPa cm^3/g, respectively.

(PACRIM-P013-2021) Mechanical and Thermal Property Investigation of NITE C/SiC Composites Reinforced by Various Carbon Fibers

S. Jung*[1]; N. Nakazato[2]; H. Kishimoto[2]; A. Kohyama[3]

1. Graduate School of Muroran Institute of Technology, Japan
2. Muroran Institute of Technology, Japan
3. NITE Corporation, Japan

The SiC matrix composites have various properties depending on the fabrication process, Nano-infiltration Transition and Eutectic (NITE) method is attractive for the fabrication materials aiming to structural components. NITE SiC matrix composites are being expected to be applied to high temperature part of aerospace system such as thruster. In this case, the material must have sufficient high temperature strength, fuel-to-product compatibility, light weight and thermal conductivity to reduce thermal stress. Inner part where the combustion products contact to the SiC composites should employ SiC fiber reinforcement because of the high oxidation resistance and good compatibility with products. Outer part of the thruster will not contact with the combustion product, therefore, lower cost and lighter weight carbon fibers may be attractive. One of the advantages of NITE method is that C and SiC fibers are able to be used in the same plate. At least it will be lighter and the thermal conductivity will be better than the usual Hi-Nicalon SiC fiber reinforcement composites. For this purpose, some pan and pitch type carbon fibers having different CTE and thermal conductivity were selected and NITE C/SiC composites were fabricated with them. This research discusses the potential of the idea with the mechanical, thermal and microstructural data of fabricated C/SiC composites.

(PACRIM-P014-2021) Fabrication Technique Developments of Complex Shaped NITE SiC/SiC Composites for Aero-space System Components

H. Kishimoto*[1]; N. Nakazato[1]; D. Nakata[1]; S. Jung[1]; H. Jung[2]; M. Onoi[2]

1. Muroran Institute of Technology, Japan
2. Metal Technology Co. Ltd., Japan

SiC or C fiber reinforced SiC matrix composites are desired materials especially for aerospace and high efficiency energy systems. The SiC matrix composites have various properties depending on the fabrication process, Nano-infiltration Transition and Eutectic (NITE) method is attractive due to capability to form dense, crystallized SiC matrix with relative lower cost, and this method suites to produce structural materials. One issue of the NITE SiC/SiC composites is the point that pressurizing at very high temperature is necessary to sinter the nano powders. It is therefore very easy to produce test specimens, however, the component having complex shapes is very difficult be shaped. Muroran Institute Technology and Metal Technology Co. Ltd. are corroborating to develop fabrication techniques of spacecraft thrusters. Ordinally, a hot isostatic pressing (HIP) is attractive to form 3D products, but the thruster has normally very complex shape, quasi-HIP method which use powders as a pressure medium may be useful to form a complex shape parts and join them each other. This research introduces the current status of complex shaping technique of NITE method and also some equipment and techniques which Metal Technology Co. Ltd. is possessing.

(PACRIM-P015-2021) High-throughput syntheses and electrochemical characterizations of complex transition metal oxides

Y. Okazaki*[1]; I. Yamada[1]; S. Yagi[2]

1. Osaka Prefecture University, Japan
2. University of Tokyo, Japan

Complex oxides consisting multiple transition metal elements such as $Ba_{0.5}Sr_{0.5}Co_{0.8}Fe_{0.2}O_{3-\delta}$ have been investigated as promising oxygen evolution reaction (OER) catalysts, especially perovskite oxides because of the wide variety of chemical compositions and electronic states. The detailed relationship between OER catalytic activities and chemical compositions provides essential insights into the design principle of highly active OER catalysts. In this study, we design the high-throughput synthesis and the electrochemical characterization system for complex metal oxide catalysts to optimize the chemical compositions for the highest activity. We synthesized simultaneously 121 samples of perovskite quaternary metal oxides [e.g., $(La_{1-x}Sr_x)(Fe_{1-y}Co_y)O_{3-\delta}$] on glass plates and Pt films from mixtures of metal nitrate solutions by using an automated dispensing device. A multichannel electrochemical measurement cell equipped with 25 distinct catalyst-loaded electrodes was constructed using a 3D printer. In this presentation, we compare catalytic activities for each composition to discuss possible factors to realize the highest activity.

(PACRIM-P016-2021) Oxygen evolution reaction catalysis of M-type hexaferrites $BaFe_{12-x}Co_xO_{19}$

F. Toda*[1]; I. Yamada[1]; S. Kawaguchi[2]; S. Yagi[3]

1. Osaka Prefecture University, Japan
2. Japan Synchrotron Radiation Research Institute, Japan
3. University of Tokyo, Japan

The reduction of the intrinsically large overpotentials in oxygen evolution reaction (OER) are vigorously pursued to achieve efficient energy conversions by adopting appropriate catalysts. Currently, catalysts using earth-abundant and low-cost transition metal oxides have been extensively investigated. An M-type hexaferrite $BaFe_{12}O_{19}$ is commonly used as a low-cost magnetic material with excellent chemical stability and corrosion resistance. Nevertheless, its performance as an OER catalyst is not suitable because of its low electric conductivity. Chemical substitution of Fe by Co in $BaFe_{12}O_{19}$ is an effective way to increase electric conductivity and probably catalytic performance. In this study, we examined the catalytic activity of $BaFe_{12-x}Co_xO_{19}$ ($0 \le x \le 12$) synthesized using high-pressure method, and investigated the relationship between Co content x and catalytic activity. OER activity drastically changed in dependence upon x. As x increased from 0 to 6, the catalytic activity increased monotonically, leading to the highest activity in this system, which may be attributed to the improved electric conductivity. On the other hand, the catalytic activities in the range of $7 \le x \le 11$ retained lower than that at $x = 6$, suggesting probable negative factors competing electric conductivity.

(PACRIM-P017-2021) NITE SiC/SiC Composites Fabricated with Higher Pressure using Improved HIP Technique

N. Nakazato*[1]; H. Kishimoto[1]; S. Jung[2]; H. Jung[3]; M. Onoi[3]

1. Muroran Institute of Technology, Japan
2. Muroran Institute of Technology, Graduate School, Japan
3. Metal Technology Co. Ltd., Japan

SiC/SiC composite is a promising structural material for the aerospace and the advanced energy systems. The Nano-Infiltration and Transient Eutectic-phase (NITE) method is expected to be one of the fabrication processes for SiC/SiC composites because of its ability to produce high-density products. Since the NITE method is performed the liquid-phase sintering under pressure, the hot isostatic pressing (HIP) process is used for the fabrication of complex-shaped products. In the HIP process using an inert gas for pressuring medium, the capsuling technique of preform is essential. Originally, the capsule material has been quartz for the NITE method. The traditional method of the sintering equipment using quartz combined with carbon has limited the shape and sintering conditions. For instance, the pressure for the NITE method has been limited below about 50 MPa for long time. Muroran Institute of Technology and Metal Technology Co. Ltd. are developing new capsuling and HIPing techniques to establish complex shape components aiming to the aerospace systems. In this research, we focused on the sintering behavior of the NITE SiC/SiC composite, especially, effects of pressure which is higher than past researches, established by developed capsuling technique using refractory metal. It is expected that the metal capsule can be fabricated by the additive manufacturing technology in the future.

(PACRIM-P018-2021) Mechanical, Thermal, and Electrical Properties of Porous SiC Ceramics Containing Excess C or Si

G. Kim*[1]; Y. Kim[1]; I. Song[2]

1. University of Seoul, Republic of Korea
2. Korea Institute of Material Science, Republic of Korea

Porous SiC ceramics containing C or Si were prepared using a pressureless sintering route without templates. The electrical, thermal, and mechanical properties of the porous SiC ceramics were investigated as functions of Si or C content and porosity. The addition of C to SiC decreased the electrical resistivity, thermal conductivity, and compressive strength of the porous SiC ceramics for an increase in C content from 0 to 50 wt%. In contrast, the addition of Si to SiC increased both electrical resistivity and compressive strength and decreased thermal conductivity for an increase in Si content from 0 to 30 wt%. The current results suggest that both the electrical and thermal conductivities of porous SiC ceramics can be adjusted in the range of 9.0×10^{-2}–3.9×10^{1}Ωcm and 1.3–31.4 W/(mK), respectively, by controlling the Si or C content.

(PACRIM-P019-2021) The influence of mineral additives on firing behaviour and properties of clay ceramics

G. Wie-Addo*[1]; A. H. Jones[1]; J. Renshaw[2]; S. Palmer[2]; A. Scrimshire[1]; P. A. Bingham[1]

1. Sheffield Hallam University, Materials and Engineering Research Institute, United Kingdom
2. Wienerberger Ltd, United Kingdom

The effect of replacing 4wt% colemanite ($2CaO.3B_2O_3.5H_2O$) and nepheline syenite ($K_{1\cdot37}Na_6(AlSiO_4)_8$) in heavy clay ceramics fired to 950°C and 1040°C and compared to the control material has been studied. Eight (8) analytical techniques have been used to explain the influences of the two inorganic additives on firing behaviour and ceramic properties. ^{57}Fe Mössbauer spectroscopy reveals that iron is present as Fe^{3+}. The differences in Fe^{3+} partitioning and site populations have been measured and are consistent with observed differences in XRD patterns and ceramic colour. Linear shrinkage tests reveal that the inorganic mineral additions can substantially affect shrinkage temperature and the extent of shrinkage. This is related to the microstructural studies observed using scanning electron microscopy. Samples containing nepheline syenite had a lower volumetric shrinkage but also a marginally lower water absorption than the control sample. Further optimisation of these or similar additives could potentially provide energy saving opportunities and reductions in CO_2 emissions for the clay ceramics used in, for example, brick manufacturers.

(PACRIM-P020-2021) Synthesis and Structural Characterization of Model Hydration Products

W. Mozgawa*[1]

1. AGH University of Science and Technology, Faculty of Materials Science and Ceramic, Poland

Sodium aluminosilicate hydrate (N-A-S-H) gel, as the main reaction product of the alkali-activated aluminosilicates, differs from the better-known aluminum-modified calcium silicate hydrate (C-A-S-H) gel. The description of structure and microstructure of the phase, which is mainly responsible for the properties of the geopolymer matrix, seems to be extremely important. Therefore, the aim of this work has been the synthesis and detailed structural analysis of systems that are equivalent to the hydration products of aluminosilicate raw materials in different conditions – the so-called N-A-S-H phase). Comparative tests will also be performed on quite well-researched systems containing calcium (C-S-H and C-A-S-H phase) Model C-S-H, C-A-S-H and N-A-S-H gels will be prepared by precipitation from solution. Resulting materials will be subjected to analysis of chemical composition using XRF and phase composition using XRD. The structure and microstructure of all resulting materials will be analyzed in detail. In order to determine the structural properties such methods as FT-IR and Raman spectroscopy will be primarily useful. The results of structural analysis will be supported by observations using the SEM microscope. This work was financially supported by the National Science Centre in Poland under grant no. 2018/31/B/ST8/03109.

(PACRIM-P021-2021) Gyrotropic birefringence via electromagnon resonance in multiferroic manganite

M. Ogino*[1]; Y. Kaneko[2]; Y. Tokura[2]; Y. Takahashi[1]

1. University of Tokyo, Applied Physics, Japan
2. RIKEN CEMS, Japan

In the multiferroics, the strong magnetoelectric (ME) coupling, which is the cross-coupling effect between electricity and magnetism, is realized. Since the ME coupling modifies the Maxwell's equations, such ME materials give rise to the exotic electromagnetic phenomena. In particular, the light-matter interaction can induce the unconventional optical gyrotropy, which is referred to as gyrotropic birefringence (GB). The strong dynamical ME coupling of electrically active spin excitation, i.e. electromagnon, potentially shows the enhanced GB. Here we report the first observation of GB on the spin-cycloidal multiferroic, perovkite manganite $Eu_{0.55}Y_{0.45}MnO_3$, by use of the time-domain terahertz polarimetry. The enhanced optical rotation arising from the GB was found on the electromagnon resonance. It is demonstrated that the GB is scaled by the bilinear coupling of ferroic order parameters $P \cdot M$ (P: Electric polarization, M: magnetization) through the spin-flop transition in the magnetic field. In addition, the quantitative spectral analysis reveals that the electromagnon and antiferromagnetic resonance cooperatively produce the enhanced GB, indicating the important role of inter-mode coupling for the enhanced dynamical ME effect.

(PACRIM-P022-2021) Luminescence Thermometry – a Fad or a Challenge?

M. Sójka*[1]; M. Runowski[2]; P. Wozny[2]; L. Carlos[3]; E. Zych[1]

1. University of Wroclaw, Poland
2. Adam Mickiewicz University, Poland
3. University of Aveiro, Portugal

Apart from such applications as modern persistent phosphors or scintillators, novel lighting solutions, and others, luminescence thermometry belongs to the most intensely developing research areas of luminescent materials. Luminescence thermometry offers a continuous real time temperature readout with high spatial resolution, universality of application and tolerance to electromagnetic field or harsh conditions. This opens the pathway for applications in a wide spectrum of fields from modern machine technologies to sophisticated biomedical imaging. In luminescence thermometry, the required temperature operating range is entirely defined by the application considered. In bio- or medical uses, only a narrow range of physiological temperatures is of interest. On the other hand, aerospace, or surface temperature distribution may easily require measuring temperature over wide ranges, at least several hundred degrees. In this presentation, we will show that luminescence thermometers working in unprecedentedly broad operating range can be effectively constructed. Supported by the National Science Center (#UMO-2020/37/N/ST5/02507)

(PACRIM-P023-2021) Effects of co-doping on transition metal oxide pigments with trigonal bipyramidal coordination units

M. Oshita*[1]; I. Yamada[1]

1. Osaka Prefecture University, Japan

The narrow hue range of inorganic pigments makes it difficult to control the chromaticity in a single phase. Mn^{3+}-doped $YInO_3$ exhibits a bright blue color, which is predominated by d-d transition of Mn^{3+} in trigonal bipyramidal coordination. Doping of other chromophore transition metals also induces another color: yellow by Fe and green by Cu/Ti, in latter of which Cu^{2+} and Ti^{4+} ions are co-doped. Co-doping also serves to adjust the color between the original chromophore elements. In this study, we aim to precisely control the chromaticity by co-doping of chromophore transition metals. A solution parallel synthesis method was adopted to adjust the chemical compositions of precursors. A mixture of solutions containing chromophore transition metals was systematically added to Y nitrate and In nitrate, followed by calcination at 1473 K in air. X-ray diffraction patterns, chromaticity, and absorption spectra were collected using the obtained oxide powders. The chromaticity shifted from yellow to blue in $YIn_{1-x-y}Mn_xFe_yO_3$ (x + y = 0.1) as the composition changed from Fe-rich (x = 0, y = 0.1) to Mn-rich (x = 0.1, y = 0), proposing that the chromaticity in (Mn, Fe)-doped $YInO_3$ could be systematically controlled between blue and yellow.

(PACRIM-P024-2021) Electromagnetic simulation of microwave and mm-wave transmission lines for 5G applications

M. Yuan*[1]; T. Brown[1]; C. Scalea[1]; E. Schwarz[1]; S. Perini[1]; T. Neuberger[1]; M. Lanagan[1]

1. Penn State University, USA

The drive toward faster data transmission for 5G has deepened interest in microwave property characterization of ceramics, polymers and metals. Transmission lines are part of the high frequency circuitry and they must have low attenuation loss. Major contributions to transmission line loss include dielectric loss, metal resistivity and surface roughness. In this presentation three types of transmission lines were investigated (microstrip, coplanar waveguide and dielectric waveguide) between 1 and 100 GHz. Recently, many finite element analysis electromagnetic simulation software packages (Comsol, Remcom) have become available to provide a "user-friendly" approach to visualizing and designing devices with high frequency material properties. These EM simulation tools will be used to design and validate microwave transmission lines up to 100 GHz. Microstrip, coplanar waveguide and dielectric waveguide transmission line geometries will be initially designed through analytical models with lossless materials. The transmission line designs will be introduced to EM simulation models with substrate permittivity values of 3-10 and dielectric loss values of 10^{-4} to 10^{-2}, which includes the liquid crystal polymers, low temperature co-fired ceramic (LTCC), alumina and glass dielectrics. How the dielectric and metal loss affect the attenuation loss of the transmission line will be addressed.

(PACRIM-P025-2021) Elaboration of tetravalent chromium-doped Yttrium Aluminum Garnet (Cr^{4+}:YAG) transparent ceramics for laser applications: Role of sintering additives

C. Perrière[1]; R. Boulesteix*[1]; A. Maitre[1]; A. Jalocha[2]; B. Forestier[2]; A. Brenier[3]

1. University of Limoges, IRCER, UMR CNRS 7315, France
2. CILAS, France
3. ILM, UMR CNRS 5306, France

The purpose of this work is to determine the influence of different additives (e.g. Ca^{2+}, Mg^{2+} and $Cr^{3+/4+}$) on the microstructural features (i.e. porosity, grain size, secondary phase, grain boundary composition) of Cr:YAG ceramics in order to elucidate the corresponding sintering mechanisms during pressureless reactive sintering. Different shaping processes (liquid and dry ways) were used to homogeneously incorporate the additives in controlled quantities. Then, chemical compositions and the location of secondary phases that are issued from the reaction of additives with YAG phase at high temperature have been identified by using Transmission Electron Microscopy (TEM) and high resolution Secondary Ion Mass Spectrometry (NanoSIMS). Special attention has been paid on the characterization of grain boundaries where atomic segregation, precipitation and/or liquid phase formation can occur. Finally, correlations between process parameters (shaping method, temperatures and composition), Cr^{4+}:YAG ceramics microstructure and optical properties (i.e. transmittance, saturation fluence) have been established.

(PACRIM-P026-2021) Residential Cogeneration System with Protonic Ceramic Fuel Cell and its societal implementation

F. Suito*[1]; K. Li[1]; R. Takatera[1]; A. Ota[1]; T. Araki[1]; H. Shimada[3]; M. MORI[2]

1. Yokohama National University, Japan
2. Central Research Institute of Electric Power Industry, Japan
3. National Institute of Advanced Industrial Science and Technology (AIST), Japan

Building hydrogen economy has been proceeding for the aim of reducing CO2 emission. Solid Oxide Fuel Cells (SOFCs) are spreading as residential cogeneration systems for the future hydrogen economy. In recent years, Protonic Ceramic Fuel Cells (PCFCs) have been focused on because of its possibility of the higher performance than that of SOFCs. In the case H2 is used as fuel for SOFCs, fuel dilution is caused due to the generated steam at the fuel electrode side. On the other hand, PCFCs show less fuel dilution because steam is generated at the air electrode side. In addition, for both fuel cells, internal steam reforming reaction occurs with CH4 fuel, and generated steam by electrochemical reaction affect their performance. Although the steam can help the reforming reaction at the fuel side in SOFCs, there is a problem of carbon deposit in PCFCs due to lack of H2O. The proper operating temperature of PCFCs is lower than conventional SOFCs, and the efficiency is influenced by the operating temperature. In this present study, the performance of the residential cogeneration system with PCFCs, SOFCs and PEFCs are compared, and operating temperature is supposed to be different from each system, 500°C for PCFCs, 750°C for SOFCs, 80°C for PEFCs.

(PACRIM-P027-2021) Improving the Stability of Series-connected Solid Oxide Fuel Cells by Modifying YSZ Electrolyte Composition

Y. Kim*[1]; H. Lim[1]

1. Changwon National University, Republic of Korea

YSZ based anode supported solid oxide fuel cells (SOFCs) were prepared, and two cells with different electrolyte thicknesses were connected in series for the simulation of a cell-imbalanced fuel cell stack. Pure YSZ cells in a series connection exhibited a rapid degradation when a thick electrolyte cell was operated under a negative voltage. On the other hand, ceria added-YSZ cells in a series connection were stable under similar operating conditions, and the power density and impedance were about the same as those before tests. The improved stability was due to the reduction of internal partial pressure in the electrolyte by locally increasing the electronic conduction. Thus, we propose a new protection method, i.e., the local addition of ceria in the YSZ electrolyte, to extend the lifetime of a cell-imbalanced SOFC stack.

(PACRIM-P028-2021) Optimizing Co-ionic Composite Electrolytes Configuration for Durable Solid Oxide Fuel Cell

A. K. Niaz*[1]; H. Lim[1]

1. Changwon National University, Republic of Korea

A bilayer comprising BCY ($BaCe_{0.85}Y_{0.15}O_{3-d}$) - GDC ($Gd_{0.2}Ce_{0.8}O_{2-d}$) at anode side and BZY ($BaZr_{0.85}Y_{0.15}O_{3-d}$) - NDC ($Nd_{0.1}Ce_{0.9}O_{2-d}$) at cathode side, was designed for improving the stability for intermediate solid oxide fuel cells. Two Pt probes were embedded in each electrolyte layer close to anode and cathode sides of electrolyte for the estimation of the local electronic conduction, which is correlated to the cell stability under negative voltage operation caused by cell imbalance. Mixed ionic and n-type electronic conduction was locally developed in the BCY-GDC region near the anode side while mixed ionic and p-type electronic conduction was locally developed in the BZY-NDC near the cathode side. On the other hand, the middle electrolyte region exhibited predominantly ionic conduction which was an effective electrolyte region. Contrary to single layer structured composite electrolytes, the bi-layered composite electrolyte cell was stably operated without electrode/electrolyte delamination under negative voltage conditions, by means of local electronic conduction. Hence, we suggest the composite electrolytes configuration, such as cathode/ BZY-NDC / BCY-GDC / anode, as a protection method against fuel cell stack degradation due to negative voltage operation.

(PACRIM-P029-2021) Validation of thermomagnetic and galvanomagnetic transport measurements (method of four coefficients) using NIST SRM 3451

A. Buchanan[1]; C. Dutra[1]; A. Jarymowycz[1]; E. Laney[1]; C. Posadas[1]; K. Thomson*[1]; M. Beekman[1]

1. California Polytechnic State University, Physics, USA

In the method of four coefficients, electrical resistivity, Seebeck coefficient, Hall coefficient, and Nernst coefficient of a material are measured in order to glean experimental insights into features of electronic structure and charge carrier scattering responsible for enhanced thermoelectric performance. While this powerful experimental technique remains underutilized in the thermoelectric materials research community, its use has increased in recent years. Like all transport coefficients, thermomagnetic properties like the Nernst coefficient require careful measurement and validation of measurement accuracy using standards or through inter-laboratory comparison. Unlike electrical resistivity and Seebeck measurements, well-defined standards are not available for Hall or Nernst coefficient. Furthermore, Nernst coefficient measurements are not commonly performed in most research labs, making inter-laboratory comparison nontrivial. Here we describe a method for validating measurement of all four of the above coefficients using only the NIST SRM 3451, without the need for inter-laboratory collaboration. Data for all four coefficients in the temperature range 80 K to 400 K are reported, which will also be of use for researchers wanting to validate new or existing measurement systems in their own laboratory.

(PACRIM-P030-2021) Improved Thermoelectric Performance of GeTe via Efficient Yttrium Doping

W. Gao*[1]; Z. Liu[1]; W. Zhang[1]; n. Sato[1]; Q. Guo[1]; T. Mori[1]

1. National Institute for Materials Science (NIMS), Japan

Optimization of the carrier concentration by doping/alloying is a key prerequisite to improve thermoelectric performance. In reality, it is, however, challenging to identify effective dopants in some specific systems. Herein, we discovered that Y doping significantly reduced carrier concentration, enhanced power factor, and therefore boosted thermoelectric performance. The high Y-doping efficiency had origin in the fact that the compensating effect due to Ge vacancies was weaker in comparison to other dopants in GeTe, because the slightly increased rhombohedral interaxial angle suggested that Ge-vacancy concentration would not be altered dramatically. In addition, first-principles calculations revealed that Y doping did not induce significant modifications on the band structure, consistent with electrical transport measurement results. Moreover, Bi doping was found to be more powerful than Sb doping to further increase electrical resistivity and to enhance performance, in which the highest peak ZT ~1.8 together with a high average ZT ~1.1 from 300 K to 723 K were achieved. Our work not only highlighted the efficacy of Y doping in enhancing the thermoelectric performance of GeTe system but also developed advanced GeTe based materials for power generation applications.

(PACRIM-P031-2021) Classical Molecular Dynamics Simulations of Lithium Thiophosphate Solid Electrolytes

S. Ariga*[1]; T. Ohkubo[1]; Y. Imamura[2]; S. Urata[2]

1. Chiba University, Graduate School of Science and Engineering, Japan
2. AGC Inc., Innovative Technology Laboratories, Japan

Ceramic and glass solid electrolytes are the key materials to success the next generation all-solid-state battery. Lithium thiophosphates are well-known compounds because of the high ionic conductivity. As a theoretical approach, the Li transport characteristics in bulk state have been well investigated using ab initio molecular dynamics (AIMD) simulations. However, due to the high computational costs of AIMD, it is still challenging to elucidate the ion conduction mechanisms in the interface between crystalline and amorphous phases. To achieve this, we developed a novel force field covering a wide range of lithium thiophosphate materials toward implementation of classical molecular dynamics (CMD) following AIMD. The training force, stress, and energy data was collected for four crystals (γ-Li_3PS_4, $Li_4P_2S_6$, $Li_7P_3S_{11}$, Li_7PS_6) and $70Li_2S$-$30P_2S_5$ glass using Vienna Ab initio Simulation Package (VASP). The force field parameters were optimized by minimizing the difference between CMD and AIMD derived from 20 configurations on each structure (total 100 systems). Class II (COMPASS) potential with bond, angle, and dihedral terms was adopted to perform the melt-quench method to make a glass structure with the fixed fractions of anionic units observed in experiments. To ensure the transferability, the developed parameters were applied to glassy and crystalline lithium thiophosphate systems.

(PACRIM-P032-2021) Nanostructured NVP and NVPF - carbon composites synthesized via Pickering emulsion method as high-performance cathodes for Na-ion batteries

A. Cymann-Sachajdak[1]; M. Wilamowska-Zawlocka*[1]

1. Gdansk University of Technology, Department of Energy Conversion and Storage, Poland

The global environmental aspects lead to increasing need of energy storage. This caused expansion of Li-ion batteries (LIBs) from portal electronics to hybrid/electric vehicles and grid systems. On the hand, limited geographical localization and increasing prices of lithium cause a huge need of reliable, alternative to LIBs energy storage devices. An increasing attention is paid to highly-abundant and cheap sodium-ion systems. Development of electrode materials of improved capacity and stability is of crucial importance for the next generation post-lithium batteries. Polyanionic compounds, such as $Na_3V_2(PO_4)_3$ (NVP) and $Na_3V_2(PO_4)_2F_3$ (NVPF), are promising candidates for Na-ion cathodes due to their high thermal stability, flat sodiation/desodiation potential plateaus and unique NASICON (Na Super Ionic CONductor)-type structure allowing for fast Na-ion migration. However, their main disadvantage is the intrinsically low conductivity, which can be improved by addition of carbon component. In this work we show an innovative Pickering emulsion assisted sol-gel synthesis method for highly-conductive NVP/carbon and NVPF/carbon nanostructures. Novel nanocomposites exhibit improved electrochemical performance compared to NVP and NVPF synthesized in a conventional way.

(PACRIM-P033-2021) Strategies for pressureless sintering of crack-free fully ceramic microencapsulated fuels

E. Kang*[1]; Y. Kim[1]; K. Lim[2]; S. Lee[2]

1. The University of Seoul, Department of Materials Science and Engineering, Republic of Korea
2. KEPCO Nuclear Fuel, Materials Development Section, Republic of Korea

A new strategy was introduced to achieve high volume fraction of tristructural isotropic (TRISO) particles (> 35 vol%) in fully ceramic microencapsulated (FCM) fuels. The proposed strategy requires (1) applying a controlled coating of a SiC matrix on the TRISO particles, (2) forming the coated TRISO particles using cold isostatic pressing, and (3) sintering the formed sample without applied pressure. The strategy was very effective for preventing both the rupture of TRISO particles and matrix cracking during sintering. The thinner the coating layer, the higher the volume fraction of the TRISO particles obtained in the FCM pellets. The maximum TRISO volume fraction (~35.3 %) was obtained when the coating thickness was ~215 μm and the TRISO pellets had no cracks in the SiC matrix. The thermal conductivity of the pellet was 45.0 and 28.5 $W(mK)^{-1}$ at 25 °C and 500 °C, respectively.

(PACRIM-P034-2021) Manufacturing hot-pressed fully ceramic microencapsulated fuels with a small amount of additives

G. Kim*[1]; Y. Kim[1]

1. University of Seoul, Republic of Korea

The effects of the processing parameters (applied pressure and sintering time) on the sintered density, thermal conductivity, and mechanical properties of the SiC matrix and fully ceramic microencapsulated (FCM) fuel pellets were investigated with a decreased amount of additives by hot pressing. The FCM pellet sintered with 2.71 wt% $AlN-Y_2O_3-Sc_2O_3-MgO$ additives at 1870 °C for 4 h under 30 MPa in argon exhibited an excellent thermal conductivity (61.7 $Wm^{-1}K^{-1}$ at a tristructural isotropic (TRISO) particle content of 43.3%), owing to the (i) reduced amount of additives, (ii) high sintered density (residual porosity ~ 0.4%), and (iii) highly thermally conductive SiC matrix (105.0 $Wm^{-1}K^{-1}$).

(PACRIM-P035-2021) Impact of dopants on diffusion in crystalline and amorphous zirconia

M. W. Owen*[3]; M. J. Rushton[3]; L. J. Evitts[3]; A. J. Claisse[1]; M. Puide[1]; W. E. Lee[2]; S. C. Middleburgh[3]

1. Westinghouse Electric Sweden AB, Sweden
2. Imperial College London, United Kingdom
3. Bangor University, United Kingdom

Grain boundaries in ZrO_2 may act as favourable pathways for corrosion species in comparison to diffusion of species through the bulk material. Impurity and alloying element segregation to grain boundaries has been observed experimentally (high Fe contents for example). These highly doped grain boundaries may become amorphous in nature and the impact of this structural change is investigated here and compared to the crystalline doped material. Atomic scale modelling methods have been conducted to analyse diffusion of all components in trivalent doped and un-doped crystalline and amorphous zirconia, ZrO_2, at a range of concentrations between 2.6-11 at. % and temperatures. Diffusion coefficients, pre-exponential factors and activation energies have been calculated for all structures at all concentrations and a comparison has been made - clear differences were observed indicating the potential benefits of amorphous grain boundary formation in retarding diffusion compared to the bulk at reactor operating temperatures. Relating these coefficients gives a description of crystalline and amorphous structures and their differences in diffusive behaviour. These values can be exploited by mechanistic higher-level fuel performance codes and models to better predict the corrosion rate of Zr-alloys.

(PACRIM-P036-2021) Optical Characterization of Core-shell Structures Based on $NaYF_4$ Converters and TiO_2 Based Nanostructures for Photocatalytic Processes

I. Lee da Silva[2]; R. F. Falci*[1]; T. M. Machado[3]; V. d. dos Anjos[2]; M. V. Bell[2]

1. Université Laval, Physics and Engineering, Canada
2. Federal University of Juiz de Fora, Physics Department, Brazil
3. Pontifical Catholic University of Rio de Janeiro, Department of Chemical and Materials Engineering, Brazil

Sodium yttrium fluoride ($NaYF_4$) has been vastly studied due to its physical and chemical properties, such as high refractive index (1.430-1.470 m^2W^{-1}), low phonon energy (< 400cm^{-1}), high transparency in the UV and visible spectral regions, high photochemical stability and low light scattering, among others. Due to its solubility for rare-earth ions, yttrium fluoride is used as a host matrix for co-doping aiming for emission processes by upconversion. $NaYF_4$:Yb, Er based upconversion nanoparticles were obtained through unassisted coprecipitation method by organic agents, in order to eliminate the presence of OH-ions. The synthesis method by coprecipitation proved to be efficient in obtaining $NaYF_4$:Yb, Er based upconversion particles. Although it was not possible to determine the morphology and particles size distribution, the synthesized material has a nanometric scale. $NaYF_4$:Yb,Er nanoparticles were characterized using X-ray diffraction, thermogravimetric analysis, differential scanning calorimetry, scanning microscopy, optical absorption, Raman spectroscopy and photoluminescence spectroscopy. After this procedure, the nanostructures will be covered with the TiO_2 semiconductor coupled to the acetylacetonate molecule, repeating the characterization procedures described above and evaluating the photocatalytic activity of the resulting hybrid material.

(PACRIM-P037-2021) Effect of groundwater composition on dissolution of simulant nuclear waste glass

J. Ayling*[1]; A. J. Fisher[1]; M. T. Harrison[2]; C. L. Corkhill[1]

1. The University of Sheffield, United Kingdom
2. National Nuclear Laboratory, United Kingdom

The UK's safety case for geological disposal of nuclear waste is dependent on understanding how the composition of contacting water affects the dissolution behaviour of the glassy waste form over millennial timescales. The geology surrounding the disposal facility and the subsequent groundwater chemistry must be taken into consideration before a site suitable for geological disposal can be selected. Experiments have been conducted to investigate the aqueous durability of simulant nuclear waste glass in three simulant groundwater solutions as well as ultra-high purity water. Accelerated dissolution tests (following the PCT-B methodology and lasting up to 112 days) were conducted on MW25 glass (an inactive simulant of the UK's nuclear waste glass with 25% Magnox waste loading), and a calcium-zinc modified version of MW28 glass. Dissolution tests were conducted with powdered glass in simulant granitic, saline, and clay groundwaters, as well as ultra high purity water (UHQ). Time dependent rates are calculated from ICP-OES analysis of the sampled leachate solutions. The structure and composition of any resultant surface alteration layers was investigated SEM-EDX and XRD analysis. The effect of the increased calcium and zinc content is discussed as well as comparison of variation in the dissolution rates and alterations layers in each of the four leachate solutions.

(PACRIM-P038-2021) Comparative study of the Effectiveness of Interatomic Potentials for the Molecular Dynamics Simulations of Alumino-borosilicate glasses

M. I. Tuheen*[1]; L. Deng[1]; J. Du[1]

1. University of North Texas, Materials Science and Engineering, USA

Molecular dynamics (MD) simulation is an effective tool to analyze the structures of glassy and amorphous materials. Success of MD simulations depend on the appropriate choice of interatomic potentials. Boron containing glass systems are challenging to model due to the lack of well-established empirical potential that can produce the correct boron coordination state depending on the glass composition and thermal history. This work aims to establish a comparative study of the several recently developed interatomic potentials in alumino-borosilicate systems. A comparative assessment was established on the basis of the short- and medium range structure features such as boron N_4 values, total correlation functions, bond angle distribution, and distribution of Q^n species. The mechanical properties of the simulated glasses were also measured and compared. This assessment provides an insight on the choices of empirical potentials for the MD simulations of alumino-borosilicate glass systems.

(PACRIM-P039-2021) 3D Printing of Nanofiber Reinforced Tissue Engineering Scaffolds

J. Lai[1]; M. Wang*[1]

1. The University of Hong Kong, Department of Mechanical Engineering, Hong Kong

3D printing provides a power platform for fabricating complex porous scaffolds for tissue engineering. Many scaffolds are made via extrusion-based 3D printing, in which inks with or without growth factors are extruded from printing heads to construct scaffolds in a layer by layer manner. Hydrogels are common biomaterials for 3D printing of scaffolds but most hydrogels have poor mechanical properties, which have severely limited their tissue engineering applications. Adding polymer nanofibers into hydrogels can lead to improved mechanical properties that may be comparable to the body tissues. On the other hand, it is widely shown that incorporating growth factors into scaffolds and controlling their release in vivo can accelerate tissue regeneration. In this study, PLGA nanofibers (PLGAf) were firstly electrospun. Short PLGAf were then mixed with a fibroblast growth factor (FGF)-containing alginate (Alg) hydrogel. Subsequently, the hydrogel mixture was extruded from printing heads at room temperature to form FGF/Alg/PLGAf scaffolds. The nanofiber dispersion in Alg hydrogels was studied and viscosity of reinforced hydrogels was assessed. The mechanical properties under compression were determined for printed scaffolds, and the in vitro FGF release behaviour was investigated. The 3D printed nanofiber reinforced scaffolds exhibited improved mechanical strength and also sustained FGF release when Alg was crosslinked.

(PACRIM-P040-2021) 3D Printing for Highly Elastic Composite Scaffolds with Shape Morphing Ability for Blood Vessel Regeneration

S. Chen[1]; M. Wang*[1]

1. The University of Hong Kong, Department of Mechanical Engineering, Hong Kong

3D printing of tissue engineering scaffolds has been investigated for blood vessel regeneration. But the lack of high elasticity still exists for the scaffolds. Also, programmed shape morphing of scaffolds to achieve tubular structures can be considered, with the scaffolds having a hierarchical structure mimicking the anatomical structure of blood vessels. In this study, highly elastic thermoplastic urethane (TPU) and shape memory poly(L-lactide-co-trimethylene carbonate) (PLATMC) were blended at high temperature and then 3D printed for making composite scaffolds with hierarchical structure. A layer of electrospun PLGA/PDA (polydopamine) fibers was then deposited on TPU/PLATMC scaffolds, with human basic fibroblast growth factor (FGF) being loaded in PDA nanospheres. Finally, a layer of GelMA/Gel encapsulated with human umbilical vein endothelial cells was printed on the scaffolds. The shape morphing and self-healing ability of scaffolds thus made was investigated at 37 C. The elastic properties of printed scaffolds were studied using tensile tests, and the in vitro FGF release behaviour was investigated. The in vitro biological behavior, including cell viability and cell distribution and proliferation, was also studied for the scaffolds. The TPU/PLATMC composite scaffolds with biomimicking structures have high potential for blood vessel regeneration.

Monday, December 13, 2021

GOMD Award Lectures

Opening Remarks and Otto Schott Award Presentation and Lectures (2021)

Room: Saturna Island

Session Chairs: Doris Möncke, Alfred University; Mathieu Hubert, Corning Incorporated

10:15 AM

(GOMD-191-2021) Viscoelastic Behavior of Molecular vs. Network Glass-Formers: Are They Fundamentally Different?

S. Sen*[1]

1. University of California, Davis, USA

The viscoelastic behavior of a wide range of inorganic and organic glass-forming liquids are investigated using rheological measurements in the linear regime. While the behavior of the network liquids is observed to follow the classical Maxwell model of spring-dashpot combination with a relatively narrow distribution of relaxation times, the molecular liquids display a strong departure from this behavior. Besides a violation of Maxwell scaling in the terminal regime, a purely viscous response and a Newtonian regime are not recovered for these molecular liquids, even at the lowest frequencies of measurement. Direct dynamical measurements using NMR and X-ray Photon Correlation spectroscopic techniques indicate that these liquids are characterized by a rather broad distribution of relaxation times. We propose that their rheological behavior is better described by the fractional Maxwell model with "springpot" elements replacing the conventional spring-dashpot combination.

10:45 AM

(GOMD-192-2021) The Interaction between Stress, Light, and Chemistry in Glass

J. Zwanziger*[1]

1. Dalhousie University, Chemistry, Canada

I review our progress in understanding the chemical basis for the stress-optic and photoelastic response of glass. By correlating the basic chemical bonding motifs in glass formers, glass modifiers, and their stress-optic response, we were able to develop a simple predictive design rule for how glass chemistry drives photoelastic response. In simple cases, like silicates, it is straightforward to apply, while in more complex systems like borates and tellurites, more detail about short range order needs to be supplied. On-going work will be briefly mentioned, including our efforts to understand the wavelength dispersion of the photoelastic response and the relation of the underlying structure to the individual photoelastic tensor elements.

Tuesday, December 14, 2021

GOMD Award Lectures

George W. Morey Award Lecture (2020)

Room: Saturna Island
Session Chair: Gang Chen, Ohio University

8:40 AM

(GOMD-048-2021) Glass: We love it but it breaks

W. Kob*[1]

1. University of Montpellier, France

Glasses have many outstanding properties that are exploited in a multitude of applications. Sadly glasses have the flaw that they break, often in a catastrophic manner, thus preventing us to use them in certain applications. The fact that on the microscopic level glasses are disordered is not a good reason for them to be fragile since the bonding between the atoms is strong. This leaves the hope that the strength of glasses can be improved by considering special compositions and a clever manufacturing process for the glass. A further step in the direction of stronger glasses is to obtain a better understanding of the fracture process on the microscopic level. In this talk I will therefore present some recent results of large scale computer simulations of sodium-silicate glasses in which we investigated how the fracture process depends on composition, temperature and other relevant parameters. The insight gained from these simulations should help to improve in the future the mechanical behavior of real glasses.

Wednesday, December 15, 2021

GOMD Award Lectures

The Norbert J. Kreidl Award for Young Scholars Lecture (2021)

Room: Saturna Island
Session Chair: Mathieu Bauchy, University of California, Los Angeles

12:10 PM

(GOMD-135-2021) Confirming Classical Nucleation Theory with Novel Energy Landscape Methods

C. Wilkinson*[1]; J. C. Mauro[1]

1. Pennsylvania State University, USA

Energy landscapes have been recognized as a powerful tool for deciphering the very nature of materials. They have already been used to understand glass relaxation, protein folding, and catalytic performance but a landscape has the potential to fully describe the kinetics and thermodynamics of any system or process. In order to enable this description for the purpose of relaxation and crystallization, we present a novel software created to map energy landscapes, energy landscapes of common glass-ceramic systems, and new fundamental insights into the crystallization of liquid systems. This is achieved by using a combination of statistical mechanics and classical nucleation theory to find an effective nucleation curve of the system BaO $2SiO_2$. The nucleation curve predicted is found to be within 1 order of magnitude of the experimentally measured values. Further insights derived from the energy landscape will be discussed.

Thursday, December 16, 2021

GOMD Award Lectures

Varshneya Glass Technology Award Lecture (2021)

Room: Saturna Island
Session Chair: Gang Chen, Ohio University

8:40 AM

(GOMD-136-2021) Nanocrystal doped glass and fibers: fabrication challenges and opportunities for novel photonics applications

H. Ebendorff-Heidepriem*[1]

1. University of Adelaide, Australia

An intriguing approach to imbue glass with unique photonic properties is the creation of luminescent nanocrystals in glass and fiber. This is achieved either by direct doping of ex-situ prepared nanocrystals into glass or by in-situ growth of nanocrystals in the glass. This paper will present recent progress on direct doping of upconversion nanocrystals for fundamental investigations and diamond nanoparticles for magnetic field sensing as well as precipitation of gold nanoparticles for plasmonic-based applications. The ex-situ technique of direct doping of nanocrystals has the challenge of balancing the undesired dissolution of the nanocrystals with the required dispersion of the nanocrystals in the glass matrix during glass processing at elevated temperature. The advantage is the unprecedented control of the nanocrystal type and location as demonstrated for fibers with intrinsic magnetic field sensitivity. The in-situ growth of nanocrystals has the challenge of engineering the glass composition to achieve controlled precipitation of the nanocrystals at elevated temperature. A novel method of controlling the size and concentration of gold nanoparticles in tellurite glass paved the way to extraordinary tuning of plasmonic effects.

Monday, December 13, 2021

GOMD S1: Fundamentals of the Glassy State

Structural Characterization of Glass: Al B Silicate Glasses

Room: Saltspring Island C
Session Chairs: Alex Hannon, Rutherford Appleton Laboratory; Doris Möncke, Alfred University

1:40 PM

(GOMD-001-2021) Correlation between Phase Separation and Local Atomic Structure in Aluminosilicate Glasses Revealed by Small- and Wide-Angle X-ray Scattering

S. K. Wilke*[1]; C. J. Benmore[2]; V. Menon[1]; R. Weber[1]

1. Materials Development, Inc., USA
2. Argonne National Lab, X-ray Science Division, Advanced Photon Source, USA

Liquid-liquid phase separation is common in binary silicates, of which Al_2O_3-SiO_2 is particularly interesting because Al can participate as both a network former and modifier. Past structural studies have shown composition-dependent coexistence of AlO_x species with coordination of x = 4, 5, and 6. It has been proposed that phase separation is driven by Coulombic repulsions between the Al cations, which are electrostatically screened in AlO_4 coordination – preventing phase separation – but not sufficiently so in AlO_6. We interrogate this theory by probing directly the correlation between phase separation and local structure in Al_2O_3-SiO_2 glasses (0-50 mol. %

Al_2O_3) using a combined small- and wide-angle X-ray scattering method (SAXS/WAXS), which measures over a range of momentum transfer $0.05 < Q < 28$ Å^{-1}. Phase separation is assessed via the Porod slope (SAXS), while atomic bond distances and coordination numbers are extracted from the pair distribution function (WAXS). Phase separation is observed in all glasses containing at least 10 mol. % Al_2O_3, which exhibit Porod slopes of 3.0-3.6 that indicate semi-diffuse interfaces. As Al_2O_3 content increases from 10 to 50 mol. %, the Al-O bond distance increases from 1.779 to 1.796 Å, and average Al-O coordination increases from 4.10 to 4.47, consistent with the cation screening theory.

2:00 PM

(GOMD-002-2021) Spectroscopic imaging of diffusion profiles in Na_2O-CaO-Al_2O_3-SiO_2 melts

M. Jacquemin*[1]; V. Sarou-Kanian[1]; P. Simon[1]; L. Hennet[1]; C. Bessada[1]; E. Gouillart[2]; E. Burov[2]

1. CEMHTI-CNRS, France
2. Joint unit CNRS/Saint-Gobain, Surface, Glass and Interfaces, France

For the first time, solid-state Magnetic Resonance Imaging (MRI) has been used in combination with Raman imaging to gain insight into structural modifications occurring across concentration gradients in silicate melts. Diffusion couple experiments were performed on peralkaline Na_2O-CaO-Al_2O_3-SiO_2 (NCAS) melts by assembling glass pieces exhibiting 5 wt% gradients in two oxide components at temperatures ranging from 1200 to 1500°C. The spatially-resolved Raman measurements were acquired post mortem on all samples. In addition, ^{23}Na, ^{27}Al and ^{29}Si Magic-Angle Spinning (MAS) chemical shift imaging was successfully performed on a sample exhibiting Al_2O_3-Na_2O gradients. Micronic spatial resolutions were achieved with both techniques, allowing to probe the structure along the specific features described by electron microprobe concentration profiles. We show that ^{23}Na and ^{27}Al chemical shift profiles are highly correlated and follow the trend of the sodium concentration profile, which spans over distances ten times greater than that of the aluminum profile. By comparing the imaging data to spectral acquisitions on the end-member glasses and a synthesized glass of the interface composition, we provide evidence that the structure at different positions on the diffusion profiles is the same as that of a glass of equivalent composition, independently of the diffusive exchange direction.

2:20 PM

(GOMD-003-2021) Structural transformations in PTR glass studied by Raman scattering

L. Glebov[1]; P. Shirshnev*[1]; M. Klimov[2]; R. Alvarez[1]

1. University of Central Florida, CREOL, USA
2. University of Central Florida, Advanced Materials Processing and Analysis Center, USA

The photo-thermo-refractive (PTR) glass is a material for volume Bragg gratings fabrication. The main property of PTR glass is the ability to change refractive index after UV-irradiation and thermal development. This is Na_2O-ZnO-Al_2O_3-SiO_2-KBr-NaF glass doped with Sn, Sb, Ce, and Ag. During thermal development at temperatures exceeding T_g, the structure of glass is changing. The study of the structural changes in PTR glass by Raman scattering technique is presented. Four types of samples are studied - virgin, volume Bragg grating with period of 0.2 μm, volume Bragg grating with period of 0.4 μm and a sample after spontaneous crystallization of an unirradiated glass at high temperature. Raman spectra in samples were measured in the region from 200 to 1300 cm^{-1}. It was found that spontaneously crystallized sample has an additional bands at 690 and 737 cm^{-1} that other samples do not have. Also it was found that intensity of bands assigned to structural units Q3 (SiO_4 unit with a single non-bridging oxygen) decreases in sequence of samples - virgin glass, grating with long period, grating with short period, spontaneously crystallized sample.

Structural Characterization of Glass: Silicate Glass

Room: Saltspring Island C

Session Chair: John McCloy, Washington State University

3:00 PM

(GOMD-004-2021) Sodium silicate glasses: The structural roles of sodium and oxygen

A. Hannon*[1]; S. Vaishnav[2]; O. L. Alderman[1]; P. A. Bingham[3]

1. Rutherford Appleton Laboratory, ISIS Facility, United Kingdom
2. University of Sheffield, United Kingdom
3. Sheffield Hallam University, United Kingdom

It is shown that modelling the first oxygen-oxygen peak in the neutron correlation function of a glass enables structural information about other correlations to be obtained, and the method is illustrated by application to a sodium silicate glass. The first O-O coordination number can be calculated from network theory, and sodium silicate crystal structures show that the mean O-O distance can be calculated from the Si-O distance, despite the distortion of the SiO_4 tetrahedra. Modelling the O-O peak for a sodium silicate glass allows the Na-O bond length distribution to be determined. For a binary glass with 42.5 mol% Na_2O, it is found that the Na-O coordination number is 4.8(2) with an average bond length of 2.45 Å, and the Na-O bond lengths are more widely distributed than in sodium silicate crystal structures. Sodium ions are bonded mostly to non-bridging oxygens (NBOs), and the Na-NBO coordination number may be four as in crystals. Sodium ions are also bonded to a smaller number of bridging oxygens (BOs). Contrary to previous reports, it is not concluded that Na-NBO bonds are shorter than Na-BO bonds, but instead that the Na-BO distribution is relatively narrow, whilst the Na-NBO distribution extends to both shorter and longer distance. The broad distribution of Na-O bond lengths arises from a relatively broad distribution of Na-NBO bond valences, subject to the overall requirement of charge balance.

3:20 PM

(GOMD-005-2021) Viscosity of sodium silicate melts with a variation of nitrogen and fluorine concentration

S. Sukenaga*[1]; M. Ogawa[2]; Y. Yanaba[3]; M. Ando[4]; H. Shibata[1]

1. Tohoku University, Institute of Multidisciplinary Research for Advanced Materials (IMRAM), Japan
2. AGC Inc., Glass Division, Figured & Plate Glass Department, AGC Yokohama Technical Center, Japan
3. The University of Tokyo, Institute of Industrial Science, Japan
4. Tohoku University, Graduate School of Engineering, Japan

Fluorine and nitrogen are important elements to control properties of silicate glasses and melts, however, these effect on the viscosity remains unclear. The present study measured the viscosity variations due to changes in the anionic composition of a simple sodium silicate system by rotating crucible method at temperatures above their liquidus. Si-29 magic-angle spinning nuclear magnetic resonance (MAS NMR) spectroscopy was used to characterize the structural changes in the neighborhood of silicon atoms against varying nitrogen and fluorine concentrations in the quenched glassy samples. The observed change in the MAS NMR spectra indicates that nitrogen atoms prefer to bond with silicon atoms, whereas fluorine atoms tend to exist in surrounding sodium cations. Nitrogen tends to increase the viscosity, whereas fluorine strongly decreases the viscosity of the sodium silicate melts even with the enhancing effect of the latter on the polymerization degree of silicates. Viscosity of silicate melts has been generally related to the overall polymerization degree of the melt. However, the viscosity of fluorine-containing silicate melts cannot be explained by this conventional scenario. Our data indicate that the sodium–fluorine complexes acted a strong lubricant role in the network melts.

3:40 PM

(GOMD-006-2021) Structural Investigation of Network Former Mixing Effects in Rare-Earth Doped Lead Zinc Phosphotellurite Glasses by NMR and EPR Spectroscopies

M. de Oliveira*[1]; J. Amjad[2]; A. S. de Camargo[1], H. Eckert[1]

1. University of Sao Paulo, Sao Carlos Institute of Physics, Brazil
2. COMSATS Institute of Information Technology, Pakistan

We will present the structural investigation of network former mixing effects in Yb^{3+}-doped Lead Zinc Phosphotellurite glasses by using a comprehensive strategy involving complementary NMR (^{31}P, ^{125}Te, ^{207}Pb) and EPR (of Yb^{3+} probe ions) spectroscopies. Phosphate local environments and connectivities are extracted from detailed ^{31}P MAS NMR. The results suggest a preference for P–O–Te over P–O–P and Te–O–Te connectivities for low Te concentrations, consistent with nonlinear compositional trends of glass transition temperatures. While sharing of the network modifiers PbO and ZnO between the phosphate and the tellurite networks is close to proportional, the NMR data indicate subtle deviations favoring phosphate ligation for high Te content. On a local scale, nonlinear trends in ^{207}Pb chemical shift parameters suggest the preferred formation of a ligand environment with approximately equal numbers of phosphate and tellurite ligands over a wide composition range. While electron spin echo envelope modulation (ESEEM) experiments suggest that the overall elemental distribution appears to be homogeneous on a 1 nm distance scale, the EPR lineshapes suggest significant changes (and possibly inhomogeneous distributions) of the local Yb^{3+} environments.

Atomistic Simulation and Predictive Modeling of Glass I

Room: Cortes Island

Session Chair: Andrew Antony, Corning Incorporated

1:40 PM

(GOMD-007-2021) Indentation-induced photoelastic patterns in oxide glasses: A numerical approach

G. A. Rosales-Sosa*[1]; Y. Kato[1]; E. Barthel[2]; S. Yoshida[3]; A. Yamada[3]; J. Matsuoka[3]; S. Nakane[1]; H. Yamazaki[1]; C. R. Kurkjian[4]

1. Nippon Electric Glass, Fundamental Technology Division, Japan
2. ESPCI, Sciences et Ingénierie de la Matière Molle, France
3. University of Shiga Prefecture, Center for Glass Science and Technology, Japan
4. Rutgers University, Materials Science and Engineering, USA

Understanding the development of stress during mechanical indentation is of great importance to predict crack generation in oxide glass systems. Analytical solutions as well as in-situ / ex-situ photoelastic measurements have given quantitative access to stress distribution around the indented area. Recently, Finite Element Analysis (FEA) calculations have been proposed to investigate plasticity in silica glass under different mechanical tests with great success. In this work, we propose a numerical approach to obtain photoelastic patterns from the stress results of FEA which can be directly compared with experimental data. The validity of the method was confirmed for the elastic contact of a sphere by comparing our results based on FEA with results based on analytical solutions. The effect of plastic model parameters on the calculated photoelastic patterns for conical indentation is discussed. It was observed that the yield stress under pure shear plays a significant role on the calculated retardation pattern compared with the effect of permanent densification. Finally, indentation experiments were performed in silica and soda-lime-silicate glasses and the resulting birefringence patterns were measured. The calculations were able to reproduce the experimental photoelastic patterns, but differences remain suggesting that available plasticity models need to be further improved.

2:00 PM

(GOMD-008-2021) Understanding brittle to ductile transition of disordered materials using energy landscape approach

L. Tang*[1]; M. Bauchy[1]

1. University of California, Los Angeles, Civil & Environmental Engineering, USA

When subjected to load, non-crystalline materials can exhibit strikingly different responses. Understanding the nature of the atomic-scale mechanism that governs the propensity for a disordered phase to break in a brittle or ductile fashion has been a longstanding interest in condensed matter. Here, we investigate numerically the fracture of several archetypal disordered materials. We reveal that the degree of ductility is quantitively encoded in the topography of the energy landscape. Our results shed light on intriguing commonalities in the fracture of disordered phases over different material families, microscopic interactions, and scales.

2:20 PM

(GOMD-009-2021) Effects of surface orientation and termination plane on glass → crystal transformation of lithium disilicate by molecular dynamics simulations

W. Sun*[1]; V. Dierolf[1]; H. Jain[1]

1. Lehigh University, Material Science and Engineering, USA

Understanding of crystal growth in glass melts is needed for processing single crystal growth via solid → solid transformation (SCGST) of $Li_2O{\bullet}SiO_2$ glass. The interface structures between $Li_2Si_2O_5$ glass and $Li_2Si_2O_5$ crystalline phases with different orientations are studied by using molecular dynamics simulation. Atomic number density and the distribution of Q_n species (Si tetrahedra with n bridging oxygen) are shown to be effective ways in characterizing crystal and glass phases. The early stages of crystal growth are significantly affected by the orientation and termination of the crystal surface, as indicated by calculated atomic density, partial ordering, atomic segregation, and change of Q_n concentration.

2:40 PM

(GOMD-010-2021) Revisiting the Atomic Structure of Glassy Silica by Force-Enhanced Atomic Refinement

Q. Zhou*[1]; M. Bauchy[1]; T. Du[1]; Y. Shi[2]

1. University of California, Los Angeles, USA
2. Corning Incorporated, USA

Despite the archetypical nature of glassy silica, its atomic structure remains only partially understood, especially at the medium-range order (e.g., its ring size distribution). Although traditional molecular dynamics simulations based on the melt-quench method can offer a direct access to the glass structure, they are limited to very high cooling rates—which is a key limitation since silicate glasses' medium-range order can be strongly influenced by the cooling rate. As an alternative route, reverse Monte Carlo simulations can be used to "invert" experimental data into three-dimensional structures, but can yield fairly unrealistic structures that are thermodynamically unstable. Here, to overcome these limitations, we adopt force-enhanced atomic refinement—which simultaneously leverages the knowledge of experimental data and interatomic potential—to revisit the structure of glassy silica. We demonstrate that this method yields a glass configuration that simultaneously exhibits higher thermodynamic stability and enhanced agreement with experimental diffraction data as compared to those produced by molecular dynamics and reverse Monte Carlo. Based on these results, we show that the medium-range order structure of glassy silica is more ordered than previously suggested.

Atomistic Simulation and Predictive Modeling of Glass II

Room: Cortes Island

Session Chair: Andrew Antony, Corning Incorporated

3:20 PM

(GOMD-011-2021) A high-throughput ab-initio approach to study optical properties of glass materials (Invited)

V. Botu[1]; B. Aitken[1]; S. Lee*[1]

1. Corning Incorporated, USA

A method to predict the spectral optical properties of glasses from a purely ab-initio first principles approach is presented. Here we generate atomic structures for 24 different and diverse oxide glass chemistries - silicates, phosphates, borates, tellurites and others. We find that the calculated refractive indices are in good agreement with experimentally measured values, with an average error of <0.2%. Overall, due to the consistent underestimation of the electronic band gap with standard ab-initio approaches, the refractive index is over-estimated. Nevertheless, the observed trends remain consistent with those seen experimentally. As an example, we delve deeper into transition metal phosphate glasses and explore the relationship between structural, electronic, and optical properties. Representative compositions were synthesized, and their optical data is then compared with simulated values.

3:50 PM

(GOMD-012-2021) Multi-Scale Modeling of Fracture in Phase-Separated Glasses

L. Tang[1]; M. Bauchy*[1]

1. University of California, Los Angeles, Civil and Environmental Engineering Department, USA

Silicate glasses exhibiting a controlled extent of nanoscale phase separation could feature higher resistance to fracture while retaining their transparency. Here, based on high-throughput molecular dynamics simulations of calcium aluminosilicate glasses, we develop a physics-informed machine learning model aiming to pinpoint promising phase-separated silicate glasses. The machine learning model predicts the thermodynamical propensity for silicate glasses to exhibit phase separation, as well as the chemical composition of the forming phases. This information is then used as input for peridynamic simulations to assess the effect of nanoscale phase separation on the fracture toughness of silicate glasses. This multi-scale modeling approach can guide the discovery of ultratough, yet transparent glasses.

4:10 PM

(GOMD-013-2021) Evaluation of local structure reproducibility of lead borate glass model constructed by a reverse Monte Carlo method using bond valence sum constraint

M. Nagao*[1]; S. Sakida[1]; Y. Benino[1]; T. Nanba[1]; A. Mukunoki[2]; T. Chiba[2]; T. Kikuchi[2]; T. Sakuragi[3]; H. Owada[3]

1. Okayama University, Graduate School of Environmental and Life Science, Japan
2. JGC Japan Corporation, Japan
3. Radioactive Waste Management Funding and Research Center, Japan

In general, lead borate glasses have low melting temperatures and high chemical durability, and they are hence suitable for immobilizing highly volatile elements such as iodine. The authors have studied lead borate glasses by various techniques. In this paper, structural models of $2PbO\text{-}B_2O_3$ glass were constructed by a reverse Monte Carlo (RMC) method with a bond valence sum (BVS) constraint. Coordination state of the constituents was analyzed to evaluate the structural reproducibility. O atoms were classified by the neighboring B and Pb atoms with different coordination numbers, and contribution of the neighbors to the BVS of O atoms was investigated. As a result, a similarity between the RMC model and $(Pb_4O)Pb_2B_6O_{14}$ and $Pb_6B_{10}O_{21}$ crystals was found, suggesting that the characteristic structures in these crystals were successfully reproduced in the RMC glass model. In order to estimate asymmetry of PbO_n polyhedra, distance between Pb atom and the centroid of surrounding O atoms was evaluated. With expanding the coordination sphere, a decrease in the distance, that is, reducing tendency in asymmetry was confirmed both in the RMC model and crystals. However, the asymmetry decrease in the RMC model was faster than that in the crystals, which might be due to structural disorder characteristic of the glass.

4:30 PM

(GOMD-014-2021) Comparing the topography of melt formed and fracture formed glass surfaces

W. Kob*[1]; Z. Zhang[1]; S. Ispas[1]

1. University of Montpellier, France

Using atomistic computer simulations we determine the roughness and topographical features of melt-formed (MS) and fracture surfaces (FS) of oxide glasses. We find that the topography of the MS is described well by the frozen capillary wave theory. The FS are significant rougher than the MS and depend strongly on glass composition. The height-height correlation function for the FS shows an unexpected logarithmic dependence on distance, in contrast to the power law found in experiments. We unravel the crucial role of spatial resolution on surface measurements and conclude that on length scales less than 10~nm FS are not self-affine fractals.

Mechanical Properties of Glass I

Room: Saltspring Island A/B

Session Chair: Morten Smedskjaer, Aalborg University

1:40 PM

(GOMD-015-2021) Highly ductile amorphous oxide at room temperature and high strain rate (Invited)

E. J. Frankberg*[1]; J. Kalikka[4]; F. García Ferré[5]; L. Joly-Pottuz[6]; T. Salminen[9]; J. Hintikka[10]; M. Hokka[1]; S. Koneti[3]; T. Douillard[6]; B. Le Saint[6]; P. Kreiml[2]; M. J. Cordill[2]; T. Epicier[6]; D. Stauffer[12]; M. Vanazzi[11]; L. Roiban[6]; J. Akola[7]; F. Di Fonzo[8]; E. Levänen[1]; K. Masenelli-Varlot[6]

1. Tampere University, Materials science and environmental engineering, Finland
2. Austrian Academy of Sciences, Erich Schmid Institute of Materials Science, Austria
3. aera a/s, Denmark
4. Tampere University, Physics, Finland
5. ABB, Switzerland
6. INSA Lyon, MATEIS, France
7. Norwegian University of Science and Technology, Department of Physics, Norway
8. Italian Institute of Technology, Center for Nano Science and Technology, Italy
9. Tampere University, Microscopy Center, Finland
10. Wärtsilä Corporation, Finland
11. F3nice, Italy
12. Bruker inc., USA

Oxide glasses are an integral part of the modern world, but their usefulness can be limited by their characteristic brittleness at room temperature. We show show that amorphous aluminum oxide can permanently deform without fracture at room temperature and high strain rate by a viscous creep mechanism. These thin-films can reach flow stress at room temperature and can flow plastically up to a total elongation of 100%, provided that the material is dense and free of geometrical flaws. Our study demonstrates a much higher ductility for an amorphous oxide at low temperature than previous observations. This discovery may facilitate the realization of damage-tolerant glass materials that contribute in new ways, with the potential to improve the mechanical resistance and reliability of applications such as electronic devices and batteries (E.J. Frankberg et al. Science 2019). Follow up studies confirm that the available plasticity mechanisms can be scalable to thermodynamic bulk scale by using microcompression and large scale atomistic simulations.

2:10 PM

(GOMD-016-2021) Improving the Fracture Toughness of Oxide Glasses through Bond Switching

M. M. Smedskjaer*[1]; T. To[1]; S. S. Sørensen[1]; J. Christensen[1]; R. Christensen[1]; L. R. Jensen[2]; M. Bockowski[3]; M. Bauchy[4]

1. Aalborg University, Department of Chemistry and Bioscience, Denmark
2. Aalborg University, Department of Materials and Production, Denmark
3. Polish Academy of Sciences, Institute of High-Pressure Physics, Poland
4. University of California, Los Angeles, Department of Civil and Environmental Engineering, USA

The low fracture toughness and practical strength of oxide glasses continue to be their Achilles heel, limiting both future applications and the possibility to make thinner, more environmentally-friendly glasses. In this work, we show that through control of a bond switching mechanism, record-high values of fracture toughness for transparent bulk oxide glasses can be achieved. Specifically, this is achieved by gas-mediated permanent densification of a structurally adaptive lithium aluminoborate glass. The fracture toughness of this densified bulk glass (1.4 MPa $m^{0.5}$) is twice as high as that of standard window glass (0.7 MPa $m^{0.5}$). Classical molecular dynamics simulations reveal that the permanent densification enables more coordination number changes and bond swapping during deformation, ultimately enhancing the degree of plasticity and toughness upon fracture.

2:30 PM

(GOMD-017-2021) Structural and Mechanical properties of ZrO_2-Al_2O_3-SiO_2 glasses prepared by a levitation technique

A. Masuno*[1]; Y. Mikami[1]; Y. Yanaba[2]; S. Sasaki[1]; H. Inoue[2]

1. Hirosaki University, Graduate School of Science and Technology, Japan
2. The University of Tokyo, Institute of Industrial Science, Japan

Al_2O_3-based glasses are generally difficult to fabricate in bulk because of their low glass-forming ability when using a conventional melt-quench method. A levitation technique that promotes deeper undercooling and avoids crystallization of the melt allows the production of bulk glasses. Many of Al_2O_3-based glasses have fascinating mechanical properties such as high elastic modulus and high hardness. Al_2O_3–SiO_2 binary glasses exhibit unique crack-resistant behaviors. The indentation cracking resistance increased by approximately seven times with an increase of Al_2O_3 content. In particular, the composition of $60Al_2O_3$-$40SiO_2$ glass, which is identical to that of mullite, has extraordinary high cracking resistance with high elastic moduli and Vickers hardness. The mullite glass has very low glass forming ability and not so high elastic modulus compared to rare-earth aluminate binary glasses, which limits a wide range of applications. In this study, ZrO_2 were added to the mullite glass to increase elastic modulus because it has a large dissociation energy. It was found that ZrO_2 drastically increased elastic modulus and furthermore improved glass forming ability. However, the crack-resistant properties decreased with an increase of ZrO_2 content. Local structure analyses of Al, Si, Zr atoms in the glasses were performed using NMR, XAFS, and MD simulations.

2:50 PM

(GOMD-018-2021) Impact of pressure on the structure and properties of metaluminous sodium aluminoborosilicate glasses

N. Stone-Weiss*[1]; N. J. Smith[2]; R. Youngman[2]; M. Bockowski[3]; L. Huang[4]; A. Goel[1]

1. Rutgers University, Materials Science and Engineering, USA
2. Corning Incorporated, Science and Technology Division, USA
3. Institute of Physics Polish Academy of Sciences, Institute of High Pressure Physics, Poland
4. Rensselaer Polytechnic Institute, Materials Science and Engineering, USA

Aluminoborosilicate glasses comprise the vast majority of essential glasses used in both everyday life and for cutting edge-technological applications. These glasses show a wide range of composition-dependent properties which are exploited for applications spanning from ultra-strong cell phone touch displays and thermal resistant glasses to glasses for nuclear waste immobilization. Applying high pressure to glasses near the glass transition has been shown to significantly affect glass structure and properties, as varied pressure can induce polymorphism in the glass network due to density alterations. We aim to elucidate the effects that pressure has upon structure and mechanical / elastic properties in sodium aluminoborosilicate glasses. By measuring glass structure via NMR spectroscopy and properties of samples prepared at both ambient and 1 GPa pressures, we expect to gain a complete understanding of the role that pressure plays in inducing polymorphic changes across a broad composition space and in turn how these changes impact glass performance and characteristics.

Mechanical Properties of Glass II

Room: Saltspring Island A/B

Session Chair: Gustavo Rosales-Sosa, Nippon Electric Glass

3:40 PM

(GOMD-019-2021) Indentation cracking in silicate glasses is directed by shear flow, not by densification (Invited)

E. Barthel*[1]; G. A. Rosales-Sosa[3]; V. Keryvin[2]; G. Kermouche[4]

1. CNRS/ESPCI, SIMM, France
2. Université de Bretagne Sud, France
3. Nipppon Electric Glass, Japan
4. Mines de Saint-Etienne, France

Silicate glasses are sensitive to cracking but more or less so, depending on composition. However exact mechanisms are still elusive, preventing progress in formulation. Over the years, a common understanding has has emerged which has it that densification explains the contrasted indentation cracking behaviours found in different types of glass. However, it has recently become more and more apparent that this postulate alone cannot account for the full picture. After a review of experimental evidence, we will present the results of systematic calculations of stress and strain fields during a full indentation cycle, based on recently developped constitutive relations for various silicate glasses, including amorphous silica and sodo-lime-silicates. We find that the impact of densification actually ranges between minor and completely negligible. In contrast, the yield stress in shear (flow stress) has first order impact, as postulated by Marsh more than fifty years ago. We will then discuss the implications of these findings and some consequences for our understanding of crack resistance in relation to shear flow, instabilities, localization, damage and fracture. reference: Indentation cracking in silicate glasses is directed by shear flow, not by densification E Barthel, V Keryvin, G Rosales & G Kermouche, Acta Materialia 194 (2020) 473-481

4:10 PM

(GOMD-020-2021) Mechanism of toughening in metal-nanoparticle implanted sodalime glass – investigation by nano-indentation technique

M. Ono*[1]; S. Miyasaka[2]; Y. Takato[2]; S. Urata[2]; Y. Hayashi[2]

1. Hokkaido University, Research Institute for Electronic Science, Japan
2. AGC Inc., Japan

We recently proposed toughening brittle oxide glass by implanting copper nano-particle. It was then implied that glass is affected by the plastic deformation of metal nano-particles by molecular simulation together with Peridynamics simulation. Here, by examination of nanometer scale indentation measurement, the increase of plasticity in the nanoparticle implanted glass is clearly observed. The amount of increase in the plastic deformation depends on the condition of the treatment such as the temperature at which implantation is performed, and the duration time of the implantation. The difference of the load-displacement curves seems to be large when indentation depth is close to that of the nanoparticle implanted layer. The effect of using different metals as copper and silver for implantation is also discussed.

4:30 PM

(GOMD-021-2021) Composition-dependent Indentation Deformation in Borosilicate Glasses

K. S*[1]; N. Krishnan[2]; N. N. Gosvami[1]

1. Indian Institute of Technology Delhi, Department of Materials Science and Engineering, India
2. Indian Institute of Technology Delhi, Department of Civil Engineering, India

The family of borosilicate glasses is particularly interesting because of their unique behavior where the addition of alkali and alkaline earth oxides to these glasses acts as a charge compensator in improving the network connectivity for boron network as opposed to playing the role of network modifiers for silicate networks. Herein, we analyze the contribution of elasto-plastic response of glasses in governing the indentation deformation mechanisms for a range of borosilicate glass compositions. The series of glasses are designed by varying SiO_2/B_2O_3 molar ratio while maintaining a constant concentration of network modifying alkali and alkaline earth oxides. We employ the nanoindentation technique followed by an annealing protocol below the glass transition temperature (T_g) to quantify the contribution of densification and shear flow as a function of the glass composition. Interestingly, we notice that the uniform elastic recovery and elastic deformation index, along with the propensity for resistance to permanent deformation, are closely related to the network connectivity of glasses as quantified based on the tetrahedral species and non-bridging oxygen atoms. Overall, we show the important role of alkali and alkaline earth modifiers in governing the composition-dependent indentation behavior of the borosilicate glass series.

4:50 PM

(GOMD-022-2021) Load and Compositional Dependence of Strength in Calcium Aluminosilicate Glasses

M. Kazembeyki*[1]; J. C. Mauro[3]; M. M. Smedskjaer[4]; M. Bauchy[2]; C. G. Hoover[1]

1. Arizona State University, School of Sustainable Engineering and the Built Environment, USA
2. University of California, Los Angeles, Civil and Environmental Engineering, USA
3. Pennsylvania State University, USA
4. Aalborg University, Denmark

Knowing how calcium aluminosilicate (CAS) glasses resist permanent deformations is essential for understanding their mechanical behaviors for a wide range of applications. Here we elucidate the functional dependence of modulus, hardness, densification and shear flow in the pressure and composition spaces. We chose a family of CAS glasses with constant CaO/Al_2O_3ratio and varying SiO_2content, as well as a commercial ~100% SiO_2glass. We perform a series of microindentation tests coupled with AFM surface measurements on each indent. Using a method proposed by Yoshida et al., we reveal and quantify the densification and shear flow, responsible for the inelastic energy dissipation. The reduction of hardness, caused by indentation size effect (ISE), is shown to be compositionally dependent; glasses with highersilica content show a greater reduction in their hardness. We demonstrate that this is intimately linked with the inelastic deformation processes, as glasses with larger silica content also experience a reduction in contribution in shear flow and an increase in densification for a fixed load. Finally, we show that the compositional dependence of the inelastic behaviors are positively correlated with the atomic packing density and Poisson's ratio of the glass, confirming that these properties also governs the ISE.

5:10 PM

(GOMD-023-2021) Predicting Normal and Anomalous Glasses Directly from the Indent Images using Machine Learning

A. Agrawal[1]; M. Zaki*[1]; N. Krishnan[1]

1. Indian Institute of Technology Delhi, Department of Civil Engineering, India

Depending on their indentation response, glasses can be broadly classified into "normal" and anomalous. Normal glasses are characterized by slow crack growth under controlled crack growth experiments, while anomalous glasses exhibit abrupt failure. Understanding the failure behavior of normal and anomalous glasses from indent pattern relies on expert knowledge. Here, using machine learning, we show that normal and anomalous glasses can be identified when trained against a modest dataset. Further, using interpretable machine learning, we also identify the features associated with the failure patterns of normal and anomalous glasses, thus providing insights into the mechanisms governing glass failure. Overall, the present study suggests that interpretable machine learning can provide insights in the post-mortem analysis of indentation patterns, thereby allowing improved understanding of glass failure.

GOMD S2: Glass and Interactions with Its Environment - Fundamentals and Applications

Dissolution and Interfacial Reactions: Alternative Conditions

Room: Saturna Island
Session Chair: Jessica Rimsza, Sandia National Laboratories

1:40 PM

(GOMD-024-2021) Multi-scale investigation of the degradation of alkali silicate glasses by unsaturated humidity

F. Alloteau[2]; O. Majérus*[1]; V. Valbi[1]; I. Biron[3]; P. Lehuédé[3]; T. Charpentier[4]; D. Caurant[5]

1. PSL University, Chimie Paristech, France
2. Rathgen Forschungslabor, Germany
3. Center of Research and Restoration of French Museums, France
4. CEA, NIMBE, France
5. CNRS, IRCP, France

Unsaturated atmospheric conditions (defined by a relative humidity, RH < 100 %) are considered as mild and not fundamentally differing from the immersion conditions regarding the alteration of glass. The degradation processes induced by unsaturated humidity are nonetheless specific, as notably observed in the Cultural heritage (CH). In this work, we addressed the mechanisms of glass atmospheric alteration, by the mean of ageing experiments at 85 RH% as a function of time and temperature. Three alkali silicate glass compositions have been studied, two are representative of "unstable" glass compositions of the CH, and one is a soda-lime silicate glass representative of a "stable" glass. The two unstable compositions were characterized by a high hydration rate that even increased in the timescale of the experiments (up to 9 months at 40°C, 85 RH%), while the stable composition hydrated at a rate two orders of magnitude lower. The temperature dependency of their hydration kinetics was not arrhenian, revealing that a self-acceleration mechanism involving several processes controlled the kinetics. The multi-spectroscopic study of the hydrated layer indicates that unstable compositions retained a significant proportion of alkalis and non-bridging oxygens in the hydrated layer, possibly in relationship with their low chemical durability.

2:00 PM

(GOMD-025-2021) Atmospheric alteration of float glass surface as a function of its composition

A. Serve*[1]; O. Majérus[2]; H. Montigaud[1]; S. Papin[1]; D. Caurant[2]

1. Saint-Gobain, France
2. Chimie Paristech, France

The float glass surface can cause quality issues for coated products. Some of these issues are linked with its evolution at the different stages of the glass production and storage. The glass surface reacts with the atmosphere as soon as it is formed and its alteration depends on glass composition and production conditions. In this work, we focused on the effect of glass bulk composition, in particular the influence of the relative Na_2O, CaO and MgO concentrations should be determined. Glasses from the industrial float production with five different compositions have been studied. The samples were altered under unsaturated atmospheric conditions (35°C, 80%RH) for different durations in order to simulate harsh storage conditions. The approach implemented consists in measuring the glass composition and hydration at the extreme surface at different stages of the ageing process (fresh glass just after elaboration and glass aged in climatic chamber). Then, we looked for a correlation between this evolution and the composition of the glass. The main analytical tools were atomic force microscopy, ATR infrared and Raman spectroscopies, and secondary-ion mass spectroscopy (SIMS). The results indicate that the alteration kinetics is correlated with the alkaline-earth nature and concentration of the glass bulk, despite different production conditions.

2:20 PM

(GOMD-026-2021) Is the poison in the container? Alteration of industrial glasses at the molecular scale

M. Tarrago*[1]; C. Le Losq[1]; E. van Hullebusch[1]; D. R. Neuville[1]

1. Université de Paris, Institut de physique du globe de Paris, CNRS, France

Glass is commonly used as a container to hold (tableware, bottles) or cook (kitchenware) food. In such applications, glass is exposed to relatively high temperatures (about 220 °C) and liquids of different pH. Commercial glassware often contains potentially toxic elements (PTE) such as Co, Cr, Cu, Pb, Sb, Zn, and Zr to improve their coloring and refractive index. European regulations (REACH, Food Contact Materials) require the industry to account for the safety of their products. However, there are currently no definite agreements on legally acceptable concentration thresholds of PTE in glass. In order to establish these thresholds, it is essential to understand the leaching processes that transport PTE from glass to food. The present study focuses on the alteration of a set of industrial and natural glasses in an acidic aqueous medium (initial pH 2.3) at 22 and 70 °C (a typical temperature of warm meals). Glass dissolution causes an increase of pH, leaching of the glass components into the liquid media (analyzed by ICP-MS) and surficial structural modifications that can be tracked by Raman spectroscopy. The main changes are due to the interaction of H_3O^+ and OH^- with the silicate network. Alteration changes the speciation of silicate Q^n-groups, the polymerization of the network and the bonds to the network modifiers. The presence of large cations such as Pb^{2+} seems to favor the alteration.

Dissolution and Interfacial Reactions: Long-Term Degradation

Room: Saturna Island

Session Chairs: Louise Criscenti, Sandia National Laboratories; Jessica Rimsza, Sandia National Laboratories

3:00 PM

(GOMD-027-2021) Acoustic and electric field-effects on dissolution reactions, and assessing zeolite precipitation in (hyperalkaline) environments (Invited)

G. Sant*[1]; M. Bauchy[1]

1. University of California, Los Angeles, USA

Dissolution and precipitation affect processes from the synthesis and degradation of microstructures, to CO2 immobilization, and nuclear waste disposal. But, our ability to affect, i.e., enhance or retard, such processes has relied on stoichiometric additives. Here, we highlight the ability of acoustic- or electric-stimulation to affect dissolution by: (a) directing energy to the solid-solution interface whereby bubble cavitation enhances dissolution, or (b) the ability of electric fields to affect ion-transport into and out of the diffusion boundary layer thereby enhancing dissolution. Importantly, while acoustic energy is able to affect all manners of dissolution (i.e., transport-, or surface-limited); electric stimulation at subcritical potentials is only effective in affecting transport-limited dissolution. The ability to affect dissolution is emphasized by reductions in the activation energy of dissolution, which may reduce, by up to 50% under conditions of stimulation. We also elaborate on the development of a new thermodynamic database ("zeo19") for predicting zeolite formation and stability in hyperalkaline environments. The database harmonizes existing and newly-derived data, and predicts zeolite phase equilibria, across compositions, and their partitioning as a function of temperature, at ambient pressure.

3:30 PM

(GOMD-028-2021) Evaluating divalent cation (Zn, Ca, Mg) behaviour during the dissolution of simulant UK high level radioactive waste glass

C. L. Corkhill*[1]; A. J. Fisher[1]

1. University of Sheffield, Materials Science and Engineering, United Kingdom

A modified version of the "MW" sodium aluminoborosilicate base glass has been adopted at Sellafield. The addition of Ca favours $CaMoO_4$ formation and Zn infers improved viscosity. The Ca/Zn MW base glass is used to immobilise reprocessing wastes containing a significant proportion of Mg. We describe the results of experiments that investigate the influence of divalent cations on the mechanisms and kinetics of glass dissolution. The addition of Zn was found to enhance durability in dilute, Si-undersaturated conditions, due to an increase in glass polymerisation with increasing Zn content, as determined by ^{29}Si MAS NMR. However, under conditions of Si-saturation, the presence of Zn prevented the formation of a passivating silica gel layer, and promoted the formation of Zn-containing smectite clays, significantly enhancing the dissolution rate with increasing Zn content. Mg had a similar effect to Zn, which was additive, however glasses containing Ca generally showed lower dissolution rates. The formation of Zn- and Mg-bearing silicates was concurrent with a significant increase in the dissolution rate, similar to Stage III "rate resumption" behaviour, which has previously only been associated with the formation of zeolites.

3:50 PM

(GOMD-029-2021) The dissolution of UK Butex high-level-waste glass under conditions relevant to geological disposal

A. J. Fisher[1]; M. T. Harrison[2]; R. J. Hand[1]; N. C. Hyatt[1]; C. L. Corkhill*[1]

1. University of Sheffield, Material Science and Engineering, United Kingdom
2. National Nuclear Laboratory, Central Laboratory, United Kingdom

The first reprocessing plant in the UK for irradiated natural uranium metal fuel, used the Butex solvent extraction process, which produced liquid waste (HAL) rich in uranium. A portion of Butex HAL remains in storage, which may be co-vitrified with Post-Operational Clean Out (POCO) and Magnox:Oxide blend waste using the current borosilicate glass (MW) or using the newly formulated Ca/Zn modified version of MW. This work describes the dissolution behaviour of simulant Butex-POCO-blend MW and Butex-POCO-blend Ca/Zn MW (containing 4.3 wt.% U) as a function of redox conditions in UHQ water, using the PCT-B method (40 °C, 1,200 m^{-1}, 1-112 d). Evaluation of the dissolution acts as a comparison for previous work performed at SRNL on a wide range of glasses, allowing inferences about anoxic dissolution behaviour (relevant to geological disposal conditions) to be made. This work also provides useful insight into the oxidation state, local co-ordination and leaching behaviour of uranium from vitrified wasteforms, with the identification of secondary phases, in an attempt to underpin the post closure safety assessment.

4:10 PM

(GOMD-030-2021) Alkali-silica gel reaction in simulant radioactive waste glasses dissolved at high pH

C. L. Corkhill*[1]; C. Mann[1]; J. R. Eskelsen[2]; A. J. Fisher[1]; E. M. Pierce[2]

1. University of Sheffield, Materials Science and Engineering, United Kingdom
2. Oak Ridge National Lab, USA

In geological disposal facilities for vitrified radioactive waste, cement will be used to line vaults, in seals and plugs, as buffer material and in general construction. The interaction of cement with vitrified waste is a key issue, since the generation of a hyper-alkaline plume could accelerate glass dissolution mechanisms. We describe an investigation of the mechanisms and kinetics of sodium aluminoborosilicate glass dissolution in the presence of cement leachates containing mixed alkalis with pH ranging from 10 to 13. The surfaces of glass powders and monoliths were characterised using SEM, TEM, EDX and µ-focus XRD. The rate of degradation of the glass in these solutions was found to be the result of two key mechanisms: (i) the formation of a porous alkali- or alkaline-earth silica gel; and (ii) the precipitation of crystalline phases, including zeolites (e.g. phillipsite-K), phyllosilicates (e.g. rhodesite) and clays (e.g. smectite). Depending on the dominant alkali or alkaline-earth element within the cement dissolution media, these key mechanisms were significantly different. These results highlight the complex, but significant effects of high pH, cementitious solutions on the dissolution of nuclear waste glasses, which should be included in future models of vitrified waste disposal in a cement-containing geological disposal facility.

GOMD S3: Optical and Electronic Materials and Devices - Fundamentals and Applications

Optical and Photonic Glass and Glass-Ceramics I

Room: Pender Island
Session Chair: Laeticia Petit, Tampere University

1:40 PM

(GOMD-031-2021) Surface engineering of glass using femtosecond laser induced plasma processing (Invited)

G. Jose*[1]

1. University of Leeds, Applied Photon Science, School of Chemical and Process Engineering, United Kingdom

We report a femtosecond laser induced plasma treatment of silica and silicate glasses to produce structural modification at the surface to depths of several hundred nanometers. Such post formation treatment of silica based glasses offers the opportunity engineer their optical, spectroscopic and mechanical properties. The conventional techniques for surface modification of glasses are ion exchange, thermal diffusion and ion implantation by accelerated ions which are limited by the type of the glass or its composition and the specificity of cations. The doping of multiple and atomically dissimilar ions into silica/silicate glasses using an ultrafast laser processing method allowed the formation of new compositions of silicate glass surfaces for applications in photonic components engineering, biosensing, displays and anticounterfeiting applications. Rare earth doping achieved at record levels in silica allowed us to fabricate short compact waveguide amplifier materials. This new processing method can be extended for surface engineering of silica on silicon waveguide fabrication. Other functionalities like strength, colour and antiglare propertied can also be produced in glass, which will be reported in this talk.

2:10 PM

(GOMD-032-2021) Hydroxyl and fundamental absorption of PTR glass in IR spectral region

L. Glebov[1]; P. Shirshnev*[1]; R. Alvarez[1]

1. University of Central Florida, CREOL - The College of Optics and Photonics, USA

Photo-thermo-refractive (PTR) glass is a photosensitive material for volume holographic elements recording. This is Na_2O-ZnO-Al_2O_3-SiO_2-KBr-NaF glass doped with Sn, Sb, Ce, and Ag. Despite multiple applications of this material in IR spectral region, structure of a long-wavelength absorption of PTR glass is not clear because of overlapping vibronic modes of glass matrix and absorption bands of hydroxyl admixture. For solving this problem the PTR glasses with different water concentration were synthesized in isolated atmosphere of dry air. Spectra of long-wavelength absorption of these samples were compared. The absorption spectra of hydroxyl centers and PTR glass matrix were separated using a mathematical modelling. Spectrum of OH groups and glass matrix were decomposed to the Gaussian curves and compared with the corresponding spectra of oxide silicate glass. It was found that besides well-known absorption bands of hydroxyl beyond 2.8 µm, a number of hydroxyl absorption bands were found in the range between 1.5 and 2.8 µm. Vibronic spectrum of glass matrix was decomposed to Gaussian bands with maxima in vicinity of 3.29, 4.00, and 4.95 µm. This approach is giving a new data about structure of hydroxyl centers and matrix of PTR glass.

2:30 PM

(GOMD-033-2021) Intrinsic absorption and luminescence of Photo-Thermo-Refractive glass

R. Alvarez*[1]; P. Shirshnev[1]; L. Glebov[1]

1. University of Central Florida, CREOL - The College of Optics and Photonics, USA

PTR glass is a multicomponent silicate glass matrix (Na_2O-ZnO-Al_2O_3-SiO_2-KBr-NaF) doped with Ce, Ag, Sn and Sb. It is a holographic phase medium due to the ability of permanent refractive index change after exposure in the photosensitive band of the material (280 nm - 350 nm) and subsequent thermal development above 500°C. While spectroscopy of dopants was studied in this glass, no data on glass matrix were observed. To provide this study, glass matrix was synthesized with contaminations below 1 ppm. It was found that even such concentration of Fe^{3+} affects absorption and luminescence spectra. To mitigate its influence, glass matrix doped with Sn was prepared. Absorption spectra of both glasses were measured. A short wavelength edge described by an exponential function was ascribed to intrinsic absorption of glass matrix. A wide asymmetric band with maximum at 232 nm was assigned to Fe^{3+} impurity. This band was dramatically decreased in the Sn doped glass. It was found that under excitation from 200 to 250 nm, luminescence spectra include two Gaussian envelopes with maxima at 220 and 232 nm. These results suggest that PTR glass matrix luminescence is mainly governed by the emission of intrinsic L centers and the admixture of Fe^{3+}.

2:50 PM

(GOMD-034-2021) Investigation of linear and non-linear optical properties of Alumino-borosilicate glasses containing Indium Tin Oxide (ITO) nanocrystals

A. Ashjari*[1]; B. Eftekhari Yekta[1]; H. Rezaie[1]

1. University of Science and Technology of Iran, Metallurgy and Materials Science and engineering, Islamic Republic of Iran

Transparent Alumino-borosilicate based glass-ceramics containing Indium Tin Oxide (ITO) nanocrystals were synthesized through the melt casting method in the glass system of SiO_2-B_2O_3-Al_2O_3-Na_2O. In this study, the linear and nonlinear optical properties of these glasses were studied. The linear optical parameters (refractive index (n) and indirect bandgap energy (Eg)) of the glass were studied using UV-Vis and NIR spectroscopy after and before the heat treatment. The third-order nonlinear optical properties of these glass-ceramics were investigated using the Z-scan technique. The results showed that the nonlinearity of the glass increased after the heat treatment and the crystallization of ITO nanocrystals. The observed nonlinear optical (NLO) properties and the calculated nonlinear optical coefficients suggest that these nanocomposites could be applied for various photonic and optical limiting applications.

3:10 PM

(GOMD-035-2021) Linear and nonlinear optical properties in glasses managed at the micrometer scale by an imprinting thermo-electrical process

M. Dussauze*[1]; L. D. Karam[2]; F. Adamietz[1]; V. Rodriguez[2]; K. A. Richardson[3]; S. Murugan[4]; T. Cardinal[1]; E. Fargin[2]

1. CNRS / Université de Bordeaux, France
2. University Bordeaux, French Southern Territories
3. University of Central Florida, USA
4. ORC, United Kingdom

Combining glass chemistry and thermo-electrical processes allow to obtain new glassy structure, to access new properties and potentially to design multifunctional glasses. It also allows the fabrication of micro-structure by developing an imprinting thermo-electrical process. Such approach has been tested to control at the micrometer scale various properties: (i) surface reactivity, (ii) surface electrical potential, (iii) gradient of refractive index and (iv) second order (SO) optical properties. In the present study, we focus our attention on the control of optical properties at micrometer scale. We have demonstrated the formation of gradient of refractive index (GRIN) with large variations (up to 5.10^{-2}) which have been optimized on both oxide and chalcogenide glasses. Using the same process, SO optical properties have been patterned at the micrometer scale in glassy materials. By adapting this approach to amorphous niobate thin films, we have induced for the first time in an amorphous material a magnitude of SO optical response comparable to that of a $LiNbO_3$ single crystal ($X^{(2)}$ =29 pm/V at 1.06 μm). This last result paves the way for the future design of integrated nonlinear photonic circuits based on amorphous materials enabled by spatially-selective and highly efficient SO optical response.

Optical and Photonic Glass and Glass-Ceramics II

Room: Pender Island

Session Chairs: Hugues Francois-Saint-Cyr, Thermofisher Scientific; Leonid Glebov, University of Central Florida

3:40 PM

(GOMD-036-2021) Crystallization study of Er^{3+} doped glasses in the $NaPO_3$-CaF_2 system

L. Petit*[1]; n. ojha[1]; i. Dmitrieva[1]

1. Tampere University, Finland

The fabrication of rare-earth (RE) doped glasses has attracted lots of interests. Silica glass has proven to be very attractive because of its wide wavelength range with good optical transparency and high mechanical strength. However, RE tend to cluster in silica glasses. Therefore, phosphate glasses have been of great interest as they are good host materials due to their ability to incorporate high amount of RE. Glass–ceramics are typically prepared by conventional melting followed by a thermal treatment of the glass. The GCs can exhibit significantly enhanced optical properties compared to their glassy counterparts, if the crystalline phase of the desired nature and structure precipitates around the rare-earth ions. Indeed, due to the crystalline environment surrounding the RE ions, the RE-doped glass-ceramics combine glass properties (large flexibility of composition and geometry) with some advantages of the RE-doped single crystals (higher absorption, emission and lifetimes). In this presentation, we will present our work on the development of new transparent Er^{3+} doped oxyfluoride phosphate glass-ceramics. We explain how these materials are prepared and characterized in order to understand the impact of the glass composition on the crystallization tendency of the glasses. We discuss the challenges related to the preparation of such glasses for fiber drawing.

4:00 PM

(GOMD-037-2021) Upconversion modulation by Au NPs in glass: A combined experimental and theoretical study

Y. Wei*[1]; J. Zhao[1]; H. Ebendorff-Heidepriem[1]

1. University of Adelaide, Australia

Remarkable upconversion emission enhancement has been widely demonstrated for individual upconversion nanoparticles in the vicinity of plasmonic structures. This, however, significantly contradicts with Au/Ag NPs-doped upconversion glass, where only minor amplification or even a decrease in the upconversion signal was observed. Here, using experimental and theoretical investigation, we resolve this long-term puzzle by taking both near-field effects and far-field attenuation into account. For tellurite glass doped with Er3+ and different amount of Au NPs doped as the demonstration system, we show that: (1) the presence of Au NPs in glasses leads to a decrease of upconversion emissions compared to only Er^{3+} doped glass, (2) for the same sample thickness (>0.5 mm), a larger intensity decrease is observed for the glass with stronger LSPR band, and (3) larger sample thickness leads to a more prominent decrease of the upconversion emissions. These experimental observations are in good agreement with simulation results, which also predict the parameters (e.g., Au NPs amount and glass thickness) that are required to achieve upconversion enhancement. Overall, the plasmon-assisted upconversion modulation in glass is a net contribution by near-field effects and far-field attenuation. Hence, the upconversion can only be enhanced below a critical glass thickness that depends on the Au NPs content.

4:20 PM

(GOMD-038-2021) Enhanced green emission in Er^{3+}/CdS co-doped borosilicate glass by energy transfer from CdS quantum dots

N. Shasmal*[1]; A. Rodrigues[1]; W. G. Faria[2]; A. S. de Camargo[2]

1. Federal University of Sao Carlos, LaMaV, Brazil
2. University of São Paulo, São Carlos Institute of Physics, LEMAF – Laboratório de Espectroscopia de Materiais Funcionais, Brazil

A borosilicate glass was doped by Er^{3+} and CdS separately and together by melt quenching technique. The CdS-doped glass shows a broad absorption band in the blue region. It exhibits bright yellow and red emission upon excitation at 350 and 460 nm respectively, which has originated from the electron–hole recombination of CdS nanocrystal and defects or traps related transitions. Upon excitation at 377 nm, Er^{3+}-doped glass shows characteristic intense green emission. A huge enhancement was observed in the emission band at 520 nm, when Er^{3+}/CdS co-doped glass was excited at 377 nm. This enhancement can be attributed to the energy transfer from CdS quantum dots to Er^{3+}. When co-doped with CdS, Er^{3+} ions form isoelectronic acceptor-like electron (AE) trap states below the conduction band of the CdS quantum dots (QDs). Those trap states capture an electron, making it negatively charged, which is then bound to a hole by the resulting Coulomb potential. The recombination energy of the bound electron-hole pair is transferred to Er^{3+} ions, which bring the electrons from the ground state to the excited states in an Auger process, and thereby enormously increasing the emission intensity in the Er^{3+}/CdS co-doped glass. This kind of luminescence enhancement is indicative of their possible applications in different photonic applications.

4:40 PM

(GOMD-039-2021) Development of glasses and glass-ceramics for MIR applications

A. Lemiere*[1]; L. Kuusela[1]; M. Guidat[1]; A. Veber[1]; L. Petit[1]

1. Tampere University, Finland

Mid-Infrared (MIR) represents a wavelength range that remains less explored in many optics and photonics branches. Among the gain media for MIR lasers, Er^{3+}-doped materials have received much attention because of the intense 2.7 μm emission from the $^4I_{11/2}\rightarrow{}^4I_{13/2}$ transition of Er^{3+} ions and its convenient 976 nm pump band. Studies have been focused on developing new host materials, including glass-ceramics with low phonon energy and low non-radiative multiphonon relaxation rate, which is required for 2.7 μm laser operation. Er^{3+} -doped germanate and tellurite glasses are promising candidates for efficient MIR lasers. Such glasses are known for their low phonon energy, high rare-earth ion solubility and less complex fabrication due to low melting temperatures of these materials. As the presence of crystalline environment around the rare-earth ions can lead to higher absorption and emission cross-sections, much effort has been focused on the fabrication of glass-ceramics from these glasses to further enhance the MIR emission. In this presentation, we will review our work on the development of new Er^{3+} doped tellurite and germanate glasses and glass-ceramics which have promising Er^{3+} spectroscopic properties in the 2500 – 3000 nm range. The authors would like to acknowledge the financial support of the Academy of Finland (Flagship Programme, Photonics Research and Innovation PREIN-320165 and Academy Project -326418).

5:00 PM

(GOMD-040-2021) Photoluminescence in bioactive sol–gel-derived nanoglasses

A. Lukowiak*[1]; K. Halubek-Gluchowska[1]; W. Bodylska[1]; M. Fandzloch[1]; B. Borak[2]

1. Institute of Low Temperature and Structure Research, PAS, Poland
2. Wroclaw University of Science and Technology, Department of Mechanics, Materials Science and Engineering, Poland

Various types of nanomaterials play currently important role in different fields of medicine. For example, silica-calcia system is well-known basic composition of bioactive glasses that are used in regenerative medicine. Nanoforms of such glass, showing higher activity than similar microsized particles, give broader application possibilities. Hereby, we present studies on SiO_2–CaO and SiO_2–CaO–P_2O_5 nanoparticles (average diameter <100 nm) of bioactive glass doped with lanthanide ions. The sol–gel route was used to fabricate the particles. The optical properties (absorption and photoluminescence spectra) of the samples as well as their structural and morphological properties were examined. The introduction of europium ions allowed to obtain optically active materials exhibiting red luminescence. Co-doping with Tm^{3+} and Yb^{3+} resulted in green light emission due to the up-conversion energy transfer among active ions when excited by 980 nm. During the bioactivity tests, when particles were immersed in a simulated body fluid, ions release to the medium and hydroxyapatite formation on the surface were analyzed. Described systems can be used for monitoring structural changes of the glass immersed in biological fluids or for bioimaging. Acknowledgment This work was supported by the National Science Centre research grant No. 2016/22/E/ST5/00530.

GOMD S4: Glass Technology and Cross-Cutting Topics

Challenges in Glass Manufacturing I

Room: Moresby Island
Session Chair: Irene Peterson, Corning Incorporated

1:40 PM

(GOMD-041-2021) Integration of Generalized Cold Cap into Glass Melter Representation (Invited)

D. P. Guillen*[1]; A. Abboud[1]; P. Ferkl[2]; M. Hall[2]; S. Lee[2]; P. Hrma[3]; R. Pokorny[4]; W. Eaton[2]; D. Dixon[2]; A. A. Kruger[5]

1. Idaho National Laboratory, Materials Science and Engineering, USA
2. Pacific Northwest National Lab, USA
3. North Wind Solutions, LLC, USA
4. University of Chemistry and Technology Prague, Czechia
5. U.S. Department of Energy, USA

Melters, ranging in size from laboratory- to pilot-scale, have been developed in support of the U.S. Department of Energy's Waste Treatment and Immobilization Plant (WTP). To begin treatment, the Low-Activity Waste (LAW) fraction from the tank waste will be extracted, sent to the WTP and vitrified into borosilicate glass. A general computational fluid dynamics (CFD)-based methodology is being developed for waste glass melters of different sizes and geometries operating with various surrogate or actual High-Level Waste (HLW) and LAW feeds. The objective is to provide a predictive capability for the thermophysical processes occurring within the melters, specifically to predict the melt rate of various HLW and LAW feeds, off-gas volume and composition, as well as plenum gas flow and residence time. Physics in the plenum, cold cap, and melt pool are solved in an integrated manner. Solution of the coupled energy and kinetic equations provide the temperature and conversion profile within the cold cap. Heat transfer to the cold cap is affected by the convection in the melt pool induced by forced bubbling. Plenum simulations account for the atmosphere, thermal radiation from the glass, and air in-leakage. The goal of this work is to optimize the vitrification system, reduce plant down time or melter operational upsets, and increase processing rate by improving our understanding of melt dynamics.

2:10 PM

(GOMD-042-2021) Glass melting rate control in electric melters

R. Pokorny*[1]; S. Lee[2]; P. Ferkl[2]; J. Klouzek[1]; P. Hrma[3]; J. Marcial[2]; A. A. Kruger[4]

1. University of Chemistry and Technology Prague, Czechia
2. Pacific Northwest National Lab, USA
3. AttainX, Support Services Contractor to the Office of River Protection, U.S. Department of Energy, USA
4. U.S. Department of Energy, Office of River Protection, USA

The rate of melting in electric glass furnaces is determined by the heat delivered, predominantly by convection, to the cold cap, where the glass batch is converted to glass. The convective heat transfer is enhanced by increasing the melter operating temperature and by inducing forced convection through stirring or bubbling the melt, an option limited by melt volatilization, electrode and refractory corrosion, and the economy. The heat transfer into the cold cap can also be increased by decreasing the cold cap bottom temperature via the kinetics of the batch-to-glass conversion. Using simulated nuclear waste melter feeds, we found that the cold cap bottom temperature is a function of the glass-forming melt viscosity, which is a dominant controlling factor for primary foam formation and stability. The glass-forming melt viscosity during the final stages of glass batch melting depends on the temperature, composition, and the presence of dissolving solid particles. Experimental data showed that the foam collapse occurred in a relatively narrow viscosity range. Using a high-temperature observation furnace, we also investigated the cold cap bottom temperature experimentally.

2:30 PM

(GOMD-043-2021) Process Control and Glass Composition Variability for Hanford LAW Vitrification (Invited)

I. S. Muller*[1]; K. Gilbo[1]; K. Matlack[1]; I. Pegg[1]

1. The Catholic University of America, Vitreous State Laboratory, USA

The Hanford Tank Waste Treatment and Immobilization Plant (WTP) is designed to separate the Hanford tank waste into low- and high-activity waste fractions, which will then be vitrified respectively into Immobilized Low Activity Waste (ILAW) and Immobilized High Level Waste (IHLW) products. Acceptable LAW glass formulations must meet a variety of product quality, processability, and waste loading requirements. To address this need we have developed and tested nearly one thousand such glass formulations at scales ranging from crucibles to continuous melter tests spanning a factor of 160 in scale-up. These results have been captured in process control algorithms that we have developed for both baseline LAW formulations and advanced high waste loading LAW formulations. These algorithms permit the calculation of the glass formulation required for each particular batch of LAW delivered to the LAW vitrification facility. In this paper we will review these process control approaches and the underlying data sets. We will discuss their application to the planned first feed to the LAW facility (AP107) under the direct feed LAW (DFLAW) flow-sheet and the projected variations in glass compositions and glass properties during the processing of feed from that tank.

3:00 PM

(GOMD-044-2021) Influence of aluminum content on multicomponent diffusion in Na2O-CaO-Al2O3-SiO2 melts

M. Jacquemin[2]; E. Burov*[1]; P. Simon[2]; L. Hennet[2]; C. Bessada[2]; E. Gouillart[3]

1. Saint-Gobain, Laboartoire Mixte Saint-Gobain/CNRS, France
2. CEMTHI, France
3. Saint-Gobain, France

The structural role of aluminum in glasses and melts of various degree of peralkalinity has been widely covered in the literature. On the contrary, little is known about its influence on the dynamics and mechanisms of exchange reactions occurring at high temperature diffusion in the aluminosiliactes. We present new multicomponent diffusion data in the Na_2O-CaO-Al_2O_3-SiO_2 (NCAS) from highly peralkaline to near peraluminous compositions. Diffusion matrices were computed by simultaneously fitting concentration profiles from six diffusion couple experiments performed at 1200 to 1500°C. We identified the mechanisms and frequencies of the main exchange reactions by the eigenvectors and eigenvalues of the diffusion matrix, respectively. From the ternary Na2O-CaO-SiO2 to near peraluminous compositions, we observe that the dominant exchange reaction is the same for all compositions domains and involves the interdiffusion of sodium and calcium. Surprisingly, it appears that the population of modifiers and charge compensators does not affect the exchange mechanisms in a significant manner. Contrary, the dynamics of this exchange is highly impacted by aluminium with one exponential dependence of its eigenvalue with the Al2O3. Same behavior for the second and third exchange reactions involving network formers has been observed.

Challenges in Glass Manufacturing II

Room: Moresby Island

Session Chair: Irene Peterson, Corning Incorporated

3:40 PM

(GOMD-045-2021) Redox Chemistry and Gas Evolving Reactions in Glass Melting and Vitrification Processes (Invited)

J. Klouzek*[1]; M. Vernerova[1]; S. Lee[2]; J. Marcial[2]; R. Pokorny[1]; P. Hrma[4]; A. A. Kruger[3]

1. University of Chemistry and Technology, Prague, Laboratory of Inorganic Materials, Czechia
2. Pacific Northwest National Lab, USA
3. U.S. Department of Energy, Office of River Protection, USA
4. AttainX, US Department of Energy, USA

Oxidation-reduction reactions take place over a wide temperature range from about 200 ° C to the maximum melting temperatures. They are usually associated with the development of large amounts of gases, which significantly affect the process of conversion of batch or feed to glass, silica particle dissolution, bubble behavior and primary and secondary foaming. The course of reactions and the resulting redox state of the glass is determined primarily by the composition of the input mixture of raw materials. To ensure an efficient bubble removal process, a combination of an oxidant and oxides of polyvalent elements or also a reducing agent and sulfate is used in the melting of industrial glasses. Sucrose and other reducing agents may be added during vitrification processes to reduce foaming and increase the melting rate. Most redox reactions, together with the decomposition reactions of carbonates in particular, contribute to the formation of the primary foam. However, the evolution of gases takes place even after the end of reactions in the feed or batch. An example is oxygen evolution from the shift of iron redox equilibrium or production of oxygen and sulfur dioxide from sulphate decomposition at melting temperatures. We studied redox reactions associated with gas evolution by thermal analysis, evolved gas analysis, pellet expansion test and electrochemical methods.

4:10 PM

(GOMD-046-2021) Measurements of Batch Expansion and Foaming Using In-situ X-ray Tomography

I. Peterson*[1]; S. A. Luksic[2]; J. Wright[1]; N. LeBlond[1]

1. Corning Research and Development Corporation, Process Research, USA
2. Pacific Northwest National Laboratory, USA

Batch expansion and foaming during the batch-to-melt reaction sequence are undesirable because they can cause suboptimal process conditions and gaseous inclusions in the glass. In-situ x-ray tomography was used to monitor structure during heating to gain a better understanding of the conditions that lead to batch expansion, collapse of the foam, and release of gases. Supplementary data on phase evolution and weight loss gathered at the same heating rate were used to interpret the expansion behavior in terms of a viscoelastic body with low permeability.

4:30 PM

(GOMD-047-2021) Accurate energy demand measurement of industrial glass batches (Invited)

M. Rongen*[1]

1. CelSian Glass & Solar, Netherlands

One of the major challenges for the glass industry is to develop routes in view of the transition to CO_2 neutral glass melting. Although laboratory experimental facilities have proved to be a powerful tool to find more efficient ways of glass melting resulting in lower energy consumption and reduced CO_2 footprint, one essential measurement set-up was still lacking. Therefore, CelSian has developed and built an experimental set-up that enables users to measure the overall energy demand needed to convert industrial sized batch to a glass melt. Energy is required to melt the raw material batch. The quantity of energy required is a sum of the energy needed for: (a) bringing the raw materials up to the reaction temperature; (b) endothermic fusion reactions; and (c) further heating up of the melt and gases to the melting tank exit. The new experimental set-up measures the overall energy demand of industrial glass batches. The energy (J) to heat up batch from room temperature to glass melt temperature can be derived from the experimental procedure. Validation measurements have been performed on a ternary glass (SiO_2-Na_2O-CaO) and float glass and results are compared with data as calculated by thermodynamic models. The impact on energy consumption by using e.g. calcined raw materials, anhydrous materials, slags, cullet and alternative raw materials will be shown.

Tuesday, December 14, 2021

GOMD S1: Fundamentals of the Glassy State

Structural Characterizations of Glass: Borate and Borosilciate Glasses, Phosphates

Room: Saltspring Island C

Session Chair: Doris Möncke, Alfred University

9:40 AM

(GOMD-049-2021) Structural study of the germanium-borate and germanium-phosphate glasses by advanced solid state NMR and Raman spectroscopies (Invited)

S. H. Santagneli*[1]; M. Nalin[1]; H. Eckert[2]

1. Institute of Chemistry- São Paulo State University (UNESP), Inorganic Chemistry, Brazil
2. São Carlos Institute of Physics, University of São Carlos, Brazil

Germanate glasses contained borate and/or phosphate that have the tendency of forming strong homogeneous glasses with pronounced stability against crystallization. These glass families are equally important as model systems for fundamental studies as well as matrices for several applications, such as, photonic and electrolytes. Despite the favorable situation, very few structural studies are available in the literature, predominantly using spectroscopic tools such as vibrational spectroscopy and NMR. With regard to the latter on structural issues, borogermanate glasses do not usually present phase separation issues and thus appear to be particularly suitable for exploring network-former mixing effects within a micro- and macroscopically homogeneous medium. To date, however, the powerful inventory of modern NMR methodology has been underutilized in the study of borogermanate glasses, leaving open many unanswered questions about their structural organization. From the viewpoint of network–former mixing effects, we hereby present a comprehensive structural study on phosphate- and borogermanate glasses. The results of network-former mixing effects within the micro- and macro-structure formations will be presented by Raman spectroscopy and modern multinuclear SS-NMR single- and double- resonance experiments using ^{31}P and ^{11}B nuclei as structural probe

10:10 AM

(GOMD-050-2021) Physical and structural properties of rare-earth-rich borate glasses

S. Sasaki*[1]; A. Masuno[1]; Y. Yanaba[2]; K. Ohara[3]; H. Inoue[2]; Y. Watanabe[2]

1. Hirosaki University, Graduate School of Science and Technology, Japan
2. The University of Tokyo, Japan
3. Japan Synchrotron Radiation Research Institute, Japan

Transparent R_2O_3–B_2O_3 binary glasses (R denotes rare-earth elements or Y) were fabricated in a very wide composition region using an aerodynamic levitation furnace. The maximum R_2O_3 content of light rare-earth compounds reached 63 mol% and decreased with a decrease in the ionic radius of R^{3+}. The thermal, optical, vibrational, and structural properties were investigated, particularly for $50R_2O_3$–$50B_2O_3$ glasses. The glass transition temperature increased with a decrease in the ionic radius of R^{3+}, while the thermal stability was not affected by the glass composition. The packing density increased with a decrease in the ionic radius of R^{3+}. Raman scattering and FT-IR spectra revealed that, in the rare-earth-rich glasses, no conventional three-dimensional networks consisting of corner-sharing BO_n (n = 3 or 4) units existed because all B atoms were formed as isolated BO_3 units. The simple environment around B atoms in the glasses led to additional IR transmittance regions, irrespective of the kinds of R. The data obtained from high-energy X-ray diffraction measurements suggested that the local structure around R resembles the v-$NdBO_3$-type crystal structure, and the O coordination number of R ranged from 6.5 to 7.7, smaller than that of the crystalline phase.

10:30 AM

(GOMD-078-2021) Structure and thermophysical properties of Zn-phosphate glasses: Results from multinuclear NMR spectroscopy and calorimetry

Y. Xia[1]; H. Chen[1]; S. Sen*[1]

1. University of California, Davis, USA

The structure of Zn-phosphate glasses with ZnO content ranging between 35 and 67 mol% is investigated using ^{31}P and ultra-high field ^{67}Zn MAS NMR spectroscopy. The ^{31}P NMR results are consistent with previous reports and confirm a nearly binary Q-species distribution. On the other hand, the ^{67}Zn NMR spectra indicate that Zn atoms are tetrahedrally coordinated to oxygen in glasses with ≤ 55 mol% Zn. The progressive decrease in the connectivity of the phosphate network upon addition of ZnO in this composition range is manifested in a monotonic drop in T_g. Further increase in Zn content results in an increase in the Zn-O coordination number from four to six. The corresponding increase in the connectivity resulting from this coordination change overcomes the effect of the loss in connectivity in the P-O-P network, which is reflected in a reversal in the compositional trend in T_g of glasses with >55 mol% ZnO. In contrast to T_g, the calorimetric fragility index of Zn-phosphate glasses is nearly independent of composition until the ZnO content increases beyond 50 mol%, beyond which the fragility index rapidly increases. This trend appears to be related to the onset of shortening of phosphate chains and of Zn atoms taking up multiple coordination states, which likely increase the temperature dependence of the configurational entropy.

10:50 AM

(GOMD-075-2021) Borophosphate Glasses for Potential Tissue Engineering Applications

M. Abbasi*[1]; A. Krishnamurthy[1]; A. lu[1]; S. Kroeker[1]

1. University of Manitoba, Department of Chemistry, Canada

Although a plethora of phosphosilicate glasses has been proposed for medical applications, the 45S5 Bioglass® developed by Hench remains the most widely used inorganic biomaterial for bone and cartilage regeneration. However, this composition is less ideal for soft-tissue regeneration, as it dissolves more slowly, which impacts the kinetics of soft-tissue healing. Research on biomedical borophosphate glasses is quite new and promising candidates for soft-tissue engineering are still being explored. In this study, the effect of the B/P ratio on the phase separation and apatite formation tendencies of borophosphate glasses, their network structures and post-dissolution structural modifications were studied by solid-state NMR and x-ray diffraction. Glasses with higher phosphate fractions are phase-separated and exhibit greater durability compared to borate-rich compositions, based on ICP-OES analysis of the leachate collected as a function of time. Apatite formation is observed in all glasses but is particularly prominent in glasses bearing 5 and 10 mol% P_2O_5. Calcium uptake from the dissolution medium is significantly higher in these glasses, suggesting that phase-separated crystallites might act as nucleation sites for apatite formation. Moreover, SEM images show hexagonal-needle-shaped apatite crystals in high-P glasses, implying that P content influences apatite morphology.

11:10 AM

(GOMD-077-2021) Insight into the structure of Iron-containing glasses: A molecular dynamics study

B. M. Alhasni*[1]

1. University of technology and applied sciences, science, Oman

Classical molecular dynamics simulations (MD) have been applied to study the short and medium-range order of sodium iron phosphate glasses. The relative amounts of bridging O_b and non-bridging O_{nb} oxygen atoms change with the addition of Fe_2O_3 content. FeO_n polyhedra become closer together and start to share corners and edges with adding more iron oxide. The model indicates that the Fe-O coordination number is decreased from 5.4 in 10Fe toward 4.5 in 40Fe. the Fe-O_{nb}-Fe bonds develop in the expanse of P-O_{nb}-Fe at high Fe_2O_3 content. The bond angle distributions for the models show that the average O–P–O bond angle is tetrahedral, defined peak at 110°, with a very small feature at 85° due to five-fold coordinated P defects. The P-O-P bond angles are peaked at around 150°. A gradual reduction in the concentration of bridging P-O-P bonds with increasing Fe_2O_3 content was observed. The connectivity and hence the depolymerisation of the PO_4 network with the additions of Fe_2O_3 and Na_2O is described by the P-P coordination number (CN) and through the analysis of Q^n distribution. A good agreement was found with the ND, XRD and EXAFS results.

11:30 AM

(GOMD-076-2021) Influence of aluminum on structure and crystallization of iron-polyphosphate glasses

P. Stoch*[1]; P. Goj[1]; A. Wajda[1]; A. Stoch[2]

1. AGH University of Science and Technology, Faculty of Materials Science and Ceramics, Poland
2. Research Networ Lukasiewicz - Institute of Electron Technology Krakow Division, Poland

Phosphate glasses are materials that can be used in many technological fields. Their possible application may be limited due to the existence of easily hydrated P-O-P bonds. The addition of Al_2O_3, Fe_2O_3 strongly improves the durability of the glasses. The subject of the studies is iron-polyphosphate glasses in which iron is gradually substituted by aluminum. The glasses have been tested using FT-IR, Raman, and Mössbauer spectroscopies, DSC, XRD. The obtained experimental results were compared to the theoretical prediction of the glass network structure by ab initio molecular dynamics. It was shown that the substitution leads to an increase in the glass network polymerization. The glass network is inhomogeneous with Al/Fe-rich and poor regions. In the poor regions, Al/Fe cations are glass network modifiers whereas in the rich they can be the glass network formers. Mayer's Bond Order analysis was used to describe bonding mechanisms in the glasses. The existence of the mix-valence iron atoms has been postulated. Thermal treatment of the glasses leads to their partial crystallization and the main crystalline compounds are $AlPO_4$, $Al(PO_3)_3$, $FePO_4$, $Fe_3(P_2O_7)_2$, $Fe_4(P_2O_7)_3$. The compounds crystallize at the surface leading to enrichment of the residual glassy phase in P_2O_5. As a consequence, an inhomogeneous glass-ceramic material has been obtained.

11:50 AM

(GOMD-074-2021) Structural investigation of binary aqueous alkali silicates: Properties of solutions, powders and coatings

H. Mohsin*[1]

1. Saint-Gobain Research Paris, France

Thermal behavior of binary aqueous Na, K and Li silicates has been investigated to better understand the macro and microscopic structural evolution in terms of intumescence. Liquid and solid-state NMR spectroscopy has been utilized for the quantification of structural units defined as Q^n. Solid-state NMR along with Raman spectroscopy has indicated the extent of the local structural arrangement. Heating aqueous alkali silicates leads to structural changes due to the evolution of water where the quantity of leftover water in the system obeys an Arrhenian evolution for Na-silicates with an activation energy of 30 $kJ.mol^{-1}$, consistent with the geochemistry literature, while multiple activation energies are displayed by K and Li-silicates. Foaming is observed at temperatures above 150°C depending upon the concentration and type of alkali, resulting in the polymerization of the network with Na-silicates exhibiting a xerogel-to-glass evolution at 400°C. Furthermore, increasing the Na concentration results in a larger quantity of water retained after a pre-drying treatment, which correlates with a lower softening temperature of the material and is, macroscopically, related to foaming. This structural investigation has allowed us to have a better understanding of the foaming behavior of alkali silicates for addressing the issues being faced in the industrial sector.

Atomistic Simulation and Predictive Modeling of Glass III

Room: Cortes Island
Session Chair: Andrew Antony, Corning Incorporated

9:40 AM

(GOMD-051-2021) Theoretical analysis of Rayleigh scattering attenuation of F-doped silica; MD simulations with force-matching potential

S. Urata*[1]; N. Nakamura[2]; K. Aiba[1]; T. Tada[3]; H. Hosono[4]

1. AGC Inc., Innovative Technology Laboratories, Japan
2. AGC Inc., Materials Integration Laboratories, Japan
3. Kyushu University, Japan
4. Tokyo Institute of Technology, Japan

Further attenuation of Rayleigh scattering of silica glass is required to extend the capacity of long-distance optical fiber communication. This study devoted to answer the riddle of F-doping effect on the attenuation of density fluctuation, which is related to Rayleigh scattering. An accurate force-matching potential was optimized by machine-learning to reproduce forces and energies calculated by DFT calculations. The FMP enabled us to conduct large-scale MD simulations and to understand how fluorine remedies the disorders in silica glass. It was found that fluorine mainly replaces an oxygen in tetrahedral SiO_4 unit, but also forms five-fold Si, such as SiO_4F and SiO_3F_2 units. The former disrupts silica network, while the latter attenuates the network rigidity by loosening the five Si-O(F) bonds. Since the two local effects of fluorine spreads through the glass network, F-doping prompts silica glass to relax further even at temperature lower than the glass transition temperature. Consequently, fluorine minimizes the density fluctuation (thus, Rayleigh scattering) of silica glass at around 1 wt%, although fluorine itself is an impurity for silica glass.

10:00 AM

(GOMD-052-2021) Toughening of soda-lime-silica glass by nanoscale phase separation

J. Christensen*[1]; S. S. Sørensen[1]; T. To[1]; M. Bauchy[2]; M. M. Smedskjaer[1]

1. Aalborg University, Department of Chemistry and Bioscience, Denmark
2. University of California, Los Angeles, Department of Civil and Environmental Engineering, USA

Oxide glasses generally possess low fracture toughness (<1 MPa $m^{1/2}$), which increases their risk of failure and thus limits their applications. Controlled nanoscale phase separation can potentially toughen oxide glasses due to the formation of a heterogeneous nanostructure. Importantly, the transparency of the glass can be retained. Using molecular dynamics simulations, we have studied the fracture behavior of glassy systems that mimic the structure of phase-separated soda-lime-silica glasses, comprising silica-rich nanodroplets in a modifier-rich matrix. Compared to homogeneous glasses, the phase-separated glasses exhibit higher toughness, which we ascribe to multiple phenomena. That is, we observe direct crack-particle interactions that hinder crack propagation, as evident from observations of crack arrest, deflection, and diversion. Furthermore, the heterogeneity of the mechanical properties leads to crack propagation in a less brittle modifier-rich matrix phase, while the droplets increase the overall stiffness, ultimately toughening the glass. Finally, we reveal how the droplets affect the distribution of local stress, potentially shielding the crack.

10:20 AM

(GOMD-053-2021) Sodium Silicate Glass Structural Properties under Compression Shockwave

A. Yadav*[1]; V. Bihani[1]; N. Krishnan[1]

1. Indian Institute of Technology Delhi, Department of Civil Engineering, India

Sodium silicate glass is also known as soda glass is one of the most commonly used glasses. However, the effect of compressive shockwave on the structural properties is not adequately understood. Therefore, to better understand the effect of shockwave on structural properties, we have performed non-equilibrium molecular dynamics simulations using Teter and Reax FF potential. The results show good agreement of Hugoniot curves with the experimental trends. Also, we show shockwave propagation through the glass changes the short-range order and medium-range order as evident from coordination number, bond angle distribution, pair distribution function, and rings. Furthermore, on comparison, the Reax FF potential (reactive potential) gives better results than Teter potential.

10:40 AM

(GOMD-054-2021) Response of Glasses to Sharp Contact Loading Studied via 3-D Nanoindentation Tests in Classical Molecular Dynamics Simulation

H. Liu*[1]; L. Huang[1]; Y. Shi[1]

1. Rensselaer Polytechnic Institute, Materials Science and Engineering, USA

To understand the response of glasses to sharp contact loading, instrumented indentation tests coupled with characterization techniques can provide insights into the crack resistance and deformation mechanisms. However, the loading condition under indentation creates complex stress states and deformation modes that are challenging to quantify in experiments. To improve glass design with better damage resistance, nanoindentation tests in large-scale classical molecular dynamics (MD) simulations have been used to obtain in-depth atomistic level knowledge on deformation mechanisums. Previous simulated nanoindentation studies mostly adopted the 2.5-D setup, which, limited by its geometry and boundary conditions, could only facilitate median crack but not other crack patterns as seen in experiments and consumer products. To overcome these limitations, we have developed a 3-D nanoindentation protocol that can generate stress/strain field and cracking patterns closer to those in real-life scenarios. We will show 3-D nanoindentation studies on brittle metallic glasses and oxide glasses in classical MD simulations. These 3-D nanoindentation results reveal different pathways for crack initiation and propagation from those observed in previous 2.5-D indentation studies. Stress criteria for the formation of different crack systems will also be discussed.

11:00 AM

(GOMD-055-2021) Ion-exchange and interfacial reaction mechanisms of silicate glass corrosion from reactive molecular dynamics simulations

J. Du*[1]; L. Deng[1]

1. University of North Texas, Materials Science and Engineering, USA

Understanding the initial interfacial reactions and ion-exchange processes is critical to elucidate the aqueous corrosion behaviors of silicate glasses that find wide industrial and technological applications. Atomic and nanoscale details of the glass–water interfacial reactions were studied by using reactive potential based molecular dynamics simulations, from which the key mechanisms of the ion exchange, as well as the kinetics of associated interfacial reactions, are elucidated. Several ion-exchange mechanisms were observed and investigated in detail. It was found that water molecules can be an intermediate during the proton transfer processes and as a terminator of the clustered reactions in the interfacial reactions. Statistical analyses of reaction kinetics show that silanol formation is a more favored process than the silanol re-formation in the initial stage of interfacial reactions. The results obtained shed lights on the complex ion-exchange mechanisms and facilitate further studies of corrosion and glass–water interactions of silicate glasses.

Mechanical Properties of Glass III

Room: Saltspring Island A/B
Session Chair: Timothy Gross, Corning Incorporated

9:40 AM

(GOMD-056-2021) A Universal Strength Distribution for Modeling the Lifetime of Optical Fiber Under Stress

G. S. Glaesemann*[1]; Y. Shu[1]

1. Corning Research & Development Corporation, USA

Modeling optical fiber mechanical reliability is a discipline dating back to the invention of low loss fiber in the early 1970's. This discipline matured along with the development of the technology until the early 2000's. In recent years optical fiber can be found in increasingly complex stress scenarios. A few meters of fiber in a data center rack can be routed around sharp corners, shut in doors, and stretched from connection to connection. Further, there are applications where extremely short lengths are placed in a single high-stress bend. Determining the risk of failure from such bending scenarios in fiber is problematic because optical fiber has a multi-modal strength distribution. One cannot assume that the flaws at risk in tight bend situations can be found by testing only short lengths of fiber. In this work a complete multi-modal strength distribution was established by combining strength data from several test methods. A method for scaling this distribution to lengths of interest was developed so that it could be placed into a fatigue lifetime model. This approach can also guide one in the development of a strength distribution relevant for the intended use of the fiber.

10:00 AM

(GOMD-057-2021) Investigation of Slow Crack Growth of Sodium Silicates with AFM

K. T. Strong*[1]; C. Nakakura[2]; J. Depoy[2]; K. Stephens[2]; T. Diebold[1]; S. Grutzik[2]; J. M. Rimsza[2]

1. Sandia National Laboratories, Material Mechanics and Triboloty, USA
2. Sandia National Laboratories, USA

Delayed failure in silicate glasses is known to be caused by the slow crack growth phenomena. It is hypothesized that a stress intensity threshold exists, below which slow crack growth will no longer occur. The knowledge of such a value would be invaluable in designing a glass-containing component where long term reliability is a necessity. Optical techniques have been used to measure crack velocities down to 10^{-13} m/s; however, it is difficult to assess these crack propagation rates over short times because of the limited resolution of light optics (~0.5 μm). Atomic force microscopy (AFM) is an attractive technique for measuring crack velocities in glass because of its ability to measure surface topography with nanometer-scale resolution, as well as its compatibility with insulating materials. The AFM was explored as a means to measure very slow crack velocities in an effort to prove whether a stress intensity threshold exists. Sodium silicate glasses with 15 – 35 mol% Na_2O were synthesized. Indentation cracks were initiated into the glass with a Vicker's indenter within a dry box and aged in a humid environmental chamber. The crack tip locations were tracked as a function of time using an AFM. Evolution of the crack tip under various conditions will be presented. SNL is managed and operated by NTESS under DOE NNSA contract DE-NA0003525

10:20 AM

(GOMD-058-2021) Atomistic Fracture Mechanisms in Sodium Silicates

J. M. Rimsza*[1]; C. Nakakura[1]; S. Grutzik[1]; K. T. Strong[1]

1. Sandia National Laboratories, USA

Unique mechanistic insight into fracture can be obtained through reactive molecular dynamics (MD) simulations. In this investigation of fracture propagation in sodium silicate glasses, a slit crack was introduced into an atomistic amorphous silica model and mode I stress was applied through far-field loading until the crack propagates. Sodium contents between 10-25% were evaluated with dynamic loading up to 1.1 MPa√m. The impact of sodium content on the mechanism of crack propagation, including defect and modifier concentration in the inelastic region, surface energy, and the residual stress fields are assessed. Structural changes of the sodium silicate glass structure in the inelastic region were also evaluated. Results are compared and validated with ongoing experimental evaluation of slow crack growth in sodium silicates. This study provides the basis for future research into crack growth of multicomponent oxide glasses in a variety of environmental conditions.

10:40 AM

(GOMD-059-2021) Stress Corrosion Cracking of Sodium Borosilicate Amorphous Phase Separated Glasses

W. Feng*[1]; F. Célarié[2]; P. Houizot[2]; T. Charpentier[3]; R. Baniel[2]; D. Bonamy[1]; C. L. Rountree[1]

1. SPEC/CEA-Saclay, Université Paris-Saclay, France
2. Institute of Physics of Rennes, University of Rennes 1, France
3. NIMBE/CEA-Saclay, Université Paris-Saclay, France

Sodium borosilicate (SBN) glasses concerns an important research topic as the three oxides (SiO_2, Na_2O, and B_2O_3) correspond to the three principal oxides of many industrial glasses. For certain chemical compositions, an important feature includes amorphous phase separation (APS) which has industrial relevance for crush resistant glasses and porous glasses. Moreover, theory, simulations, and experiments evidenced two-phase APS glasses with the possibility of three-phase APS for some compositions. APS inside the glasses induces local heterogeneous structures, which alters the glasses' physical and mechanical properties (including stress corrosion cracking (SCC) behavior). However, the connection between the structure of APS glasses and their properties remains poorly understood, especially the SCC behavior. In this presentation, we will look at some SBN samples with compositions falling in the hypothesized three-phase APS zone. These results will be complemented with structural, physical, and mechanical properties for a holistic viewpoint of the glasses.

11:00 AM

(GOMD-060-2021) Kinetic model for environmentally assisted crack growth threshold

S. Grutzik*[1]

1. Sandia National Laboratories, Materials and Failure Modeling, USA

We present a crack tip kinetics model for environmentally assisted crack growth (EACG) in glass. The model is physics-based, includes explicit dependence on temperature and humidity, and is valid for all three slow crack growth regimes (reaction controlled, transport controlled, and approach to critical). We demonstrate the model with a set of EACG data for soda lime silicate glass from the literature. We extend the approach to include a simple description of viscoelasticity which enables prediction of a threshold for subcritical fracture growth with comparison to measured threshold values.

11:20 AM

(GOMD-061-2021) Change in Fatigue Resistance with Stressing Rate

G. Scannell*[1]; Y. Shu[2]

1. Corning Incorporated, Mechanics and Reliability Sciences, USA
2. Corning Incorporated, Characterization Sciences, USA

Lifetime predictions from subcritical crack growth in glass are of increasing importance with the increased use of glass in applications with built-in stresses, such as curved displays, foldable phones, or fibers in tight bends. Accurate predictions of the lifetime are made difficult by the limited feasibility to test out to sufficiently long times to be relevant to product use life. Recent dynamic fatigue experiments have shown a decrease in slope of with a decrease in stressing rate for several glass compositions. Such change in the fatigue resistance parameter is not projected by the commonly used Power Law or exponential models. Comparison of the change in slope to the Power Law and the exponential models shows how both models are conservative in their long lifetime predictions. To more accurately predict fatigue lifetimes, the hyperbolic sine model was solved for the dynamic fatigue data. This sinh model was able to accurately predict the change in failure strength with stressing rate in dynamic fatigue, and was able to predict crack velocity experiment data when converted to a k-v plot.

GOMD S2: Glass and Interactions with Its Environment - Fundamentals and Applications

Dissolution and Interfacial Reactions: Residual Rates and Gel Layers

Room: Saturna Island

Session Chairs: Louise Criscenti, Sandia National Laboratories; Jessica Rimsza, Sandia National Laboratories

9:40 AM

(GOMD-062-2021) A multi-scale investigation of the mechanisms controlling the corrosion of borosilicate glasses in hyper-alkaline media

F. Wang[1]; N. Balasubramanya[1]; Q. Qin*[1]; R. Youngman[2]; P. Mukherjee[3]; N. Stone-Weiss[1]; A. Goel[1]

1. Rutgers University, Material Science and Engineering, USA
2. Corning Incorporated, Science and Technology Division, USA
3. Michigan Technological University, Materials Science and Engineering, USA

The overarching goal of the present multi-scale investigation is to unearth the kinetics and mechanisms of corrosion of borosilicate glasses in hyper-alkaline (pH = 13) environments as a function of their chemical composition. Accordingly, a series of 3 – 6 component borosilicate glasses including ISG were synthesized and their dissolution behavior in TMAH solution were studied with a series of state-of-the-art characterization techniques including MAS-NMR, XPS, ICP-OES, ERDA, and HR-TEM. The glass compositions underwent congruent corrosion in the forward rate regime, whereas the controlling mechanism of corrosion in the residual rate regime depends on the presence/absence of Ca in the surrounding environment and can be explained on the basis of the dissolution–re-precipitation model. The dependence of corrosion kinetics and the chemistry of alteration products (in the residual rate regime) on the glass composition will be discussed in the presentation. The results presented in this contribution will ultimately supplement the scientific literature attempting to understand the fundamental science governing the aqueous corrosion of silicate-based glass chemistries and add to the growing database required to develop non-empirical predictive models for designing glasses with controlled dissolution rates.

10:00 AM

(GOMD-063-2021) Effect of alkali content on the durability of borosilicate glasses

A. Krishnamurthy*[1]; T. Nguyen[1]; L. Segato[1]; S. Kroeker[1]

1. Univeristy of Manitoba, Chemistry, Canada

Dissolution of silicate glasses targeted for high-level waste immobilization occurs in three stages: 1) ion exchange and surface hydrolysis, 2) formation and maturation of a gel alteration layer on the glass surface, and 3) precipitation of silicate minerals. The effect of the Na and Si fractions on the transitions between stages and the formation of the alteration layer has not been adequately documented. In this study, the network structure of sodium borosilicate glasses with different Na/Si ratios was characterized using solid-state nuclear magnetic resonance spectroscopy. Standard dissolution tests were carried out and elemental release was measured as a function of time. Above 25 mol% Na_2O, non-bridging oxygens dominated the borosilicate network, and the elemental profiles revealed a direct relationship between the alkali concentration and the kinetics of the initiation of different glass-dissolution stages: stage 2 was observed prematurely in glasses with a Na_2O fraction above 25 mol%. The dissolution profiles were invariant to Na concentration above 25 mol% Na_2O, suggesting a non-linear dependence of the chemical durability on sodium loading. Faujasite precipitated in high-Na glasses (40 mol%) within the 28-day timeframe of this study. We conclude that Na_2O must not exceed 20 mol% for borosilicate glasses intended to function as immobilization wasteforms.

10:20 AM

(GOMD-064-2021) Towards a comprehensive understanding of the effect of high field-strength cations on glass durability

Q. Qin*[1]; N. Stone-Weiss[1]; P. Mukherjee[2]; A. Goel[1]

1. Rutgers University, Material Science and Engineering, USA
2. Michigan Technological University, Materials Science and Engineering, USA

Understanding the impact of constituent oxides on glass corrosion behavior and its dependence on surrounding environment has attracted extensive research interest as it plays a key role in designing new glass compositions for industrial application. Nowadays, high field-strength cations (HFSC), such as Li^{+}, Ca^{2+}, and Zr^{4+}, are frequently seen in functional glass materials as property modifiers. However, their influence on the chemical durability of the resultant glasses is yet to be clear. Therefore, this study aims to gain new fundamental knowledge towards a comprehensive understanding of how HFSCs participate in the corrosion process. Five glass compositions containing different HFSCs (i.e. La3+, Zr^{4+}, Ti^{4+}, and Nb^{5+}) were designed and leached at pH 2 for up to 120 days. Advanced characterization techniques including ICP-OES, SEM-EDS, MAS NMR, TEM and XPS were utilized to characterize both the pre- and post-dissolution glasses and monitor the dissolution progress of different compositions over time. In-depth discussion on the dependece of corrosion kinetics and mechanisms on the glass composition and the effect of HFSCs on the overall glass durability will be provided in the presentation.

10:40 AM

(GOMD-065-2021) Investigation of possible intrinsic stress in the alteration layer of corroded glass

H. Kaya*[1]; D. Ngo[2]; S. Gin[3]; S. H. Kim[1]

1. Pennsylvania State University, USA
2. Avery Dennison, USA
3. CEA Marcoule DE2D/SEVT, France

Temperature, pH, and concentration of solution species are known to affect glass corrosion rates. However, given the thin-film nature with a high surface to volume ratio of the alteration layer, corrosion rates might be affected by possible intrinsic stress build-up that is commonly present in crystalline thin films. Any induced stress could alter chemical reactivity of the alteration layer and accelerate corrosion which is not considered by the current models for glass corrosion rate prediction. In this study we first investigate if the alteration layer has any intrinsic stress, and later if any present stress could accelerate glass corrosion rates.

GOMD S3: Optical and Electronic Materials and Devices - Fundamentals and Applications

Optical and Photonic Glass and Glass-Ceramics III

Room: Pender Island

Session Chair: Luiz Jacobsohn, Clemson University

9:40 AM

(GOMD-066-2021) Enlightening Optical Materials Development via a Correlative 2D-3D Microscopy Workflow (Invited)

H. Francois-Saint-Cyr*[1]; R. Passey[1]; L. Casalena[1]; T. Robine[2]; Z. Lu[2]; P. Le Coustumer[3]; S. Lacomme[3]; F. Pigeonneau[2]; W. Blanc[4]

1. ThermoFisher Scientific, USA
2. CEMEF, France
3. Universite de Bordeaux (BIC), France
4. Université Côte d'Azur, France

Optical materials have become embedded in so many aspects of our current technologies at large, from the most top-secret military projects to the most common civilian applications, that we tend to forget how complex their development remains. To match tightly defined optical requirements to the properties of the materials being developed, engineers and scientists require advanced analytical techniques to characterize their added-value materials in 2D and 3D. We aim to illustrate a workflow for fiber optics, encompassing Secondary Electron Microscopy (SEM), Focused-Ion Beam (FIB), Transmission Electron Microscopy (TEM), and the associated Electron Dispersive Spectroscopy (EDS). The electrically non-conductive nature of this silicate-based material under charged-particle beams, the extremely low concentration of rare-earth doping elements, and the presence of light halogens, create many challenges for these techniques. The broad distribution in size, shape, and concentration of the dielectric nanoparticles (DNP) amplify the complexity of the materials analysis and characterization. Applying a broad range of analytical techniques leads to a better understanding of the composition-structure-property relationships of the DNP in those fibers, as those results will guide toward the fabrication of fiber lasers with low light scattering, or toward fiber sensors with higher light scattering.

10:10 AM

(GOMD-067-2021) Development of a novel optical battery from glass-based materials

N. García Arango[1]; S. Vuori[2]; M. Lastusaari[2]; L. Petit*[1]

1. Tampere University, Faculty of Engineering and Natural Sciences, Finland
2. Turku University, Finland

Materials with persistent luminescence (PeL) have been of great interest as they can find multiple applications in optical recording, biochemistry, night vision, displays, signals or anti-counterfeiting, just to cite few applications. Persistent luminescence is an emission which can last for a long time after removing the irradiation source (from seconds to hours). Recently, composites made of Tm^{3+}, Yb^{3+} codoped crystals and $SrAl_2O_4$:Eu^{2+},Dy^{3+} particles displaying green PeL after UV charging have been successfully prepared, demonstrating that energy transfer can occur between the rare-earth (RE) ions in the crystals and the $SrAl_2O_4$:Eu^{2+},Dy^{3+} particles, showing that PeL can be obtained using NIR charging. Such materials are of great interest for applications dealing with solar cell or even in-situ drugs delivery. In this work, glass-based composites containing Tm^{3+}, Yb^{3+} and $SrAl_2O_4$:Eu^{2+},Dy^{3+} particles are prepared and characterized. Initially, it will be explained how to tailor the material composition, so the composite exhibits strong blue upconversion, comparing the emission strength between glass and glass-ceramics (containing tailored crystalline phases). Secondly, it will be shown that the blue upconversion under 980nm can be used to charge the green persistent luminescent particles confirming that the newly developed glassed-based materials are promising for photonic and biomedical applications.

10:30 AM

(GOMD-068-2021) Highly efficient ytterbium doped aluminosilicates oxyfluoride glasses and glass-ceramics for potential laser-cooling applications

T. Meyneng*[1]; Y. Messaddeq[1]; Y. Ledemi[1]; j. thomas[2]; R. Kashyap[2]

1. Laval University, COPL, Canada
2. Ecole Polytechnque de Montreal, Canada

In this work, we present the development of high efficiency ytterbium doped oxyfluoride glass-ceramics. Glasses from the composition 37,6 SiO_2 – 22,4 Al_2O_3 – 19 YF_3 – 20 LiF – 1 YbF_3 were prepared from the melt-quenching technique. Melting temperature was varied in order to affect fluorine evaporation during synthesis and affecting properties of the final glasses/glass-ceramics. Thermal analysis, transmission and refractive index were measured on the glasses. Ceramization treatments were performed from 550 to 650°C. X-Ray diffraction analysis revealed that fluorine concentration in the parent glass can lead to either $YLiF_4$ – glass ceramics (GCs), for low-fluorine samples and YF_3 – GCs for high-fluorine samples. GCs samples were characterized in terms of transparency after different treatment time. Sample microstructure was observed using SEM/TEM imagery. Finally, fluorescence intensity and quantum yield were measured on glasses and glass-ceramics. The study shows that ceramization treatments improves fluorescence properties compared to the parent glass, both with $YLiF_4$-GCs and YF_3-GCs. $YLiF_4$ GCs shows high thermal stability, opening possibility for fiber drawing and shaping, but presents lower fluorescence performance. YF_3 GCs on the other hand provide higher intensity and quantum yield, but at the cost of a lower thermal stability.

10:50 AM

(GOMD-069-2021) Transparent Yb^{3+} doped oxyfluorophosphate glass-ceramics

M. Hongisto*[1]; A. Veber[1]; N. Boetti[2]; S. Danto[3]; V. Jubera[3]; L. Petit[1]

1. Tampere University, Photonic glasses, Finland
2. Fondazione LINKS – Leading Innovation & Knowledge for Society, Italy
3. CNRS, Univ. Bordeaux, ICMCB, France

Materials doped with Yb3+ have been of great interest as Yb3+ has a very simple energy level structure, causing limited multiphonon relaxation and excited state absorption (ESA). Yb3+ ions in phosphate glasses possess high emission cross-section, broad absorption and emission band. Some oxyfluoride glasses can produce transparent glass-ceramics (GCs) containing fluorite nanocrystals after heat treatment. A GC contains at least one type of functional crystalline phase and a residual glass, produced through a thermal treatment. The heat treatment can lead to surface and/or volume precipitation of the crystals. To be considered as promising optical materials, the GCs should be composed of crystals homogeneously distributed in the volume glass matrix and contain the rare-earth ions. The GCs should also be transparent. In this presentation we present new Yb3+-doped oxyfluorophosphate glasses in the NaPO3-Na2O-NaF system that were prepared and characterized. The study on replacing Na2O with NaF and its impact on the physical, thermal, optical and structural properties of Yb3+ free glasses is detailed. We first present the spectroscopic properties of the Yb3+ glasses in this system. Then, we discuss the changes in the optical properties induced by the heat treatment of the glasses.

11:10 AM

(GOMD-070-2021) Photodegradation of organic dyes using sodium phosphate glasses containing silver nanoparticles

D. Manzani*[1]; T. I. Rubio[1]

1. University of São Paulo, São Carlos Institute of Chemistry, Brazil

Glass is a versatile material that can be applied in a great variety of uses, and its utilization in heterogeneous photocatalysis is currently a growing field of research. Adding plasmonic nanoparticles to the substrate is a method employed in the enhancement of photocatalytic activity. This work aimed to obtain silver nanoparticles (Ag-NPs) on the surface of sodium phosphate glass systems through Na+/Ag+ ion exchange and thermal treatment. The synthesized compositions contain titanium dioxide and were tested for photocatalysis activity towards methylene blue (MB), an organic dye. Two different methods were carried out. In the first, glass powder was added to the MB aqueous solution, forming a suspension, which had its photocatalytic activity assessed; the light was irradiated through a borosilicate glass reactor. In the second, the glass pieces were placed at the bottom of a cuvette, and the solution was then poured; the light was irradiated directly from above towards the cuvette. The photodegradation of MB was monitored via its absorption band in the electronic spectra every 10min for 1h. In this time window, it was possible to obtain a maximum of 57.8% of degradation for the suspension method and 68.3% for the bulk method.

GOMD S4: Glass Technology and Cross-Cutting Topics

Challenges in Glass Manufacturing III

Room: Moresby Island
Session Chair: Irene Peterson, Corning Incorporated

9:40 AM

(GOMD-071-2021) High performance 3D Scan analysis of glass furnace corrosion (Invited)

D. Boloré*[1]; D. Cetin[1]; T. Champion[2]; B. Legin[2]; E. Lopez[2]; D. Rogers[1]; M. Gaubil[2]

1. SEFPRO, Saint-Gobain Research North America, USA
2. SEFPRO, Saint-Gobain Research Provence, France

High Refractory performance is a key component for improving furnace lifetime, along with design and operating parameters. Predicting refractories' real performance from lab-scale testing can be challenging, as some corrosion mechanisms occur on large areas, in heterogeneous temperature conditions and develop after a long time. On the other hand, quantitatively assessing the corrosion of industrial melters is not an easy task as the post-mortem inspection is generally constrained by time and safety. For the last five years, the SEFPRO R&D team has developed methods to precisely and extensively measure corrosion of our refractories at very various scales. In this presentation, we will focus on the insulation fiberglass market to present recent analyses improvements. We will review the behavior of SEFPRO WOOL™ chrome-alumina refractories under different service conditions, including 3D scan measurements acquired directly from the field. Comparing the data (obtained over several measurement campaigns) allows a deeper understanding of corrosion mechanisms in the furnace. These elements are used by SEFPRO together with the glassmakers to enhance refractories selection and improve furnace lifetimes.

10:10 AM

(GOMD-072-2021) Surface modification of float glass by ion exchange for improved mechanical performances and antibacterial or antiviral properties

G. Orsi[1]; M. Biesuz[1]; V. M. Sglavo*[1]

1. University of Trento, Italy

A certain interest has arisen in the last months regarding the possible modification of the surface of many articles to improve their anti-bacterial or anti-viral properties. The introduction of Cu- or Ag-containing compounds has been revealed to be a relatively simple and efficient route in that direction. Ion exchange has been abundantly used in the past for the modification of the surface composition of silicate glasses especially for improving its mechanical resistance. In the present work, an attempt has been carried out to combine two different effects which can be induced through a single ion exchange process: (i) the reinforcing effect guaranteed by the interdiffusion between larger ions from a salt bath and smaller ion in the glass to generate a typical "stuffing" effect on the material surface and (ii) the anti-bacterial or even anti-viral functionality promoted by the diffusion of Ag or Cu ions from the salt into the glass surface. The combined process is analyzed in terms of processing conditions and final performances with respect to the base glass composition. The results revealed certain limitations in the concomitant interdiffusion among the different ions which require an optimization of the final specific functionality of the glass.

10:30 AM

(GOMD-073-2021) Influence of glass composition on chemical strengthening efficiency

A. Abdurakhmanova[1]; V. M. Sglavo*[1]

1. University of Trento, Italy

Chemical strengthening has assumed a key role in the processing of glass for many advanced applications, from "touch" electronics to pharmaceutical packaging, from automotive to constructions and safety or armor devices. It is fundamentally based on an ion exchange process involving larger ions (K or Na) present in a molten salt which diffuse substitute smaller ions (Na or Li) into the glass; the "stuffing" effect accounts for the creation of a compression on the surface of the material which mechanically shields the cracks and improves the mechanical resistance. The magnitude and the depth of the surface residual compression depend on the processing conditions (mainly temperature and duration of the ion exchange) but also on the glass. In the present work two slightly different soda lime silicate float glasses have been considered and subjected to identical chemical strengthening processes to point out how some minor differences in the chemical composition can affect the ion exchange efficiency. Different mechanical properties like residual stress, flexural strength and surface damage sensitivity are studied and correlated to the strengthening process and the evolution of the glass structure analyzed by complementary spectroscopic techniques. The results can help to design proper glasses proner to be used in chemical strengthening processes.

GOMD S1: Fundamentals of the Glassy State

Data-based Modeling and Machine Learning for Glass Science I

Room: Moresby Island
Session Chair: Adama Tandia, Corning Incorporated

1:30 PM

(GOMD-079-2021) Text extraction and Natural Language Processing in Glasses (Invited)

E. Olivetti*[1]

1. Massachusetts Institute of Technology, USA

Advances in applying natural language processing (NLP) to scientific text have been successfully applied to well-studied material systems with large amounts of data. However, we need ways to leverage literature data in materials domains without thousands of papers. Applying NLP pipelines to these types of materials science systems can be challenging due to the general schema and the noisiness of automatically extracted data. In this presentation, we demonstrate how to leverage domain knowledge to build upon existing data extraction techniques and improve extraction accuracy using examples in the pursuit of alternative cementious feedstocks fields. This presentation will describe an effort to integrate artificial intelligence with material science to support the development of low environmental impact concrete mixtures. Generative modeling approaches can be used to learn from this and other data to optimize the design of concrete mixtures.

2:00 PM

(GOMD-080-2021) Knowledge extraction from glass literature using natural language processing

V. Venugopal[1]; M. Zaki[1]; S. Bishnoi[1]; M. Agarwal[1]; N. Krishnan*[1]

1. Indian Institute of Technology Delhi, Civil Engineering, India

Most of the knowledge in the literature is in the form of text embedded in the manuscript. As such, access to this information require the user to go into each of the manuscripts and read the full text. This impedes easy access to answer questions such as: (i) what synthesis protocol should be used to make glass X?, (ii) how many studies are there on glasses containing element Y?, or (iii) are there any XRD patterns of glasses containing element X, prepared using Y protocol? To address these challenges, here we present an approach employing natural language processing and text mining. Specifically, we analyze the abstracts, figure captions, and the texts of more than 100,000 manuscripts in glass science. The analysis of this text allows direct dissemination of knowledge in the form of caption cluster plots (plots with keywords associated with each of the figure captions), topic modeling (classifying the abstracts automatically based on the topics), and elemental map (map of abstracts having selected elements). These approaches can aid the direct dissemination of knowledge that is otherwise buried in the materials science literature.

2:20 PM

(GOMD-081-2021) Combining Experimental and Simulation Datasets in Machine Learning for Glass Properties Prediction

K. Yang*[1]; Z. Yin[2]; Y. Song[1]; M. Bauchy[1]

1. UCLA, Civil and Environmental Engineering, USA
2. UCLA, Department of Mathematics, USA

Machine learning is increasingly recognized as a promising path to accelerate the discovery of novel glasses with improved properties and functionalities. However, despite some notable success stories, applying machine learning to predict glass properties still faces various challenges. In particular, machine learning models require as a prerequisite the existence of data that are (i) available, (ii) complete, (iii) consistent, (iv) accurate, and (v) numerous. Although experimental data are usually accurate, they are often not numerous enough to enable meaningful deep learning approaches. As an alternative path, synthetic data generated by high-throughput molecular dynamics simulations can offer large, consistent datasets. However, their limited accuracy does not always yield a perfect agreement with experiments—which makes it challenging to directly combine experimental and simulation data within universal, unifying datasets. Here, to address this challenge, we present a new "data fusion" approach that can simultaneously leverage the distinct advantages of experimental and simulation data—wherein experimental and simulation data mutually inform, augment, and advance each other. We demonstrate that our fused machine learning model systematically outperforms models that are solely trained based on experimental (or simulation) data.

2:40 PM

(GOMD-082-2021) Extracting the Synthesis and Processing Protocols from Materials Science Literature for Improved Property Prediction

M. Zaki*[1]; N. Krishnan[1]; J. Jayadeva[2]

1. Indian Institute of Technology Delhi, Department of Civil Engineering, India
2. Indian Institute of Technology Delhi, Department of Electrical Engineering, India

Recently, extensive studies have been carried out to use machine learning and artificial intelligence to enable computational design and discovery of novel materials. Most of these studies focus on inverse design or property prediction with focus on composition–property relationships. However, a crucial parameter that has received lesser attention is the material processing route along with the relevant parameters. Herein, combining a text-mining approach with machine learning, we present a framework that can predict the material synthesis parameters for inorganic glasses. Specifically, we analyze the text associated with more than 50,000 research articles to obtain the synthesis parameters of oxide glasses with a variety of input components (more than 30). We show that the processing parameters, when combined with the composition, provides improved prediction of the material properties. Finally, we show that the approach presented herein can outperform heuristic approaches for glass synthesis, thereby allowing scientists to choose the most appropriate synthesis route for developing glasses with targeted properties.

Data-based Modeling and Machine Learning for Glass Science II

Room: Moresby Island
Session Chair: Adama Tandia, Corning Incorporated

3:50 PM

(GOMD-083-2021) Modeling atomistic disorder using automatic differentiation and machine learning (Invited)

E. D. Cubuk*[1]

1. Google Brain, USA

With the increasing availability of materials property databases and computational resources, a data-driven approach to modelling disordered solids is becoming more attractive. Automatic differentiation and machine learning are emerging as two promising tools for materials science. I will talk about some recent work on using automatic differentiation and machine learning to model disordered solids, with applications including emprical force-fields, prediction of dynamics, and structural optimization. If time permits, I will also talk about some of the common failure modes of machine learning models in materials science, and potential ways to mitigate these failures.

4:20 PM

(GOMD-084-2021) Decoding the "Genome" of Inorganic Glasses using Interpretable Machine Learning

R. Ravinder*[1]; S. Bishnoi[2]; M. Zaki[1]; N. Krishnan[1]

1. Indian Institute of Technology Delhi, Department of Civil Engineering, India
2. Indian Institute of Technology Delhi, School of Interdisciplinary Research, India

In materials, understanding and predicting the composition–structure–property relationship is the key to developing novel materials. Such predictions are typically hindered by the complex physics happening at different length and time scales, along with the large number of structural and compositional arrangements possible. As an alternative route, data-driven approaches such as machine learning can prove key to predict structure and composition of materials. Herein, using a large database of glasses (> 450,000 compositions) with up to 232 components, we develop machine learning based models for predicting 25 relevant glass properties. Further, employing Shapley additive explanations (SHAP), we demonstrate the role of each of the input component in controlling the glass property quantitatively. Interestingly, we observe that the components may exhibit a coupled effect—presence of one additional component may reverse the performance of another component as in the case of boron anomaly. We show that SHAP can accurately capture these effects thereby providing insights to composition–property relationship and underlying physics. We believe this step towards decoding the glass genome through interpretable machine learning can significantly improve our fundamental understanding on the compositional control of glass properties.

4:40 PM

(GOMD-085-2021) Understanding the optical properties of glasses using interpretable machine learning

N. Krishnan[1]; M. Zaki*[1]

1. Indian Institute of Technology Delhi, Civil Engineering, India

Designing glasses with tailored optical properties is a major challenge for several scientific and daily-life applications such as telescopic and microscopic lenses and display screens. Recent studies suggest that machine learning (ML) methods can be used to develop high-fidelity composition–property models for glasses. However, these ML models, owing to their "black-box" nature, are highly uninterpretable and does not provide any insights into the fundamental nature of the compositional control on glasses' properties. Here, using machine learning algorithms and a database containing ~150,000 glasses, we develop models to predict refractive index and abbe number of oxide glasses with 49 and 47 distinct components, respectively. Further, employing SHapley Additive exPlanations (SHAP), we interpret the composition–property models to understand the compositional control on refractive index and abbe number. Interestingly, we demonstrate that by selecting the top 20 components, models with comparable accuracy as that of the ML models with close to 50 components can be developed. Overall, the present work shows that interpretable ML can be used to develop a fundamental understanding of the compositional control of materials' properties.

Mechanical Properties of Glass IV

Room: Saltspring Island A/B
Session Chair: Satoshi Yoshida, AGC Inc.

1:30 PM

(GOMD-086-2021) Fracture Toughness of a Metal-Organic Framework Glass

T. To[1]; S. S. Sørensen[1]; M. Stepniewska[1]; A. Qiao[1]; L. R. Jensen[2]; M. Bauchy[3]; Y. Yue[1]; M. M. Smedskjaer*[1]

1. Aalborg University, Department of Chemistry and Bioscience, Denmark
2. Aalborg university, Department of materials and production, Denmark
3. University of California, Department of civil and environmental engineering, USA

Metal-organic framework (MOF) glasses feature unique thermal, structural, and chemical properties compared to traditional metallic, organic, and oxide glasses. In practical engineering, the knowledge of their strength and toughness are important, yet the small sizes of MOF glasses limit the testing. Recently, the discovery of a zeolitic imidazolate framework (ZIF) glass with ultra-high glass-forming ability, namely ZIF-62, has enabled the preparation of >25 mm^3 samples and thus mechanical testing besides indentation. In this work, we report the first measurement of the fracture toughness (K_{Ic}) of MOF material using a self-consistent single-edge precracked beam method and find it to be ~0.1 MPa $m^{0.5}$, which is low even compared to brittle oxide glasses and in between the values of foam and elastomers. Molecular dynamics simulations and theoretical calculations confirm the low K_{Ic} value of ZIF-62 glass, ascribing the origin to the weak coordinative bonds (Zn-N), which preferentially break during crack propagation.

1:50 PM

(GOMD-087-2021) Bulk Metallic Glasses' Response to Oscillatory Stress is Governed by the Topography of the Energy Landscape

L. Tang*[1]; M. Bauchy[1]

1. University of California, Los Angeles, Civil & Environmental Engineering, USA

When subjected to oscillatory loading, bulk-metallic glasses (BMGs) tend to fail under notably low main applications of optical fibers is as distributed er stress than under monotonic deformation—a behavior known as fatigue. Despite the practical importance of fatigue in structural applications, its atomic mechanism and origin remain largely unknown. Here, for the first time to the best of our knowledge, we reveal that BMGs' response to oscillatory stress is quantitatively encoded in the topography of the energy landscape. These results highlight a close correlation between atoms' dynamics and static energy landscape.

2:10 PM

(GOMD-088-2021) Towards damage resistant glasses by tailoring heterogeneities through consolidation of glassy nanoparticles

Y. Zhang*[1]

1. Rensselaer Polytechnic Institute, Materials Science and Engineering, USA

Oxide glasses are widely used in modern day life thanks to their superior optical, thermal, mechanical and chemical properties. However, oxide glasses are generally brittle, leading to poor damage resistance and low practical strength, which seriously limits the range of their commercial applications. Here we demonstrated a universal, feasible and effective toughening strategy to prepare damage resistant oxide glasses by tailoring heterogeneities through consolidation of glassy nanoparticles with pressure in molecular dynamics simulations. Using SiO_2, Al_2O_3 and Al_2O_3-SiO_2 glasses as examples, we showed that oxide glasses can display significant ductility and even necking behavior before the final fracture due to the introduction of structural heterogeneities. In addition, we showed that oxide glasses can be simultaneously strengthened and toughened by the synergistic introduction of both structural and chemical heterogeneities.

Finally, due to a stress facilitated relaxation mechanism, consolidated oxide glasses possess a surprising work hardening ability, which has never been reported before. Our findings suggest that when processed properly, amorphous oxides can deform plastically and fracture stably like ductile metals, which could lead to a paradigm shift in designing damage resistant oxide glasses.

2:30 PM

(GOMD-089-2021) Compositional Dependency of Inelastic Dissipation Mechanisms During Scratching in CAS Glasses

M. Kazembeyki[1]; S. Soman[1]; J. C. Mauro[4]; M. M. Smedskjaer[3]; M. Bauchy[2]; C. G. Hoover*[1]

1. Arizona State University, School of Sustainable Engineering and the Built Environment, USA
2. University of California Los Angeles, CIVIL AND ENVIRONMENTAL ENGINEERING, USA
3. Aalborg University, Denmark
4. Pennsylvania State University, USA

Understanding and controlling scratch resistance in glasses is of great importance, as this can be one of the limiting factors in glass applications. Thanks in part to the wide use of glasses that humans interact with, such as the screens for portable/smart electronic devices, resistance to scratch induced damage has become a primary concern. Previous studies have shown that glasses dissipate energy on smaller scales partially through densification, shear flow and cracking. In this study, we focus on the scratch resistance of a family of Calcium Aluminosilicate (CAS) glasses whose composition follows xCaO+xAl O +(100-2x)SiO, where 'x' is in mol%. We performed a series of instrumented scratch tests on the nano scale using a conospherical probe. The scratch hardness (Hs) was obtained by dividing the work of lateral deformation by the total deformed volume. In addition, an Atomic Force Microscope was used to map the surface after the elastic deformation was recovered. Although Hs did not exhibit a significant evolution with respect to SiO content, the energy dissipated by inelastic mechanisms showed a strong compositional dependency, also evidenced by pile-up volumes and changes in the depth of the scratch trough AFM scans.

2:50 PM

(GOMD-090-2021) Scratch behavior of ion-exchange strengthened soda lime silicate and sodium borosilicate glass

A. Talimian*[1]; T. Csanádi[2]; R. Limbach[3]; J. Dusza[2]; L. Wondraczek[3]; D. Galusek[1]

1. Centre for Functional and Surface Functionalized Glass (FunGlass), VILA, Slovakia
2. Institute of Materials Research, Slovak Academy of Sciences, Slovakia
3. University of Jena, Otto Schott Institute of Materials Research, Germany

Although chemical tempering has been used to improve the strength of silicate glasses, the influence of ion-exchange strengthening on the wear and scratch-resistance of the glasses is unclear. In this work, we studied the effects of sodium-potassium ion exchange on the scratch resistance of commercial soda-lime silicate and sodium borosilicate glasses. Ion-exchange processes were carried out in molten potassium nitrate at various temperatures (400-475 °C) and times (4-24h). The generated surface compression by ion exchange was determined by instrumented indentation, the scratch behaviour of samples was measured by multi-pass scratch experiments, and the surface chemical composition of glass was analysed by Energy Dispersion X-Ray Spectroscopy (EDXS). μ-Raman analysis was used to examine the glass structure. The relationships among the surface compression, crack and scratch resistance and glass structural changes are critically discussed.

GOMD S2: Glass and Interactions with Its Environment - Fundamentals and Applications

Dissolution and Interfacial Reactions: Composition Effects

Room: Saturna Island

Session Chairs: Nicholas Smith, Corning Incorporated; Jessica Rimsza, Sandia National Laboratories

1:30 PM

(GOMD-091-2021) Structural Drivers for the Chemical Durability of Alkali Borosilicate Glasses

N. J. Smith*[1]; S. Goyal[1]; A. Tandia[1]; J. P. Icenhower[1]; E. Bakowska[1]

1. Corning Incorporated, USA

Prospective design of novel technical glass compositions for commercial applications requires detailed knowledge—and ideally the ability to predict—aqueous release kinetics from multicomponent compositions in acids, bases, and neutral solutions that may be encountered during manufacture or end-use. Such durability properties derive fundamentally from glass structure, and the consequent chemistry of various network bonds with aqueous solutions across a wide pH range. The underlying aspects of structure that govern durability response are easier to deduce in some systems than others, for example, in "well-behaved" aluminosilicate glasses, in contrast to the more structurally complex trends that occur in borosilicate glasses. In this talk, we review salient points of glass structure hypothesized to drive the durability response, and highlight a statistical-mechanics-based approach to quantitatively evaluate these structural features. We then examine results from an array of static kinetic experiments on simplified alkali borosilicate glasses in aqueous media, focusing on elemental extraction rates far from saturation, in order to offer important insights about the structural drivers for release rates.

1:50 PM

(GOMD-092-2021) Investigating the Origin of Anomalous Water Diffusion in Silica Glass at Low Temperatures

B. D. Hausmann*[1]; M. Tomozawa[1]

1. Rensselaer Polytechnic Institute, Materials Science & Engineering, USA

Water diffusion in silica glass exhibits anomalous behavior at low temperatures. Solubility of water (reacted hydroxyl) in silica glass at 650°C under a constant water vapor pressure, 355 torr, was found to increase initially with time, exceeding the equilibrium solubility, estimated by extrapolation from the higher temperature data. Then water solubility decreased with time at a constant temperature and water vapor pressure. This has been attributed to compressive stress generation from water diffusion, which promoted reaction of molecular water with the glass. After some time, stress relaxation caused a reversal of this reaction in the glass surface. In order to confirm this hypothesis, samples were successively heat treated at 355 torr water vapor pressure at temperatures between 350 and 650 °C, the same conditions employed for water diffusion. Stress measurement by birefringence in the samples was performed as a function of water diffusion depth and compared with previously obtained data of OH diffusion profiles and surface concentrations. Residual stress was found to directly correlate with the change in surface OH concentration, as well as with OH diffusion depth for residual compressive stress, confirming the hypothesis.

2:10 PM

(GOMD-093-2021) Characterizing the Dissolution Behavior and Structure of Bioactive Na_2O-CaO-B_2O_3-P_2O_5 Glasses

R. L. Blatt*[1]; R. Brow[1]

1. Missouri University of Science & Technology, Materials Science and Engineering, USA

Laboratory experiments have been conducted in which series of glasses with the compositions (mol%) (100-y)(0.5Na_2O-0.5CaO)-xB_2O_3-(y-x)P_2O_5 (40≤y≤70 and 0≤x≤y) were reacted in simulated body fluid (SBF) for 1 to 168 hours. Across all series, dissolution rates initially decreased with the addition of B_2O_3, but subsequently increased with further additions. Final solution pH increased from 2 to 8 with increasing B_2O_3 contents. ICP-OES analysis combined with tracked weight loss and XRD revealed that some glasses dissolved congruently, while others formed calcium phosphate reaction layers. Speciation of borate and phosphate structural units were determined from ^{11}B and ^{31}P MAS NMR spectra and are used to explain the compositional dependence of the dissolution rates. In particular, increasing the relative concentrations of tetrahedral borophosphate sites decreases dissolution rates.

2:30 PM

(GOMD-094-2021) Early stage dissolution of simulated UK Ca/Zn HLW glass in conditions relevant to disposal

M. T. Harrison*[1]

1. National Nuclear Laboratory, WM&D, United Kingdom

A new CaZn borosilicate glass formulation for the immobilisation of highly-active Mo-rich waste feeds has been developed in the UK. Non-active lab- and full-scale trials have provided the underpinning to allow deployment this glass on the Vitrification Plant at Sellafield. The demonstration of acceptable product quality to date has used a standard leach test in pure water at high temperatures (90 °C) to compare with the data available for existing UK HLW glass. However, these tests, whilst a useful relative comparison, provide little information about the effect of lower ambient temperatures and the effect of more complex natural groundwaters. Hence, Radioactive Waste Management (RWM) in the UK is funding a series of durability experiments in conditions more representative of a potential geological repository using the expected range of simulated CaZn glass compositions. These include tests at 40 °C in generic granitic (high strength rock, HSR) and clayey (low strength sedimentary rock, LSSR) groundwaters as well as de-ionised water for long (up to 3 years) and short durations to probe both the initial and residual dissolution rates. This presentation will summarise dissolution rate measurements from leach tests of powder samples at high surface-area-to-volume ratio (1200 m^{-1}) and short durations (up to 42 days) at 40 °C in HSR, LSSR and DIW for comparison with the results at 90 °C.

2:50 PM

(GOMD-095-2021) Revisiting the Immobilized Low-Activity Waste Glass Dissolution Model for the Hanford Site Integrated Disposal Facility

J. Neeway*[1]; C. Lonergan[1]; S. Kerisit[1]; B. Parruzot[1]; J. V. Crum[1]; J. V. Ryan[1]; M. Asmussen[1]; G. L. Smith[1]

1. Pacific Northwest National Lab, USA

Before immobilized low-activity waste glass can be disposed at the Hanford site Integrated Disposal Facility (IDF), a performance assessment (PA) that estimates the facility's long-term impacts on the public and the environment must be carried out. The model used for glass dissolution is comprised of a kinetic portion, which accounts for pH and temperature, and a thermodynamic portion, where the rate is proportional to the activity of orthosilicic acid, a(H_4SiO_4), with respect to a quasi-equilibrium value, K_g. In addition, the glass ion exchange reaction is represented by a constant release of Na into the system. Here, we discuss the model and how the determination of the model parameters has evolved through the years. Specifically, we focus on the K_g term and ion exchange term and discuss how previous experimental methods were likely overly conservative. Lastly, we present results from a validation activity where the IDF glass dissolution model with empirically determined parameters has been compared to experimental results from static corrosion experiments. Collectively, these activities have not only led to increased efficiencies in parameter determination for a wide variety of glasses but also increased confidence in glass disposal.

GOMD S3: Optical and Electronic Materials and Devices - Fundamentals and Applications

Optical Fibers and Waveguides I

Room: Cortes Island

1:30 PM

(GOMD-096-2021) Advanced glasses and optical fibers for high energy lasers

J. Ballato*[1]; P. Dragic[2]; T. Hawkins[1]

1. Clemson University, USA
2. University of Illinois at Urbana-Champaign, USA

Continued progress in the development of optical fiber-based lasers has led to the present state where further improvements in performance are limited by intrinsic optical nonlinearities. In order to manage such limitations, laser designers have largely adopted the approach of microstructuring the fiber to shift nonlinear thresholds to high optical powers. The nonlinearities are accepted as fixed and performance is enhanced through fiber geometric complexity. This talk treats a different option, which is to mitigate optical nonlinearities at their fundamental origin: the glass with which the light interacts. Specifically, this talk focuses on the formulation, fabrication, lasing, and power-scaling of a conventional core / clad optical fiber that possesses intrinsically low thermo-optic properties to mitigate transverse mode instability (TMI). The core glass composition further exhibits intrinsically low Brillouin scattering.

1:50 PM

(GOMD-097-2021) Laser cooling in a silica optical fiber at atmospheric pressure

J. Knall[2]; P. B. Vigneron[2]; M. Engholm[4]; P. Dragic[3]; N. Yu[3]; J. Ballato*[1]; M. Bernier[5]; M. Digonnet[2]

1. Clemson University, USA
2. Stanford University, USA
3. University of Illinois at Urbana-Champaign, USA
4. Mid Sweden University, Sweden
5. Universite Laval, Centre d'optique, photonique et laser, Canada

The first laser cooling in a silica optical fiber will be discussed. In particular, material trends with cooling, including fiber composition and OH^- concentration, will be the focus. Specifically, six Yb-doped silica fibers were studied with the best fiber cooling by -70 mK with only 170 mW/m of absorbed pump power at a wavelength of 1040 nm. This fiber had an extremely low OH^- loss and a higher Al concentration (2.0 wt.% Al), permitting a high Yb concentration (2.52 wt.% Yb) without incurring significant quenching. Strong correlations were found between the absorptive loss responsible for heating and the loss measured at 1380 nm due to absorption by OH^-. The measured dependencies of the temperature change on the pump power and the pump wavelength are in excellent agreement with predictions from an existing model.

2:10 PM

(GOMD-098-2021) Development of a solar concentrator coupled with optical fiber bundle for deployment of solar energy

J. Roy*[1]; S. Morency[1]; G. Dugas[1]; Y. Messaddeq[1]

1. Laval University, Center for Optics, Photonics and Lasers, Canada

The global energy demand is increasing sharply with fast global development. We need cost-effective and efficient technologies to utilize solar energy for fulfilment of world's energy requirements and also to save the environment. We have developed a concentrated solar energy delivery system using parabolic mirror based solar concentrator coupled with silica opical fiber bundle to harvest solar energy for various applications. It consists of four parts: optical reflector, optical fiber bundle, solar tracker and mechanical support. We have used 10 m long optical fiber bundle to deliver concentrated sunlight. The fiber bundle is composed with 460 pieces silica optical fiber. The diameter of the each fiber is 600 μm. The optical fibers are made with pure silica to tolerate high temperature at the focal point of the concentrated sunlight. We have achieved 96 suns concentrated solar energy using only 50 % reflection efficiency of the primary mirror. This system can be used for daylighting system in underground car parking, small office and buildings. This system can also used in laboratory as a direct sunlight source instead of solar simulator, Xenon arc lamp etc. to perform experiments like solar cell testing, photocatalysis.

2:30 PM

(GOMD-099-2021) Elaboration of chalcogenide microstructured optical fibers preforms by 3D additive manufacturing

J. Carcreff*[1]; F. Chevire[1]; E. Galdo[1]; R. Lebullenger[1]; A. GAUTIER[1]; J. Adam[1]; D. Le Coq[1]; L. Brilland[2]; R. Chahal[2]; G. Renversez[3]; J. Troles[1]

1. University of Rennes 1, France
2. Selen Optics, France
3. Institut Fresnel, France

In recent years, a growing interest has settled for optical materials and fibers for the mid infrared (mid-IR) region. This interest originates from societal needs for health and environment for instance, and also from demand for defence applications. Chalcogenide glasses are good candidates for the realization of new and innovative mid-IR systems. In this context, chalcogenide microstructured fibers have attracted great interest in the field of versatile mid-IR fiber transmission, supercontinuum generation, and sensing. Different methods exist to manufacture these fibers such as stack and draw, drilling, extrusion or molding. In our work, we have studied a new method based on 3D printing, the additive manufacturing. This method can be implemented with $Te_{20}As_{30}Se_{50}$ glass (TAS), which is very stable against crystallization. A hollow-core preform of chalcogenide glass was made from a homemade 3D printer. The printed preform has been drawn into a hollow-core fiber. An anti-resonant hollow core fiber showing numerous transmission bands in the 2-12 μm has been obtained.

2:50 PM

(GOMD-100-2021) Tunable Rayleigh scattering in low-loss nanoparticle-doped optical fibers for long-range distributed sensing applications

V. Fuertes de la Llave*[1]; N. Gregoire[1]; P. Labranche[1]; Y. Ledemi[1]; S. LaRochelle[1]; Y. Messaddeq[1]

1. Centre d'optique, photonique et laser (COPL), Canada

One of the main applications of optical fibers is as distributed optical fiber sensors (DOFS), due to the possibility of sensing different parameters by exploiting the intrinsic scattering in the fiber and provide simultaneously spatial resolution along the whole fiber under test. In particular, optical backscatter reflectometry (OBR) is one of the most popular methods to measure Rayleigh scattering in fibers, since it is characterized by high sensitivity and a sub-millimeter spatial resolution over ten-meter-level range. Owing to the encouraging features of OBR systems, the current trend in DOFS is to enhance the Rayleigh scattering in them. Nanoparticle-doped optical fibers is one of the approaches that show the best performance. Nevertheless, the high optical losses induced by that scattering enhancement restrict considerably their sensing distance to few meters. Here, it is shown that through the tailoring of nanoparticle characteristics in the fiber core, it is possible to obtain low-loss tunable Rayleigh enhanced backscattering of ~ 26-46 dB, with respect to a SMF-28 fiber, paired with two-way optical losses of ~ 0.1-9 dB/m. Therefore, distributed sensing over distances from 4 m to more than 200 m are possible, along with a tunable distributed temperature sensitivity from 18.6 pm/°C to 23.9 pm/°C, ~1.9-2.4 times larger than in a SMF-28.

Optical Fibers and Waveguides II

Room: Cortes Island

3:50 PM

(GOMD-101-2021) Low-loss lead-germanate glass fibers for mid-infrared applications: Balancing dehydration efficiency and metallic lead formation

H. Ebendorff-Heidepriem*[1]; P. Wang[2]

1. University of Adelaide, Australia
2. State Key Laboratory of Transient Optics and Photonics, Xi'an Institute of Optics and Precision Mechanics, Chinese Academy of Sciences (CAS), China

For lead-germanate glass fibers, reducing the content of hydroxyl (OH) groups and the formation of metallic Pb species is essential to pave the way for their applications as low-loss mid-IR fiber optics since OH and metallic Pb species cause intense absorption and scattering loss, respectively, in the mid-IR spectral range. Combining a dry O_2-rich atmosphere containing ultra-dry N_2 together with the use of chloride dehydration agent and nitrate oxidation agent compound was found to enable efficient dehydration effect and absence of metallic Pb scattering sources in the dehydrated glasses. Heat-treatment experiments of glass samples at preform extrusion and fibre drawing conditions revealed the co-effects of glass melting and heat treatment atmospheres on the formation of nano- and micron-scale metallic Pb species in both the as-produced and heat-treated lead-germanate glasses. Using this advance in knowledge, we successfully fabricated low-loss lead-germanate glass fibers with no presence of reduced metallic Pb particles by optimizing dehydration agent, glass melting, preform extrusion and fiber drawing conditions. The optimized fabrication conditions reduced the unstructured fiber loss by almost one order to <0.3 dB/m at 1.55 μm.

4:10 PM

(GOMD-102-2021) Rare-earth germano gallate optical fibers for Mid-InfraRed optical applications

F. Calzavara*[1]; C. Strutynski[1]; T. Guerineau[1]; L. Loi[2]; R. Laberdesque[2]; J. Rampnoux[3]; S. Morency[4]; Y. Ledemi[4]; Y. Petit[1]; M. Dussauze[5]; F. Désévédavy[6]; F. Smektala[6]; S. Danto[1]; L. Canioni[2]; V. Jubera[1]; Y. Messaddeq[4]; E. Fargin[1]; T. Cardinal[1]

1. ICMCB-CNRS, France
2. Center for Intense Lasers and Applications (CELIA), France
3. Laboratoire Ondes et Matière d'Aquitaine (LOMA), France
4. International Associated Laboratory (LIA) LuMAQ, Centre d'Optique, Photonique et Laser (COPL), Canada
5. Institute of Molecular Science (ISM), France
6. Laboratoire Interdisciplinaire Carnot de Bourgogne (ICB), France

The ability to produce robust fiber-based integrated optical systems operating over a wide spectral domain from the ultra-violet up to mid-infrared is one of today's key challenges in photonics. Over recent decades, several glass systems such as fluorides, heavy metal oxides and chalcogenides appeared to be promising alternatives to SiO_2 glasses, but have not yet been able to compete with its high thermal robustness and remarkable mechanical assets. Following an original approach, this work reports on the production of light guiding fibers from rich rare-earth gallium oxide-based glass compositions. The impact of the addition and the nature of rare earth (RE) ions to improve the delay to crystallization will be discussed. These materials exhibit relatively high vitreous transition temperatures (~ 735 °C), improved surface micro-hardness and low phonon oxide glass matrix. Those vitreous systems offer an excellent optical transmission extending from ultraviolet wavelengths (~ 0.280 μm) up to 6 μm. They also show a relatively high third order polarizability and high RE mid-infrared emitting centers solubility. This unique combination of superior thermal, mechanical and optical properties represents a promising alternative for the development of highly robust power scaled fiber devices operating from the UV up to the challenging 2 – 5 μm optical domain.

4:30 PM

(GOMD-103-2021) Chalcogenide step-index fibers for mid-IR supercontinuum generation and application of supercontinuum absorption spectroscopy

R. Bizot*[1]; A. Lemiere[2]; F. Désévédavy[1]; B. Kibler[1]; P. Mathey[1]; G. Gadret[1]; F. Smektala[1]

1. Laboratoire ICB, Photonic, France
2. Tampere University, Finland

Broadband supercontinuum sources are largely studied for decades since it is implied with a large field of applications such as spectroscopy, metrology, medical diagnostics and others. Chalcogenide glasses are good candidates for these applications as these materials offer a wide transmission far in the infrared region and also have a high non-linear refractive index. The majority of organic molecules absorb within the infrared region and some compositions of chalcogenide are known to transmit light until 20 μm. Furthermore, atmospheric windows 3-5 μm and 8-14 μm are part of the spectral range covered by chalcogenide glasses. We report in this work a step-index chalcogenide fiber respecting the European Registration, Evaluation, Authorization and Restriction of Chemicals (REACH). The fiber synthetized is made of germanium, selenium and tellurium and it is indeed free of notoriously toxic elements such as arsenic or antimony. Here we present a spectral broadening extending in the mid-infrared from 1.5 to 17.5 μm. An experiment of gas detection with nitrous oxide (N_2O) was therefore done around 8 μm by using Supercontinuum Absorption Spectroscopy (SAS) technique.

4:50 PM

(GOMD-104-2021) TeO_2 – ZnO - La_2O_3 tellurite glass system investigation and purification for robust mid infrared optical fibers

M. Evrard*[1]; A. MALDONADO[1]; F. Désévédavy[1]; C. Strutynski[1]; G. Gadret[1]; J. Jules[1]; C. Brachais[1]; T. Mansuryan[4]; V. Couderc[4]; M. Dutreilh-Colas[3]; S. Danto[2]; T. Cardinal[2]; P. Thomas[3]; F. Smektala[1]

1. Laboratoire ICB, Photonics, France
2. ICMCB-CNRS, France
3. IRCER, France
4. XLIM, France

Ternary diagram TeO_2-ZnO-La_2O_3 (TZL) was investigated in order to determine the whole vitreous domain and to select the most appropriate formulations for step-index optical fibers manufacturing. Optical characterizations on bulk samples were performed to explain the evolution of the refractive index as a function of the composition. Then, thermal and structural analyzes permitted to identify core and cladding formulations for fibers drawing. After the demonstration of successful step-index optical fibers manufacturing, core and cladding glasses were then purified to capture and remove OH species from the matrix. Dehydrating reagents, such as ZnF_2 and LaF_3, were then introduced in the glass matrix by cation substitution under controlled atmosphere into a glovebox and characterizations were conducted in order to determine the efficiency of each fluoride. Optical losses were measured and a minimum of 0.43 dB/m at 2.6 μm is observed for single-index purified fibers while non-purified fiber presents an attenuation higher than 1.5 dB/m for the same wavelength. Finally, step-index preforms were synthetized by the Built-in-Casting method by combining purified and non-purified formulations, attenuation measurements were performed on the resulting fibers and exceptional transmission properties up to 4 μm were reported.

5:10 PM

(GOMD-105-2021) Hybrid Glass-Metal Optical Fibers: Elaboration, Simulation And Electro-Optical Characterization

A. R. Maldonado*[1]; W. Correr[1]; M. Evrard[2]; R. Bizot[2]; F. Désévédavy[2]; J. Jules[2]; G. Gadret[2]; C. Strutynski[2]; C. Brachais[2]; Y. Messaddeq[1]; F. Smektala[2]

1. Laval University, Centre d'Optique, Photonique et Laser, Canada
2. Université de Bourgogne Franche-Comté, Laboratoire Interdisciplinaire Carnot de Bourgogne, France

The emergence of multimaterials optical fibers is of tremendous technological interest in photonics to combine the remarkable properties of glasses with those of other materials such as metals or polymers in order to form a fully integrated fiber optical system with multiple functionalities. Among these hybrid fibers, the development of fibers combining both optical signal and simultaneous electrical transport function could bring alternative interesting solution in many fields such as telecommunications, medicine or sensing. During the last few years, many designs and techniques have been implemented, mostly on silica-based fibers, to get continuous metallic electrodes along an optical fiber: fiber capillary twin holes filling with molten metal, preform feeding with a wire coil during the drawing, stack-and-draw technique and manual insertion of solid wires. Here, we report the engineering of two different glass-metal hybrid fibers based on tellurite glasses then chalcogenide glasses with a rod-in-tube and co-drawing approach. Tellurite glasses and chalcogenides glasses have been chosen for their wide transmission window in the mid-infrared range, their strong non-linear optical properties and their low T_g compatible with many metals and polymers compared to silica glasses.

GOMD S4: Glass Technology and Cross-Cutting Topics

3-D Printing of Glass

Room: Saltspring Island A/B

3:50 PM

(GOMD-106-2021) 3D-printing of chalcogenide glasses an original way for fabricating mid-infrared optical components

J. Troles*[1]; J. Carcreff[1]; J. Chevire[1]; R. Lebullenger[1]; A. GAUTIER[1]; R. Chahal[2]; J. Adam[1]; L. Calvez[1]; C. Boussard-Pledel[1]; L. Brilland[2]; F. Charpentier[3]; H. Tariel[3]; G. Renversez[4]

1. University of Rennes 1, France
2. SelenOptics, France
3. Diafir, France
4. Aix Marseille Univ, France

In recent years, a growing interest has settled for optical materials and fibers for the mid infrared (mid-IR) region. This interest originates from societal needs for health and environment for instance, and also from demand for defence applications. Indeed, the mid-IR spectral region contains the atmospheric transparent windows (3-5 μm) and (8-12 μm) where thermal imaging (military and civilian) can take place. In this context, we have investigated an alternative way for fabricating Mid-IR chalcogenide optical components such as preforms, optical fibers, sensors and lenses by using an original 3D printing process. In our work, the 3D-printing set-up is based on a customized commercial RepRap-style 3D printer upgraded for soft glasses. Especially, the feeding mechanism is customized for brittle materials. By using this additive manufacturing method, preforms with complex designs can be fabricated in a single step within a couple of hours, with a high degree of repeatability and of precision of the geometry. In addition, we have shown that an infrared sensor can be realized with our 3D printer. This original 3D printing method, open the way to many applications involving chalcogenide fiber manufacturing but also many other chalcogenide glasses optical devices.

4:10 PM

(GOMD-107-2021) Development of functionalized phosphate materials for extrusion-based 3D-printing

S. Kaser*[1]; R. Zaki[1]; C. Strutynski[1]; S. Danto[1]; S. H. Santagneli[2]; M. Dussauze[3]; D. Bernard[1]; M. Faessel[4]; J. Sabatier[5]; S. H. Messaddeq[6]; Y. Messaddeq[6]; T. Cardinal[1]

1. ICMCB-CNRS, France
2. UNESP, Instituto de Quimica, Brazil
3. ISM-CNRS, France
4. IUT Bordeaux, Technoshop Coh@bit, France
5. IMS-CNRS, France
6. COPL, Canada

3D-printing of inorganic vitreous materials has seen a growing interest worldwide for multiple applications. Phosphate glasses are known for their high solubility for rare earth and metallic ions, good chemical stability and good thermal properties making them compatible with processing at moderate temperature (<500°C). Here, phosphate glass compositions and thermal properties were adapted and optimized for extrusion-based 3D-printing via Fused Deposition Modeling. Both aluminum phosphate glass compositions and novel transparent zinc hydrated phosphate materials were investigated. The structure and properties of the synthetized materials were studied using vibrational and solid-state NMR spectroscopies. The 3D-printed materials were characterized by SEM, vibrational and luminescence spectroscopy. The residual porosity was quantified by X-ray tomography. These glass matrices offer new opportunities for 3D-printing of inorganic vitreous materials and composites.

Wednesday, December 15, 2021

GOMD S1: Fundamentals of the Glassy State

Glass Formation and Structural Relaxation I

Room: Saltspring Island C

Session Chair: Daniel Neuville, IPGP-CNRS-USPC

8:30 AM

(GOMD-108-2021) Influence of iron redox on viscosity of silicate melts (Invited)

D. R. Neuville*[1]

1. IPGP-CNRS-Université de Paris, Géomatériaux, France

Viscosity of silicate melts is one of the most important physical properties to understand the high-temperature phenomena with regard to magmatic system. Effects of composition and temperature on the viscosity have been elucidated for a long time. Although iron is a main component of magmatic system, its influence on the viscosity remains unclear because the iron behavior is complicated: iron has two redox states, i.e. Fe^{3+} and Fe^{2+}. We elucidated the viscosity of the iron-sodium-silicate system with a variety of iron redox states at temperatures close to their glass transition (T_g). Near the glass transition temperature, viscosity increases for more than 3 order of magnitude with increasing the ratio Fe^{3+}/Fe^{tot}. The tendencies in the viscosity as well as structural variation against Fe^{3+}/Fe^{tot} ratio support the consensus for structural roles of Fe^{3+} and Fe^{2+} from the previous studies: Fe^{3+} ions have stronger tendency to behave as a network former than Fe^{2+} ions.

9:00 AM

(GOMD-109-2021) Structural Control on the Rheological Behavior of Binary P-Se Supercooled Liquids

B. Yuan*[1]; B. Aitken[2]; S. Sen[1]

1. University of California, Davis, Materials Science and Engineering, USA
2. Corning Incorporated, USA

The structural evolution of P_xSe_{100-x} ($0 \leq x \leq 25$) supercooled liquids and its effects on their shear relaxation behavior are investigated using Raman spectroscopy and oscillatory parallel plate rheometry. The P atoms in these glasses form $Se{=}P(Se_{1/2})_3$, $P(Se_{1/2})_3$ and $_{2/2}SeP\text{-}PSe_{2/2}$ units, which serve as cross-linkers of the $[Se]_n$ chain segments. The frequency dependence of the storage and loss moduli as well as the relaxation spectra $H(\tau)$ indicate that liquids with >85% Se are characterized by the coexistence of a slow and a fast relaxation process corresponding to bond scission/renewal and Se chain segmental motion, respectively. On the other hand, liquids with ≤85% Se display only the bond scission/renewal process as the length of the Se chain fragments becomes too short to sustain the segmental dynamics. The $H(\tau)$ spectra of the P_xSe_{100-x} liquids with >85% Se suggest that the relaxation-time τ distribution for Se chain dynamics is a sensitive function of the chain length and hence, of the associated conformational entropy. On the other hand, the distribution of τ for bond scission/renewal is controlled by the variety of bond types and structural moieties involved. The strong temporal coupling between these two relaxation processes is likely indicative of a cause-and-effect relationship between the unique interconversion of the P-Se polyhedra and the rearrangement of the Se chain segments.

9:20 AM

(GOMD-110-2021) Glass Transition Temperatures of Sodium Borovanadate Glasses

M. Goeks[2]; N. Leonard*[1]; S. Feller[1]

1. Coe College, Physics, USA
2. Northern Michigan University, Physics, USA

The glass transition temperatures of sodium borovanadates are reported and compared to those of a variety of other glass systems including borovanadates. The work on sodium borovanadates in the present work as well as the work on sodium borates from past work are compositionally described by varying R, the molar ratio of sodium oxide to boron oxide (Na_2O/B_2O_3), for fixed families of K, the molar ratio of vanadium pentoxide to boron oxide (V_2O_5/B_2O_3). R varies between 0 and 2 while the values of K are 0, 0.1, 0.3, 0.6, and 0.9. The results of the present work match well with a number of systems trend-wise including sodium borosilicates, lithium borovanadates, lead borovanadates, potassium borovanadates, and the sodium borovanadates taken from the literature. Some discrepancy of glass transition temperatures between the present work and others are noted and possible explanations will be discussed. The sharing of sodium cations between the borate and vanadate part of the glass was modeled and compared with a cation sharing model developed for cesium borovanadates. The authors acknowledge NSF for their support of this research (NSF-DMR-1746230) and (NSF-REU-1659581).

Glass Formation and Structural Relaxation II

Room: Saltspring Island C
Session Chair: Daniel Neuville, IPGP-CNRS-USPC

10:20 AM

(GOMD-111 2021) Rheological Behavior of Molecular vs. Network Chalcogenide Supercooled Liquids

S. Sen*[1]; W. Zhu[1]; B. Yuan[1]; B. Aitken[2]

1. University of California, Davis, USA
2. Corning Incorporated, USA

The viscoelastic behavior of supercooled glass-forming liquids along the binary join As_4S_3-GeS_2 with structures varying from predominantly molecular to a three-dimensional tetrahedral network is studied using small-amplitude oscillatory shear parallel plate rheometry. The storage shear modulus G' shows a scaling behavior of $G'(w) \sim w^n$ in the terminal (low-frequency) regime where n varies between 1 and 2 and shows increasingly anomalous departure from the expected value of 2 (Maxwell scaling) with increasing molecule content. A concomitant departure from the Maxwell scaling is also observed for the loss modulus G'' at frequencies above the G'-G'' crossover. On the other hand, the variation of the phase angle δ with the complex modulus indicates that the molecular liquid does not display a purely viscous response even at the lowest frequencies. These results, combined with an analysis of the relaxation spectra of these liquids, suggest that the anomalous behavior of molecular liquids may be linked to their rather broad relaxation spectrum and the presence of slow relaxation processes associated with molecular clusters. Such fundamental differences between the rheological behavior of molecular and network liquids may explain the significantly higher fragility of the former.

10:40 AM

(GOMD-112-2021) Origin of Nearly Constant Loss in Metallic Glass

L. Zella*[1]; J. Moon[2]; D. Keffer[1]; T. Egami[1]

1. University of Tennessee, Materials Science and Engineering, USA
2. Oak Ridge National Laboratory, Materials Science and Technology Division, USA

The Nearly Constant Loss (NCL) relaxation is a subtle phenomenon that lays between fast beta relaxation and the Boson peak in time scale. Despite its direct relevance to important dynamical phenomena including aging, deformation, and atomic diffusion, the mechanism behind the NCL remains largely unknown. Using Dynamic Mechanical Analysis (DMA) and atomic level stress calculations in molecular dynamics (MD), we attempt to quantify the atomic level mechanisms responsible for NCL in $Cu_{64.5}Zr_{35.5}$ metallic glass, a prototypical metallic glass. Decomposition of system stress into individual atomic level stress enables the analysis of the dynamic response of a partial stress removing atoms of interest. Removing the liquid-like atoms characterized by their nearly $\pi/2$ phase shift in DMA shows strong correlations between liquid-like atoms and NCL in the metallic glass and further analysis leads to the conclusion that more diverse liquid-like dynamic behaviors are responsible for NCL than previously reported exclusively rattling or caging motions. *This work was supported by the U.S. Department of Energy (DOE), Office of Science, Office of Basic Energy Sciences, Materials and Science and Engineering Division.

GOMD S3: Optical and Electronic Materials and Devices - Fundamentals and Applications

Charge and Energy Transport I

Room: Saltspring Island A/B
Session Chair: Ana Candida M Rodrigues, Federal University of Sao Carlos

8:30 AM

(GOMD-113-2021) Thin-Film Glassy Solid Electrolytes: A New Functionality for Glass Enabling High Energy Density Li and Na Batteries (Invited)

S. W. Martin*[1]

1. Iowa State University, Materials Science & Engineering, USA

Fast ion conducting glasses have long been considered as alternatives to flammable liquid electrolytes in Li and Na batteries. However, to date, there has never been before the unique combination of required electrochemical properties for use as a solid electrolyte with the equally important requirements of viscoelastic behavior to form them into thin films suitable for high ion conductivity separators. In this first ever report of thin film fast ion conducting glasses, we will summarize our efforts to produce thin films of Li and Na ion conducting glasses and test in them in symmetric and asymmetric cells.

9:00 AM

(GOMD-114-2021) Lithium ion sites and their contribution to the ionic conductivity of R Li_2O-B_2O_3 glasses with R ≤ 1.85

A. Ruckmann[1]; G. Beckler[1]; W. Guthrie[1]; M. R. Jesuit[1]; M. Boyd[1]; N. Barrow[2]; N. Tagiara[3]; S. Feller[1]; E. I. Kamitsos[3]; C. B. Bragatto*[1]

1. Coe College, Physics Department, USA
2. Johnson Matthey Technology Centre, United Kingdom
3. Theoretical and Physical Chemistry Institute, Greece

According to the Nernst-Einstein equation, the electric conductivity (σ) in glasses is related to the number of mobile species (n) and their speed (μ). In the case of lithium borate glasses RLi_2O-B_2O_3, the electric conductivity is known to be due the movement of lithium ions, therefore its concentration should directly affect n and therefore σ. Using the topological constraint theory (TCT), a recent work published in the literature established the existance of two distinguishing lithium species in this glass system, according to their bond rigitidy with the glass network, separating which Li ions actually contribute to n and which are immobile. To test this new proposed approach, high content lithium oxide glasses were prepared (up to R = 1.85) without any signs of heterogeneity by using a fast quenching technique. Results show that instead of being related to the number of loose lithium ions, the ionic conductivity is mostly likely related to the different boron species which act as lithium sites, given that the number of non-bridging oxygens increases proportionally to the lithium to boron ratio when R ≥ 0.5, the region where the plateau in electrical properties is observed. This result indicate that n is related not only to the Li ions, but to the type of Li-O pairs.

9:20 AM

(GOMD-115-2021) Structure, strong liquid behavior, and electrochemical characterization of mixed anion glasses in the series $Na_4P_2S_{7-6x}O_{4.62x}N_{0.92x}$; $0 \le x \le 0.5$

M. Olson*[1]

1. Iowa State University, MSE, USA

Optimizing glassy solid electrolyte materials requires balancing high ionic conductivity achieved in sulfide systems with thermal stability achieved in oxy-nitride systems. Glasses in the oxy-sulfide-nitride series $Na_4P_2S_{7-6x}O_{4.62x}N_{0.92x}$; $0 \le x \le 0.5$ were studied to understand the effect of oxygen and nitrogen on a sulfide composition. The conductivity decreased with increasing x, but DSC experiments revealed an increase in the thermal stability. The short-range order structures of these materials were studied using NMR spectroscopy to understand the structure-property relationship in these materials. The x=0.2 chemistry showed the best compromise between conductivity and resistance to crystallization, demonstrating stable thermal cycling to temperatures > 100 °C past its T_g, making it a viable candidate for thin film glass processing. The x=0.2 glass was drawn successfully into a thin film using the redraw method inside of a N_2 glovebox, achieving thicknesses ranging from 60-100um. The ionic conductivity and interfacial stability of the thin film were probed using Electrochemical Impedance Spectroscopy (EIS). Cyclic voltammetry was performed in order to determine the oxidative stability window for the thin film. Cycling experiments were used to determine the stability of the film against an applied voltage and sustained current.

9:40 AM

(GOMD-116-2021) Glass-forming ability and ionic conductivity of glasses and glass-ceramics from the new $Na_{3.4}Al_xSc_{2-x}Si_{0.4}P_{2.6}O_{12}$ ($1.0 \leq x \leq 1.7$) Nasicon series

J. Ortiz-Mosquera[2]; A. Nieto-Muñoz[2]; A. Rodrigues*[1]

1. Federal University of Sao Carlos, Materials Engineering, Brazil
2. Federal University of Sao Carlos, Programa de Pós-graduação em Ciência e Engenharia de Materiais, Brazil

The synthesis of Nasicon-structured compounds via the glass-ceramics route has been widely investigated. Among the advantages of the glass-ceramics route, as compared to the solid-state reaction followed by sintering, it may be mentioned the reduced porosity, microstructure control, and easiness of shaping. However, to obtain glass-ceramics with a controlled microstructure, it is important to start from a precursor glass with a good glass-forming ability and stability against crystallization. To allow the synthesis of Sc-Nasicon glass-ceramics, we have added aluminum oxide to the base glass to improve the glass-forming ability, thus obtaining the $Na_{3.4}Al_xSc_{2-x}Si_{0.4}P_{2.6}O_{12}$ ($1.0 \leq x \leq 1.7$) Nasicon series. In this series, Al^{3+} replaces the isovalent Sc^{3+}, maintaining the molar ratio of mobile Na^+ ions constant. It is observed that the ionic conductivity decreases when Sc^{3+} is replaced by Al^{3+}, which has a smaller ionic radius. This behavior is attributed to the decreases in the unit cell volume, which leads to a decrease in the bottleneck size, through which the Na^+ ion must move.

Charge and Energy Transport II

Room: Saltspring Island A/B

Session Chair: Ana Candida M Rodrigues, Federal University of Sao Carlos

10:20 AM

(GOMD-117-2021) On the correlation between atomic-level structure and ionic transport of lithium phosphorous oxynitride amorphous electrolyte (Invited)

R. B. Nuernberg*[1]; A. Landry[1]; R. Bayzou[2]; O. Lafon[2]; F. Le Cras[3]; B. Pecquenard[1]

1. ICMCB-CNRS, France
2. University Lille, CNRS, France
3. University Grenoble, CEA, LITEN, DEHT, France

Lithium phosphorous oxynitride (LIPON), having an amorphous structure, was first synthesized at the Oak Ridge National Laboratory in the 1990s. Its development has arisen from the pursuit of solid electrolytes stable against lithium metal (Li) to fabricate micro-batteries employing radiofrequency sputtering. Strictly speaking, LIPON is not thermodynamically stable against Li. However, a passivating solid electrolyte interphase emerging between Li and LIPON comprises decomposition products (Li_3N and Li_2O) thermodynamically stable against Li. Along with the electrochemical stability, the low electronic ($<10^{-10}\Omega^{-1}.cm^{-1}$) and moderate ionic ($>10^{-6}\Omega^{-1}.cm^{-1}$) conductivity have been also refereed as key properties for its applications as thin-film electrolytes. The relationships between sputtering conditions, chemical composition, local structure, and ionic conductivity of LIPON thin films have been intensively explored and reported in literature. Most studies converge on the fact that the ionic conductivity can be enhanced by increasing the nitrogen concentration up to a saturation point. The controversial debated question is centered on the resulting structure once nitrogen is incorporated and on its structural role regarding the ionic transport.

10:50 AM

(GOMD-118-2021) Patterning of the Surface Electrical Potential on Chalcogenide Glasses by a Thermoelectrical Imprinting Process

M. Dussauze*[1]; L. D. Karam[1]; R. Alvarado Meza[1]; K. A. Richardson[3]; A. Pradel[2]; A. Piarristeguy[2]

1. CNRS / Université de Bordeaux, France
2. University Montpellier, France
3. University of Central Florida, USA

The development of novel sensing systems requires breakthroughs in the conception of multifunctional materials. In this sense, while extensive research has been dedicated to the individual tuning of the electrical or optical properties of different materials, the combination of both features would result in a promising field of research that would further extend opportunities for engineering novel function in sensor geometries. In the present work, we employed chalcogenide glasses and focused on the spatial control of their surface electrical potential via a thermoelectrical imprinting process. Glass compositions based on the system Ge-Sb-S-Na were prepared by varying the sulfur stoichiometry and the sodium content. Each glass was thermally poled using electrodes with specific patterns, and subsequent structural modifications and surface electrical potential were then evaluated. Raman mapping show structural modifications attributed to alkali depletion following the patterns of the electrodes used for the imprinting process. Furthermore, KPFM measurements show clearly defined motifs on the electrical potential which can vary in sign within an amplitude range of 10V and exhibit patterning at the micrometer scale. We observed that the efficiency of the surface potential imprinted was dramatically impacted by the glass' sulfur and sodium content.

11:10 AM

(GOMD-119-2021) The incorporation of LiPON in the mixed oxy-sulfide-nitride solid electrolyte: 0.58 Li2S + 0.315 SiS2 + 0.105 [(1-y) $Li_{0.67}PO_{2.83}$ + y $LiPO_{2.53}N_{0.31}$]

V. M. Torres*[1]; P. Philipp[1]; S. Kmiec[1]; S. W. Martin[1]

1. Iowa State University, Materials Science and Engineering, USA

Glassy solid electrolytes are a promising solution to the development of solid-state batteries. Mixed Oxy-Sulfide Nitride (MOSN) glasses combine the high ionic conductivity of sulfide glasses, the excellent chemical stability of oxide glasses, and the electrochemical stability of nitride glasses. The systematic incorporation of LiPON during glass synthesis was used to explore the effects of nitrogen and oxygen on the behavior of mixed anion electrolytes. The MOSN series $0.58Li_2S + 0.315SiS_2 + 0.105[(1-x)Li_{0.67}PO_{2.83} + xLiPON]$ (x = 0.0, 0.1, 0.2, 0.3, 0.4, 0.5) was prepared via melt-quench synthesis. Electrochemical impedance spectroscopy (EIS) was used to determine the temperature-dependent ionic conductivity of the glass series and Differential scanning calorimetry (DSC) was used to determine the thermal properties of these MOSN glasses. Fourier transformation-infrared spectroscopy (FTIR) and magic angle spinning nuclear magnetic resonance spectroscopy (MAS NMR) were used to understand the structure-property relationships in these mixed anion materials. It was found that x = 0.2 displayed the highest ionic conductivity suggesting that nitrogen incorporation can be used to optimize MOS glasses. NSF Grant NSF-DMR-1936913 is acknowledged.

11:30 AM

(GOMD-120-2021) Structure and electrochemical properties of 0.6 Li_2S + 0.3 SiS_2 + $LiP_{1-x}Al_xO_{3-x}$ glassy solid-state electrolytes

J. Wheaton*[1]; S. Leland[1]; S. Kmiec[1]; S. W. Martin[1]

1. Iowa State University, Materials Science and Engineering, USA

Lithium silico-oxy-thio-phosphate glasses have been shown to be desirable for use as solid-state electrolytes due to their electrochemical stability, high intrinsic ionic conductivities approaching 1 mS/cm, and their ease of processability. Alumina doping of some glassy-solid-electrolyte (GSE) systems has been shown to further increase the conductivity, while also increasing the connectivity of the glass forming network and potentially further suppressing the crystallization of the supercooled liquid. The aluminum doped mixed oxysulfide series, 0.6 Li_2S + 0.3 SiS_2 + $LiP_{1-x}Al_xO_{3-x}$ (x = 0, 0.025, 0.05, 0.10) was synthesized through a melt-quench method and the short-range order structures of the glasses were characterized using Raman and infrared spectroscopies. Electrochemical impedance spectroscopy (EIS) was used to determine temperature dependent ionic conductivity of the bulk glasses. In addition, electrochemical characterization was conducted in symmetric Li metal cells to determine the stability of the glass in contact with Li through time-dependent EIS. Galvanostatic cycling experiments were also conducted to further investigate the Li cycling stability of the glass against lithium metal.

Glass-based Optical Devices I

Room: Pender Island

Session Chair: Tushar Sanjay Karnik, Massachusetts Institute of Technology

8:30 AM

(GOMD-121-2021) Progress in Mid-Infrared Supercontinuum Generation in Chalcogenide Glass Fibers (Invited)

S. Dai*[1]

1. Ningbo University, Laboratory of Infared materials and devices, China

Chalcogenide glasses have the advantages of a wide transparency window and high optical non-linearity, making them good candidates for mid-infrared supercontinuum generation. We describe both the history and recent developments in mid-infrared supercontinuum generation from chalcogenide fibers according to three kinds of fiber structures: step-index, microstructured and tapered fibers. We also review the coherence properties of mid-infrared supercontinuum generation and all-fiber supercontinuum sources based on chalcogenide fibers.

9:00 AM

(GOMD-122-2021) Monolithic chalcogenide glass waveguide integrated interband cascaded laser

H. Lin*[2]; C. Kim[3]; L. Li[4]; M. Kim[3]; W. Bewley[3]; C. D. Merritt[3]; C. L. Canedy[3]; I. Vurgaftman[3]; A. Agarwal[1]; K. A. Richardson[5]; J. Hu[1]; J. R. Meyer[3]

1. Massachusetts Institute of Technology, Materials Science and Engineering, USA
2. Zhejiang University, College of Information Science & Electronic Engineering, China
3. Naval Research Laboratory, Code 5613, USA
4. Westlake University, China
5. University of Central Florida, The College of Optics & Photonics, USA

Mid-infrared photonic integrated circuits (PICs) that combine on-chip light sources with other optical components constitute a key enabler for applications such as chemical sensing, light detection, ranging, and free-space communications. In this paper, we report the monolithic integration of interband cascade lasers emitting at 3.24 μm with passive, high-index-contrast waveguides made of chalcogenide glasses. Output from the chalcogenide waveguides exhibits pulsed peak power up to 150 mW (without roll-over), threshold current density 280 A/cm^2, and slope efficiency 100 mW/A at 300 K, with a lower bound of 38% efficiency for coupling between the two waveguides. These results represent an important step toward the realization of fully-integrated mid-infrared PICs.

9:20 AM

(GOMD-123-2021) Photonic on-chip chalcogenide long-wave infrared gas sensor

T. Karnik*[1]; Q. Du[1]; K. Chen[1]; M. Kang[2]; K. A. Richardson[2]; J. Hu[1]

1. Massachusetts Institute of Technology, Materials Science and Engineering, USA
2. University of Central Florida, College of Optics and Photonics, USA

For the past few decades, on-chip photonic sensors have been widely developed for various applications like industrial process monitoring, pollution detection and medical diagnoses. Long-wave infrared (LWIR), typically spectrally spanning from 8-12 μm are critical for gas sensing and identification as it coincides with the vibrational bands (the "fingerprint" region) of most molecules. However, there is limited work on photonic LWIR photonic gas sensing due to large material absorption in silicon based materials. In this talk, we demonstrate a LWIR suspended chalcogenide glass gas sensor by taking advantage of the large IR transparency window of As_2Se_3 and $Ge_{33}As_{12}Se_{55}$ glass. The chalcogenide glass waveguides are fabricated on a silicon chip and then suspended from the silicon substrate to avoid excessive material absorption. A gas chamber is also incorporated into the system to allow controlled gas flow and uniform gas concentration along the gas sensor. The sensing capability is also quantified by monitoring the optical absorption peak of SF_6 gas at 10-11 μm. Our results has presented a prototype LWIR sensor device that can potentially be applied to various gas molecule "fingerprint" detection and sensing.

9:40 AM

(GOMD-124-2021) Chalcogenide glass photonic chip-based tunable ultrafast Raman soliton source

Q. Du*[1]; z. li[2]; C. Wang[2]; J. Zou[2]; K. A. Richardson[3]; Z. Cai[2]; J. Hu[1]; Z. Luo[2]

1. Massachusetts Institute of Technology, Materials Science and Engineering, USA
2. Xiamen University, China
3. University of Central Florida, USA

The advantages of low cost, compact size, and reduced power consumption make photonic chip-based ultrafast laser sources an appealing technology for diverse applications. In this talk, we report what we believe to be the first on-chip ultrafast soliton laser source based on Raman soliton self-frequency shift. By capitalizing on strong optical nonlinearity and versatile dispersion control in $Ge_{28}Sb_{12}Se_{60}$ chalcogenide glass waveguides, we have demonstrated 185 fs duration Raman soliton generation, possessing continuous wavelength tunability from 1589 nm to 1807 nm with signal-to-noise ratios consistently exceeding 65 dB. The source operates with pump pulse energies as low as 1.08 pJ, representing over three orders of magnitude reduction compared to those of fiber Raman lasers. Our work therefore enables a new class of broadly tunable, energy-efficient, compact, and potentially cost-effective on-chip ultrafast laser sources.

Glass-based Optical Devices II

Room: Pender Island
Session Chair: Myungkoo Kang, University of Central Florida

10:20 AM

(GOMD-125-2021) Next-generation phase change materials for programmable integrated photonics (Invited)

O. Muskens*[1]

1. University of Southampton, United Kingdom

In this presentation I will give an overview of our experimental results on the use of new ultralow loss phase change materials for programmable integrated photonic circuits. The antimony-based chalcogenides Sb_2S_3 and Sb_2Se_3 have been recently demonstrated as promising alternatives to conventional phase change materials that have been used in electronics but are too lossy for applications in the visible and near-infrared part of the spectrum. With a bandgap in the visible, these materials offer capabilities in the near-infrared and telecom range. The excellent matching of their refractive index with that of high-index semiconductors facilitates their integrated into conventional opto-electronic devices. Recently we have demonstrated application of Sb2Se3 thin films on top of silicon photonic integrated circuits and have achieved very large optical phase shifts by reversibly switching the material between its states.

10:50 AM

(GOMD-126-2021) Phase-change-material metasurface for nonlinear mid-infrared frequency conversion

M. Shalaginov*[1]; F. Yue[2]; R. Piccoli[2]; T. Gu[1]; K. A. Richardson[3]; R. Morandotti[2]; J. Hu[1]; L. Razzari[2]

1. Massachusetts Institute of Technology, USA
2. INRS-EMT, Canada
3. University of Central Florida, USA

Mid-infrared is a frequency band of strategic importance for a number of medical, industrial, and security-related applications. A key stumble block for mid-IR engineers is a lack of affordable high-performance optical equipment for collecting and detecting mid-IR signals. In contrast, the near-IR systems are significantly more mature than their mid-IR counterparts. As a plausible solution, the mid-IR light sources can be bridged together with near-IR detectors via upconversion mechanisms, such as third-harmonic generation (THG). Here, we address this challenge by studying nonlinear optical properties of a novel chalcogenide alloy $Ge_2Sb_2Se_4Te_1$ (GSST), which features broadband transparency for both amorphous and crystalline phases. We retrieved its effective third-order susceptibilities, which turned out to be much higher in comparison to other transparent nonlinear materials. Additionally, GSST demonstrates drastic contrast in the susceptibilities, which allows tuning of the nonlinear responses in a wide range. To enhance THG conversion efficiency, we designed and implemented a metasurface composed of cylindrical GSST resonators supporting magnetic dipole resonant modes. Due to tight confinement of fundamental mid-IR fields and low losses, we observed nearly a 30-fold increase in THG conversion efficiency for the crystalline-GSST metasurface.

11:10 AM

(GOMD-127-2021) Electro-thermal switching of optical phase-change materials using graphene

C. Ríos*[1]; Y. Zhang[1]; M. Shalaginov[1]; M. Kang[2]; K. A. Richardson[2]; T. Gu[1]; J. Hu[1]

1. Massachusetts Institute of Technology, Materials Science and Engineering, USA
2. University of Central Florida, CREOL, USA

The integration of Phase-Change Materials (PCMs) to photonic devices has enabled nonvolatile reconfiguration. This functionality allows photonic systems to be active, i.e., access multiple optical responses, while using zero "standby" power. This exceptional combination of properties is possible because PCMs exhibit large and stable optical properties modulation upon a fast and controlled solid-state transition. Current devices based on PCMs for free-space-optical applications use metal or tin-oxide microheaters, which are not transparent in the IR and require considerable operational powers. In the on-chip integrated counterpart, PCMs are switched using non-scalable optical pulses, or doped-silicon micro-heaters, which only work in Si platforms. To overcome these issues, a substrate-agnostic and nonperturbative platform is demonstrated using an optoelectronic framework that employs undoped-graphene microheaters to thermally and reversibly switch the bistable optical phase-change material $Ge_2Sb_2Se_4Te_1$. An in-situ Raman spectroscopy method is utilized to demonstrate reversible switching between four different levels of crystallinity. Additionally, active and broadband Si_3N_4 integrated photonic modulators and reconfigurable neural-network-designed metasurfaces are proposed. These results position single-layer graphene as the lowest power consumption approach for switching PCMs of comparable size.

11:30 AM

(GOMD-128-2021) All-optically reconfigurable frequency-selective resonance splitting in chalcogenide microring resonators

B. Shen*[1]; H. Lin[2]; S. Sharif Azadeh[3]; J. Nojic[1]; M. Kang[4]; F. Merget[1]; K. A. Richardson[4]; J. Hu[5]; J. Witzens[1]

1. Institute of Integrated Photonics, RWTH Aachen University, Germany
2. College of Information Science & Electronic Engineering, Zhejiang University, China
3. Max Planck Institute of Microstructure Physics, NINT Department, Germany
4. College of Optics and Photonics, University of Central Florida, USA
5. Massachusetts Institute of Technology, Department of Materials Science & Engineering, USA

This work reports a method to enable all-optically reconfigurable control of resonance splitting of one or multiple arbitrarily selected azimuthal orders in a microring resonator. This is accomplished by inscribing Bragg gratings in photosensitive $Ge_{23}Sb_7S_{70}$ chalcogenide microring resonators via a novel cavity-enhanced photo-inscription process, in which injection of light at the targeted C-band resonance frequency induces a spatially-varying refractive index change. The so-formed Bragg grating precisely matches the selected resonance order without introducing optical losses. Long-term room temperature stability of the photo-inscribed Bragg gratings has been verified in darkness and during operation with reduced optical power levels. The Bragg gratings can be reconfigured by first erasure with flood illumination of visible light at 561 nm and subsequent re-inscription. We also report controlled splitting of multiple resonances by inscribing superimposed Bragg gratings.

11:50 AM

(GOMD-129-2021) Unveiling True Three-Dimensional Microstructural Evolution in Chalcogenide Nanocomposites as a Route to Advanced Infrared Functionality

M. Kang*[1]; I. Martin[2]; R. Sharma[1]; C. Blanco[1]; M. Shalaginov[3]; S. Antonov[4]; T. Prosa[2]; D. Larson[2]; J. Hu[3]; H. Francois-Saint-Cyr[5]; K. A. Richardson[1]

1. University of Central Florida, CREOL, College of Optics & Photonics, USA
2. CAMECA Instruments Inc, USA
3. Massachusetts of Institute of Technology, USA
4. Max-Planck-Institut für Eisenforschung GmbH, Germany
5. Thermo Fisher Scientific, USA

As a promising solution for infrared (IR) gradient refractive index (GRIN) applications envisioned to realize an arbitrarily-shaped, single optical component with minimal chromatic aberration, Ge-As-Pb-Se chalcogenide glass and glass-ceramic nanocomposites have received interest due to their ability to modulate refractive index through the formation of high index nanocrystalline phases.

We present these multi-phased nanocomposites as a testbed optical material to validate that atom probe tomography as a suitable three-dimensional (3D) metrology method can provide critical morphological and chemical information where the 2D perspective of transmission electron microscopy falls short. We demonstrate that such knowledge in a 3D space allows us to predict optical properties which would closely match experimental data. This first-ever finding establishes the novel material's quantitative process-structure-property relationship, which serves as a toolbox enabling optical designers to benefit from inputs appropriate for use in effective media predictions. The approach validated in this study, blending glass science, metrology, and IR GRIN optics, results in a new material design paradigm that can be applied to a wide range of multi-phase optical nanocomposites where knowledge of phase formation, geometry, and chemistry across a 3D space is required.

GOMD S5: Glass Education

Education and Professional Development

Room: Saturna Island

Session Chair: Mathieu Hubert, Corning Incorporated

8:30 AM

(GOMD-130-2021) How to get the most out of your ACerS Membership (Invited)

M. Mecklenborg*[1]

1. The American Ceramic Society, USA

The American Ceramics Society (ACerS) is a premier materials societies, with a strong focus on connecting glass scientists and ceramicists around the world. ACerS includes numerous organizations to support involvement and volunteer opportunities within ACerS Divisions, U.S. and International Sections, and Technical Interest groups. Mark Mecklenborg, ACerS Executive Director, will highlight the different opportunities for increasing your involvement in ACerS for early, mid, late career scientists and engineers. For every career stage, industry, and geographic location, there is a way to be involved with ACerS.

9:00 AM

(GOMD-131-2021) Engineering Superheroes: A Multi-Platform Education Initiative (Invited)

R. Castro*[1]

1. University of California, Davis, Material Science & Engineering, USA

Here we discuss a new educational initative designed to inspire a generation of future materials engineers. The program leverages the universe superheroes to introduce science and engineering concepts at different educational levels, from K to freshman college. The K-12 portion is based on producted videos which can be used as support for teachers to introduce concepts of materials science and engineering. For example, the video introduces composites by discussing the properties of the fantastic Captain America's shield. The program has lesson plans made available to teachers targetting the different age groups. At the college level, a first-year course was created to excite students about MSE majors. The course is project based and the students learn basic and some advanced concepts of the 'science of superheroes'. Concepts of energy generation, batteries, nuclear power, and many others can be introduced to different levels of complexities by captivating the students' attention through superheroes characters. At the end of the course, the students use creativity and what they learned in the course to propose a novel super-material.

9:30 AM

(GOMD-132-2021) Challenges and opportunities for glass courses at Corning

M. Hubert*[1]; D. McEnroe[1]; T. St Clair[1]; C. Hogue[1]

1. Corning Incorporated, USA

With the Covid pandemic, 2020 has been a challenging year on many levels, including education activities. In this presentation, the challenges encountered by Corning scientists engaged in teaching glass science and technology classes at the local Community College when switching to fully remote learning, as well as some of the solutions implemented in response, will be discussed. We will also present how the "work-from-home" mandates established in 2020 were turned into an opportunity to teach an internal glass course to Corning employees, reaching out to a broad audience and promoting exchanges while working remotely.

9:50 AM

(GOMD-133-2021) North American Summer School on Photonic Materials (NASSPM): An experiment on incorporating laboratory experience in short courses

H. Jain*[1]; K. A. Richardson[3]; Y. Messaddeq[2]

1. Lehigh University, Institute for Functional Materials and Devices, USA
2. Laval University, Canada
3. University of Central Florida, USA

NSF's Int'l Materials Institute for New Functionality in Glass (IMI-NFG) pioneered the concept of weeklong international school, which brought together subject experts from various countries to teach glass science and engineering to graduate students and post-doctoral researchers. The goal was to attract bright young scientists to glass research, introduce them to the latest scientific and technological advancements, while building an international community with lasting professional friendships. The idea sparked broad interest and was significantly expanded in the form of ICG Summer School, which has been held for the past 14 years at University of Montpellier. The success of Montpellier School has been emulated by a similar school in Wuhan, China. These forums have served to teach formation, structure, and properties of glass to graduate students/postdocs from all over the world. Although they have offered exposure to cutting edge topics of glass science and engineering, they did lacked hands-on lab experience. This gap was filled by NASSPM held at Laval U., which focused on optics and photonic materials. It was highly successful based on feedback from 80 students from 12 countries. In this presentation, we describe the challenges and benefits of this mode of advanced training of graduate level workforce in materials research.

10:10 AM

(GOMD-134-2021) Pasteur Partners PhD (P3) - an alternative to traditional doctorate in USA

H. Jain*[1]; V. Dierolf[1]; A. Jagota[2]; L. Columba[2]; K. Zimmerman[2]

1. Lehigh University, Institute for Functional Materials and Devices, USA
2. Lehigh University, USA

We describe the initial experience with a new model of doctoral training, which is in contrast to traditional curiosity-driven research, and is especially valuable to those who wish to see the impact of their doctoral dissertation in real life within a reasonable timeframe. It provides graduate training through active partnership with companies that are familiar with potential applications and the needs of the society at large. Its features include: 1. The student is actively involved in defining the scope of her/his dissertation. P3 program begins with a pre-program summer internship at a company to get the 'big picture' of the problem of interest through discussions with industry researchers. 2. With a broad perspective of research problem, the student selects appropriate courses in consultation with the academic adviser. Overall, each scholar's program is personalized with co-advisers from both the university and a partner company. 3. A one to two semester company residency is required to ensure use-inspired, industrial perspective as well as the rigor of a doctoral degree. 4. In addition to meeting Departmental requirements, the student is required to take modular, graduate-level professional development courses, co-taught by faculty and industry researchers. P3 is developed with support from NSF's Innovation in Graduate Education program.

Thursday, December 16, 2021

GOMD S1: Fundamentals of the Glassy State

Topology and Rigidity I

Room: Saltspring Island C

10:20 AM

(GOMD-137-2021) Topological Origin of the Low Thermal Expansion of Glassy Silica

Q. Zhou*[1]; M. Bauchy[1]; Y. Shi[2]

1. University of California, Los Angeles, Civil and Environmental Engineering, USA
2. Corning Incorporated, USA

Unlike other multi-component silicate glasses, glassy silica exhibits an extremely low coefficient of thermal expansion below the glass transition temperature. Interestingly, recent in situ neutron scattering results (Shi et al., JNCS 2019) have showed that the short-range order of glassy silica (e.g., interatomic distances) exhibits a coefficient of thermal expansion that is notably higher than that observed at the macroscale—so that the origin of the low thermal expansion of glassy silica remains unclear. To address this question, we adopt the force-enhanced atomic refinement (FEAR) simulation method to explore the atomic structure of silica glasses at varying temperature. Based on these simulations, we demonstrate that silica's sub-Tg coefficient of thermal expansion is encoded in the topology of its medium-range order.

10:40 AM

(GOMD-138-2021) Energy Landscape Signature of the Flexible-to-Rigid Transition in Network Glasses

Z. Zhao*[1]; H. Liu[1]; L. Tang[1]; M. Bauchy[1]

1. University of California, Los Angeles, Civil and Environmental Engineering, USA

Based on their connectivity, network glasses can be classified as flexible, stressed–rigid, or isostatic, if the number of topological constraints is lower, larger, or equal to the number of atomic degrees of freedom, respectively. Thanks to the absence of any stoichiometric requirement, the rigidity of glasses can be continuously tuned (e.g., from flexible to stressed–rigid) by changing their chemical composition. Interestingly, optimally-constrained isostatic glasses have been noted to exhibit unusual properties (e.g., nearly-reversible glass transition, low relaxation, desirable mechanical properties, etc.). However, the physical origin of the unusual behaviors and properties of isostatic glasses remain unclear. Here, based on molecular dynamics simulations, we investigate how the flexible-to-rigid transition in network glasses is encoded in their energy landscape. To this end, we introduce a simplified, yet realistic model of network glasses with varying connectivity. We characterize the topography of these glasses by adopting the activation-relaxation technique (ART), which enables a systematic search of saddle points and transition pathways in the energy landscape surface. We demonstrate that the flexible-to-rigid transition arises from an interplay between low-energy saddle points (in flexible glasses) and topological frustration (in stressed–rigid glasses).

11:00 AM

(GOMD-139-2021) Structural analysis for high-temperature levitated melts

S. Kohara*[1]

1. National Institute for Materials Science (NIMS), Japan

Investigating high temperature melts is very challenging, because chemical reactions with containers are difficult to avoid. In addition, the diffraction signal from a crystalline container makes it difficult to obtain high quality data from melts. However, the use of levitation methods can overcome these problems and, in many cases, enables deep undercooling and enhanced glass formation, owing to the elimination of extrinsic heterogeneous nucleation. Recently, studies on melts and the formation of glasses from supercooled melts using levitation techniques have been of particular interest in understanding the structure of these materials and the process of glass formation. We have performed synchrotron X-ray diffraction measurements on high-temperature melts by utilizing the aerodynamic levitation technique. Moreover, structural modeling was introduced to discuss the structure of Al_2O_3 and Er_2O_3 melts, in comparison with that of ZrO_2 melt, in terms of topology.

Glass under Extreme Conditions I

Room: Saltspring Island A/B

Session Chair: Nadja Lonnroth, Corning Incorporated

10:20 AM

(GOMD-140-2021) Atomistic Origins behind the Temporal Optical Evolution of Ge-Sb-S Chalcogenide Glasses upon Gamma Irradiation for Use in Extreme Environment

M. Kang*[1]; B. Sohn[2]; Q. Du[3]; D. Ma[3]; R. Pujari[3]; L. Sisken[1]; C. Blanco[1]; C. Goncalves[1]; C. Arias[1]; A. Yadav[1]; S. Novak[1]; C. Schwarz[4]; R. M. Gaume[1]; J. Hu[3]; A. Agarwal[3]; D. T. Tan[2]; K. A. Richardson[1]

1. University of Central Florida, CREOL, College of Optics & Photonics, USA
2. Singapore University of Technology and Design, Singapore
3. Massachusetts Institute of Technology, USA
4. Ursinus College, USA

Chalcogenide glasses are often employed in applications where their linear and nonlinear optical responses make them a unique choice. However, the response of these properties is often not measured under such representative conditions to assess their time-dependent behavior. We present a combined theoretical-experimental method to correlate the time-dependent linear and nonlinear optical responses of bulk Ge-Sb-S glasses to their metastable coordination topological defects (CTDs) created upon gamma irradiation. The evolution of the CTDs is evidenced by an irradiation-induced increase in the ratio of edge-shared to corner-shared GeS_4 tetrahedra which decays at room temperature during recovery, demonstrating the glass' metastable behavior. Through this structural transition, the mass density of the glass undergoes a momentary increase and subsequent decrease, giving rise to a change and subsequent relaxation of linear refractive index and bandgap energy. Concurrent with this evolution in linear optical properties, the nonlinear response of the glass is found to be unaffected, likely due to a counter effect associated with the network's free electrons created upon irradiation. We expect the material system to be deployable as robust, reversible radiation sensors in extreme environments such as space and ground-based radioactive facilities.

10:40 AM

(GOMD-141-2021) Effect of ion radiation on the indentation response of oxide glasses

X. Ren*[1]; M. M. Smedskjaer[1]; H. Peng[2]; T. Du[1]

1. Aalborg University, Denmark, Department of Chemistry and Bioscience, Denmark
2. Lanzhou Uinversity, China

Radiation is known to affect the surface structure and properties of glasses. In this work, we investigate the effect of ion irradiation (Xe ions) treatment on the mechanical properties of two series of oxide glasses. First, a series of sodium borosilicate (NBS) glasses with varying Si/B and Na/B ratios, finding that radiation reduces the hardness and increases the crack initiation resistance. Second, a series of calcium aluminoborosilicate (CABS) glasses that feature high crack initiation resistance prior to any treatment. Again, we find a decrease in hardness and increase in crack initiation resistance upon irradiation. In order to explore the mechanism of the irradiation effects, we also perform surface structural analysis (Raman, infrared spectroscopy) as well as atomistic simulations. Finally, in order to achieve more damage-resistant oxide glasses, we also explore the possibility to improve the mechanical properties through combining radiation treatment with other extrinsic post-treatment methods (hot compression, humid aging).

GOMD S2: Glass and Interactions with Its Environment - Fundamentals and Applications

Nuclear Waste Mobilization I

Room: Moresby Island

Session Chairs: Ming Tang, Clemson University; Natalie Smith-Gray, Washington State University

10:20 AM

(GOMD-142-2021) Predicting Zeolites' Stability during the Corrosion of Nuclear Waste Immobilization Glasses

B. Zhen-Wu[1]; J. V. Ryan[2]; G. Sant[1]; M. Bauchy*[1]

1. University of California, Los Angeles, Civil and Environmental Engineering Department, USA
2. Pacific Northwest National Lab, USA

During the long-term corrosion of nuclear waste glasses under nuclear waste disposal conditions, the precipitation of zeolitic phases has been linked to a delayed acceleration in glass corrosion (known as "Stage III"). Hence, predicting the thermodynamic propensity for zeolites to form upon the dissolution of nuclear waste glasses is key to ensure their long-term performance. Here, we compile a unified, internally-consistent thermodynamic database to estimate the stability of secondary phases (including zeolites, clays, and calcium–silicate–hydrate phases). Based on this, we report a geochemical modeling method allowing us to predict the stability of secondary phases upon the dissolution of nuclear waste immobilization glasses. We show that this approach offers a realistic description of the stability of the secondary phases forming during the dissolution of the International Simple Glass as a function of temperature and pH. We find that the formation of silica and clay secondary phases is thermodynamically favored at low pH ($pH < 10$), whereas, in contrast, zeolite (analcime) and calcium–silicate–hydrate phases are favored at high pH ($pH > 10.5$). This suggests that thermodynamics (i.e., not solely kinetics) plays a key role in determining the range of solution pH wherein stage III corrosion may occur, i.e., when zeolite formation is favored.

10:40 AM

(GOMD-143-2021) Machine learning-based prediction of the corrosion behavior of nuclear waste glasses

Y. Song*[1]; J. Vienna[2]; J. Ryan[2]; M. Bauchy[1]

1. University of California, Los Angeles, Civil and Environmental Engineering, USA
2. Pacific Northwest National Laboratory, USA

Vitrification represents one of the most efficient approaches for safely confining resultant fission products from nuclear power plants/legacy waste. However, it has been challenging to determine the long-term corrosion behavior of different nuclear waste glasses. In that regard, vapor hydration test (VHT) is an efficient test method to accelerate the corrosion process for estimating the corrosion rate. Herein, we train a neural network model based on a VHT dataset that comprises a large selection of waste glasses. The model. Our model achieves state-of-the-art accuracy for mapping the oxide compositions to their VHT performance (i.e., long-term corrosion potential). With several techniques for interpreting the trained neural network model, we further obtained valuable insights on how each of the oxides impacts the corrosion behavior. Based on this study, we demonstrate that machine learning-based analysis can serve as a powerful tool for designing new glasses that are more corrosion resistant.

11:00 AM

(GOMD-144-2021) Predicting nepheline precipitation in waste glasses using ternary submixture model and machine learning

X. Lu*[1]; I. Sargin[2]; J. Vienna[1]

1. Pacific Northwest National Lab, Energy and Environment Directorate, USA
2. Washington State University, USA

Nepheline precipitation in nuclear waste glasses during vitrification can be detrimental due to the negative effect on chemical durability often associated with its formation. Developing models to accurately predict nepheline precipitation from compositions is important for increasing waste loading since existing models can be overly conservative. In this study, a dataset of 955 glasses, including 352 high-level waste glasses, was compiled from literature data. Previously developed submixture models were refitted, where a misclassification rate of 7.8% was achieved. In addition, nine machine learning (ML) algorithms were applied to evaluate their ability to predict nepheline precipitation from glass compositions. Model accuracy, precision, recall/sensitivity, and F1 scores were compared between different ML algorithms and modeling protocols. Model prediction with an accuracy of ~0.9 (misclassification rate of ~10%) was observed for different algorithms under certain protocols. This study evaluated various ML models to predict nepheline precipitation in waste glasses, highlighting the importance of data preparation and modeling protocol, and their effect on model stability and reproducibility. The results provide insights into applying ML to predict glass properties and suggest areas for future research on modeling nepheline precipitation.

Surfaces and Coatings I

Room: Cortes Island
Session Chair: Joy Banerjee, Corning Incorporated

10:20 AM

(GOMD-145-2021) Surface analysis with vibrational spectroscopy – Correlations between vibrational spectral features and silicate glass network structures (Invited)

S. H. Kim*[1]

1. Pennsylvania State University, Chemical Engineering & Materials Science, USA

Infrared and Raman spectroscopy can be used for structural analysis of glass surfaces. Specular reflection infrared (SR-IR), infrared spectroscopic ellipsometry (IR-SE), and confocal Raman spectroscopy are good examples of such vibrational spectroscopy techniques. Recently developed techniques combining atomic force microscopy and infrared spectroscopy (AFM-IR) allow hyperspectral imaging in the silicate stretching region of the subsurface structure of glass. However, all these technical advancements are not fully or properly utilized for structural analysis of surface / subsurface glass network due to inaccurate (or still controversial) interpretation rules for stretch and deformation vibrational modes of silicate glass networks. This talk will briefly review vibrational spectroscopic analysis methods for optically-flat monolithic glass surfaces and then discuss how changes in vibrational spectral features can be related to the structures of glass network, especially distributions of Si-O bond lengths and angles. The comparison of experimental spectra silica and silicate glasses with the network structures simulated with molecular dynamics questions the validity of conventional spectral interpretation methods and suggests alternative interpretation schemes.

10:50 AM

(GOMD-146-2021) Water adsorption isotherm on glass and its impact on interfacial adhesion

Y. Lin*[1]; J. Banerjee[2]; N. J. Smith[2]; G. Agnello[2]; W. Walczak[2]; R. G. Manley[2]; S. H. Kim[1]

1. Pennsylvania State University, Chemical Engineering, USA
2. Corning Incorporated, USA

Particulates adhering to glass surfaces are contaminants that can cause electrical breakdown or malfunction when applied to device displays. This adhesion is highly sensitive to the water meniscus formed in humid environments and can be modeled by measuring the pull-off force as a function of relative humidity between either a spherical or cone-shaped cantilever and flat glass using atomic force microscopy (AFM). In this work, the effective thickness and hydrogen bonding interaction of water layers adsorbed on monolithic glass surfaces with different treatments were probed using transmission FTIR at Brewster's angle with p-polarized light. The effective thickness of water layers on treated glass surfaces was then utilized to estimate the adhesive forces originating from van der Waals and capillary forces for the particulate-glass interface, which then are compared with the pull-off force measured with AFM. Correlations between the adsorbed water layer, the elemental surface composition, and the adhesion force for glass with different treatments are discussed in this work, which provides insight into controlling the particulate contamination during industrial glass processing.

11:10 AM

(GOMD-147-2021) Liquid Entrenched Smooth Surface (LESS) - A New Class of Slippery Surface on Glass and Ceramics (Invited)

T. Wong*[1]

1. The Pennsylvania State University, Department of Mechanical Engineering, USA

The slippery rim of Nepenthes pitcher plants has inspired a new class of multi-functional slippery surface coatings since 2011. From the early development of slippery liquid-infused porous surfaces (SLIPS) to the latest liquid entrenched smooth surface (LESS), these slippery surface coatings have demonstrated a number of unique features that outperform state-of-the-art liquid repellent surface coating technologies. These slippery surfaces have successfully been translated in commerical space and applied in a number of impactful applications ranging from anti-fouling medical devices to marine anti-fouling to global sanitation. In this talk, I will describe our recent development of LESS, which is a sprayable non-fouling surface coating that exhibit extreme repellency towards liquids, bacteria, and viscoelastic solids. Importantly, LESS-coated surfaces can repel viscoelastic solds with dynamic viscosities spanning over nine orders of magnitude, which is three orders of magnitude higher than has previously been reported for other repellent materials. The use of LESS coating to address global sanitation and water scarcity issues will be discussed as an application example.

GOMD S3: Optical and Electronic Materials and Devices - Fundamentals and Applications

Laser Interactions with Glass I

Room: Saturna Island
Session Chair: Gözden Torun, EPFL

10:20 AM

(GOMD-148-2021) Thermal erasure of nanopores induced by femtosecond laser in silicate glasses

M. Cavillon*[1]; Y. Wang[1]; B. Poumellec[1]; M. Lancry[1]

1. University Paris Saclay, France

Femtosecond (fs) lasers enable high peak powers and 3D materials processing making them attractive tools for a variety of applications in glass science, including long lifetime data storage or high temperature sensing (>800°C). In this work, we focus on fs laser-induced modifications of the type II variety, corresponding to the formation of sub-wavelength nanoporous. The latter are known to be thermally stable for few hundreds of hours at ~1000°C in silica. However, to date, the influence of glass composition on the thermal stability of such laser-induced structures is not fully understood. To investigate this, a variety of silicate glasses (including aluminosilicate, germanosilicate, commercial silica (with and without OH, Cl) and silicate (Borofloat33, ULE) glasses) were irradiated with a fs laser under various experimental conditions (e.g., laser polarization orientation, pulse energy). The samples were subsequently heat-treated up to ~1200°C and the birefringence strength induced by the formed nanostructures was monitored. The experimental data were compared with a viscoelastic model (Rayleigh-Plesset), and it was found that the viscous dynamic of the glass between its annealing and softening temperatures is a key parameter to predict thermal aging of such nanostructures. Others parameters, including nanopore size or glass surface tension, will also be presented at the conference.

10:40 AM

(GOMD-149-2021) Erasure mechanisms of Type II nanogratings in silica

Y. Wang*[1]; M. Cavillon[1]; B. Poumellec[1]; M. Lancry[1]

1. university of paris saclay, France

Since the discovery of femtosecond laser-induced nanogratings in 2003, there is a growing interest in high-temperature applications like Structural Health Monitoring (SHM) or 5D optical data storage. The thermal stability is an important factor to be considered to ensure reliable functionality during the optical component lifetime for these kinds of devices. Nanogratings exhibit excellent thermal stability at high temperatures (typically they can be stabilized for 100 hours at 1000°C in silica). From an optical view, they can easily be detected and monitored optically through their birefringent response. The latter arises from the sum of several contributions including form birefringence, stress-induced birefringence, densification/expansion, and point defects. In this work, we discussed these contributions to the total birefringence response in silica, as well as their corresponding thermal erasure during a step isochronal annealing process using optical measurements, SEM, Raman spectroscopy, and UV absorption analyses.

11:00 AM

(GOMD-150-2021) Controlled orientation of $LiNbO_3$ nanocrystals induced in Li_2O – Nb_2O_5 – SiO_2 – B_2O_3 glasses by femtosecond laser irradiation

E. Muzi[1]; M. Cavillon*[1]; M. Lancry[1]; B. Poumellec[1]; D. Janner[2]

1. Université Paris-Saclay, ICCMO, France
2. Politecnico di Torino, DISAT, Italy

Crystallization in glass can be employed to design various "smart optical functions" (including waveguides, frequency doubling, acousto-optic modulators) in future nano/micro-scaled integrated photonic circuits. In this work, we investigate the crystallization of $LiNbO_3$ nanocrystals induced by femtosecond laser inside Li_2O–Nb_2O_5–SiO_2–B_2O_3 glasses (B_2O_3 ranging from 0% to 21 mole%), fabricated by the melt-quenching technique. In this work, anisotropy of the second harmonic generation (SHG) response is investigated in both static and scanning (writing speed at least up to 100 μm/s) fs-laser irradiation conditions. Among the results to highlight, SHG is detected much faster in high content B_2O_3 glass relative to B_2O_3-free glass (1 and 20 seconds at 100 mW, respectively). Additionally, B_2O_3-rich glasses promote faster fabrication of birefringent structures with larger birefringent response (>150 nm at λ=550 nm), as well lower glass-making temperatures. We reveal that $LiNbO_3$ crystallization is controlled by varying fs-laser parameters (polarization, pulse energy, repetition rate, scanning speed), and the polar axis of $LiNbO_3$ nanocrystals can be oriented on demand through writing laser polarization. This light-enabled orientation ability leads, for instance, to modulation of $\chi^{(2)}$ properties.

11:20 AM

(GOMD-151-2021) Dependence of lattice rotation rate on the direction of single crystal crystal growth in glass

E. J. Musterman*[1]; V. Dierolf[2]; H. Jain[1]

1. Lehigh University, Materials Science and Engineering, USA
2. Lehigh University, Physics, USA

Rotating lattice single (RLS) crystals are a special class of solid, which are fabricated in glass by local heating through laser irradiation. These metamaterials with lattice engineered using special crystal growth conditions offer the opportunity to create novel optical devices. By scanning the laser along the glass surface, anisotropic forces are imposed at the crystal growth front causing the lattice to rotate about an axis parallel to the glass surface and perpendicular to the growth direction. The rate of this rotation has been found to depend qualitatively on a number of laser parameters and the growth's crystallographic direction. In this presentation, we explore this crystal growth direction dependence of rotation rate for a model system of Sb_2S_3 RLS crystals fabricated in $84Sb_2S_3$-$16SbI_3$ glass. These rotation rates are measured by electron backscatter diffraction (EBSD) and are interpreted using a dislocation-based model for rotation and the characteristics of dislocations in this model system. A thorough understanding of the crystallographic dependence of rotation rate presented in this talk would help engineer the lattice of RLS crystals for specific applications.

GOMD S1: Fundamentals of the Glassy State

Glass Crystallization and Glass-Ceramics I

Room: Pender Island

Session Chair: Matthew McKenzie, Corning Incorporated

1:30 PM

(GOMD-152-2021) Phase-field modeling of crystal nucleation in undercooled liquids (Invited)

L. Granasy*[1]; F. Podmaniczky[1]; T. Pusztai[1]

1. Wigner Research Centre for Physics, Institute for Solid State Physics and Optics, Hungary

It will be reviewed how phase-field models contribute to a better the understanding of various aspects of crystal nucleation, including homogeneous and heterogeneous processes, and their role in microstructure evolution. Results obtained both by conventional phase-field approaches that rely on spatially averaged order parameters and by recent molecular phase-field approaches will be presented. We present simple cases of homogeneous and heterogeneous nucleation, nucleation in eutectic/phase separating systems, phase selection via competing nucleation processes, and exotic nucleation modes such as precursor-assisted two-step nucleation and growth front nucleation, a process, in which grains of new orientations form at the solidification front, yielding crystal sheaves, spherulites and biomorphic crystallization.

2:00 PM

(GOMD-153-2021) Nanoscale microstructure and chemistry of transparent gahnite glass-ceramics revealed by atom probe tomography

A. Mitchell*[1]; D. Perea[2]; M. Wirth[2]; J. Ryan[2]; R. Youngman[3]; A. Rezikyan[3]; A. Fahey[3]; D. Schreiber[2]

1. Corning Incorporated, Glass Research, USA
2. Pacific Northwest National Lab, USA
3. Corning Incorporated, Characterization Sciences, USA

Transparent gahnite glass-ceramics are composite materials comprised of nanocrystallites and residual glass. In this system, uniform, internal nucleation of gahnite ($ZnAl_2O_4$) is aided by the nucleating agent, tetragonal zirconia. Due to small (<20 nm), overlapping crystallites, it is difficult to characterize these materials using conventional techniques such as TEM. Here, atom probe tomography (APT) data is presented that captures the 3-D distributions of elements at the nanometer scale within transparent gahnite glass-ceramics for the first time. The APT data shows clustering of ZrO and the proximal location of the nucleating agents relative to the main crystalline phase, gahnite. These measurements show that gahnite does not crystallize in a core-shell relationship around ZrO_2. Residual glass compositions were directly measured, where previously they had only been estimated using mass balance approaches. Finally, no evidence of enrichment of residual glass components or depletion of gahnite components was found near the crystallites. Based on conventional understanding of the nucleation and crystallization of glass-ceramics and this system, the lack of chemical gradients near the glass-crystal boundaries is surprising. For nanocrystalline, transparent gahnite glass-ceramics, APT has proven to be a powerful tool to answer many longstanding fundamental questions.

2:20 PM

(GOMD-154-2021) Low Temperature Nucleation Anomaly in Silicate Glasses - Is it Real? (Invited)

K. F. Kelton*[1]; X. Xia[3]; M. E. McKenzie[2]; R. Youngman[2]

1. Washington University, Physics, USA
2. Corning Incorporated, USA
3. Washington University, USA

Understanding and controlling nucleation in silicate glasses is essential for glass formation and the production of glass ceramics. For over 40 years there have been reports that in many silicate glasses the commonly used Classical Theory of Nucleation fails to predict the measured nucleation rates for temperatures below the temperature of the maximum nucleation rate. The data indicate that at these temperatures the work of critical cluster formation ceases to decrease with decreasing temperature as expected, but instead plateaus and begins to increase. Many explanations for this anomalous behavior have been offered, but none have provided a satisfactory answer. Here, new time-dependent nucleation measurements are presented for a $5BaO$-$8SiO_2$ glass that was reported to show the anomalous behavior, with nucleation heating times that were over 100 times longer than previously used. The data obtained at a temperature 50K below the peak nucleation temperature are consistent with predictions from the Classical Theory. They show that the anomalous behavior previously reported is an experimental artifact arising from an inadequate amount of heating time at low temperatures. These results call into question the reports of anomalous behavior at low temperature in the nucleation rates of other silicate glasses.

Glass Crystallization and Glass-Ceramics II

Room: Pender Island

Session Chair: Matthew McKenzie, Corning Incorporated

3:10 PM

(GOMD-155-2021) In-situ and ex-situ characterization of crystallization in silicate mold fluxes by the Hot Thermocouple Technique

P. Porter*[1]; R. J. O'Malley[1]; R. Brow[1]

1. Missouri University of Science & Technology, Materials Science and Engineering, USA

Molten Na-Ca-aluminosilicate fluxes are used in continuous steel casting processes to mediate thermal transport between the molten steel and the water-cooled copper mold shell. The flux layer experiences shear stresses and thermal gradients of approx. 1000°C over a 1-5mm thickness, and must partially crystallize to achieve optimal thermal transport properties. In this work, we used the Hot Thermocouple Technique (HTT) as a tool for both in-situ characterization of crystallization behavior, and to produce samples with known thermal histories for ex-situ structural characterization and phase identification. The HTT uses a thermocouple as both a resistive heating element and temperature sensor to control the thermal history of samples. The present device uses a Type-B Pt/Rh thermocouple to heat at temperatures up to 1650°C and at heating (or quench) rates up to 100°C/s. Time-temperature-transformation diagrams for the formation of cuspidine and nepheline from commercial and simplified mold-flux materials with varying basicity were determined from both in situ and ex-situ experiments, the latter of which included data from powder X-ray diffraction, Raman spectroscopy, and cathodoluminescence microscopy.

3:30 PM

(GOMD-156-2021) Advances in nucleation modeling: A case study of soda lime silicate

M. E. McKenzie*[1]; R. Welch[2]; F. Chen[3]; X. Xia[3]; J. C. Mauro[2]; R. Youngman[1]; K. F. Kelton[3]

1. Corning Incorporated, Science & Technology, USA
2. Pennsylvania State University, USA
3. Washington University in St Louis, USA

Modeling nucleation in silicate glass is a difficult due to the disparate time and length scales that are associated with the rare event. Recently, the Toy Landscape Model (TLM) and Grand Canonical Monte Carlo (GCMC) techniques have advanced our knowledge in this area. TLM provides an approximation to the energy landscape using Tg, fragilty, and molecular dynamics (MD) to predict the nucleation rate. GCMC focuses solely on the thermodynamic aspect, allowing one to peer deeper into the nucleation reaction mechanisms, competing pathways, and micro-separations. Additionally, we employ a seeding method to study the development of the crystal-glass interface within MD. Coupled with experiment and the modeling techniques, we study the complicated ternary composition: soda-lime-silicate. TLM can reproduce the experimental nucleation rate within 8 orders of magnitude and correctly predicts the maximum nucleation temperature. GCMC can reproduce the thermodynamic barrier accurately and the preferred sodium enriched pathway. Further analysis shows how the developing interface forms and the ability to probe into competing nucleation pathways (combeite vs. devitrite). This work serves as first step to fully atomistic model of a multi-component nucleation process.

3:50 PM

(GOMD-157-2021) Understanding Crystallization Though Atomistic Simulations (Invited)

C. Wilkinson*[1]; J. C. Mauro[1]

1. Pennsylvania State University, USA

It must be frankly admitted that predicting the nucleation curves of industrial glass ceramics that have not been well characterized is of high commercial value and of intense difficulty. Despite this difficulty several methods have appeared in recent literature to enable a possible predictive method. These methods use Energy Landscapes, Molecular Dynamics, or Monte-Carlo techniques to achieve different parameterizations of nucleation models. Each approach has strengths and weaknesses but assessing which method will work for an arbitrary system is a non-trivial task. In this talk, we will review each method and discuss the relative strengths and weaknesses. The models will be discussed in the framework of two common glass ceramics: BaO 2SiO2 and Li2O 2SiO2. The data presented will be a combination of that taken from literature and novel results. Additionally, a roadmap for the future of each method is presented and discussed.

4:20 PM

(GOMD-158-2021) Compositional Effects on the Crystallization of BCZT Perovskite Glass-Ceramics

D. K. Dobesh*[1]; M. Cicconi[1]; D. de Ligny[1]

1. Friedrich-Alexander-Universityet Erlangen-Neurnberg, Glass, Germany

[$(Ba,Ca)(Zr,Ti)O_3$] (BCZT) ceramics are amongst lead-free perovskite alternatives for replicating and exceeding lead-based materials in their ferro and piezoelectric response. Glass-ceramics can provide a scalable and machinable material that can attain the electromechanical response and develop devices with advantageous properties. We present our study of the structural dependence of Ba- and Ti-rich boroaluminosilicate glasses to nucleation and crystallization mechanisms. The glass-ceramic system was investigated to determine the thermal kinetics for producing perovskite crystals within a glass matrix. The compositional space and glass stability were explored by the Na/Ca ratio to suppress surface crystallization. The incorporation of various amounts of Zr established the solubility limit and effect on the glass structure. Nucleation and crystal evolutions were analyzed with differential scanning calorimetry. We established the correlations between heating rates and crystallization temperatures with the activation energies. Classification of crystal phases was accomplished using X-ray powder diffraction and Raman Spectroscopy. Raman and Brillouin spectroscopy correlated structural dependencies of the glass to thermal treatment stages and the crystal to residual glass interface. The dielectric response of the glass-ceramics has been investigated through relative permittivity and dissipation loss.

Topology and Rigidity II

Room: Saltspring Island C

1:30 PM

(GOMD-159-2021) Topology-Informed Machine Learning for Predicting Glass Stiffness

K. Yang*[1]; B. Yang[2]; N. Krishnan[3]; C. G. Hoover[4]; M. M. Smedskjaer[5]; M. Bauchy[1]

1. UCLA, Civil and Environmental Engineering, USA
2. UCLA, Computer Science, USA
3. Indian Institute of Technology Delhi, Civil Engineering, India
4. Arizona State University, School of Sustainable Engineering and the Built Environment, USA
5. Aalborg University, Department of Chemistry and Bioscience, Denmark

Discovering novel glasses with enhanced properties and functionalities is essential to address many energy, communications, and infrastructure challenges. To this end, data-driven machine learning techniques offer a promising approach to accelerate glass discovery. However, although traditional machine learning models excel in interpolating existing datasets, they usually fail at extrapolating predictions far from the training set. This is a serious drawback since it limits their ability to identify new promising glasses in compositional domains that are yet to be explored—and such glasses are the most likely to feature properties that are very different from those of present glasses. Here, we present a new "topology-informed" machine learning framework that overcomes this limitation. We show that embedding some information about glass topology in machine learning models greatly enhances their ability to offer robust predictions far from their training sets.

1:50 PM

(GOMD-160-2021) Understanding Medium-Range Order Structure of Glasses using Persistent Homology

S. S. Sørensen*[2]; C. A. Biscio[3]; M. Bauchy[1]; L. Fajstrup[3]; M. M. Smedskjaer[2]

1. University of California, Los Angeles, Department of Civil and Environmental Engineering, USA
2. Aalborg University, Department of Chemistry and Bioscience, Denmark
3. Aalborg University, Department of Mathematical Sciences, Denmark

The medium-range order (MRO) structure of glasses can significantly influence their properties, yet the understanding of MRO, particularly in relation to the origin of the first sharp diffraction peak (FSDP) in the structure factor, remains elusive. Recent developments in persistent homology (PH), a type of topological data analysis, provide a new path for analyzing MRO structure in amorphous materials. In this work, we employ PH to analyze the MRO structure of sodium silicate glasses, which in turn has been obtained by classical molecular dynamics simulations. By studying the size of certain topological features, we find striking similarities to the length scales associated with the FSDP. In addition, we find that PH captures the compositional dependence of the FSDP in the sodium silicates. Overall, we expect the developed methodology to be easily extendable to understand MRO in other glassy systems, whenever the atomic positions can be reliably obtained.

2:10 PM

(GOMD-161-2021) Machine learning-based rapid fly ash screening for greener concrete production

Y. Song*[1]; G. Sant[1]; M. Bauchy[1]

1. University of California, Los Angeles, Civil and Environmental Engineering, USA

The use of cement in concrete production involves heavy carbon footprints. Industrial waste ashes such as fly ash are often used as Supplementary Cementitious Materials (SCMs) for partial replacement of cement in concrete. However, due to the heterogeneous nature of this material, the empirical identification method of strength-reactive fly ash is still largely empirical. The actual strength potential of fly ash in concrete is further complicated by its interaction with cement during the hydration process. In that regard, the current limitations of conventional knowledge-driven research can be made up by the emerging data-driven machine learning methods. Our studies have investigated different approaches to leverage the cutting-edge machine learning techniques to (i) enable rapid screening of reactive fly ash from the raw materials, and (ii) predict the broadly accepted strength activity index (SAI) to assess the real influence of various fly ashes to cement strength. With our trained machine learning models, these analyses can be carried out with the sole inputs of bulk material compositions and properties that are easily measurable. Our studies demonstrate that proper use of machine learning techniques can remarkably facilitate the investigation of SCMs, as well as promote the optimal strategy of using SCMs in concrete production.

Glass under Extreme Conditions II

Room: Saltspring Island A/B

Session Chair: Dominique de Ligny, University Erlangen-Nürnberg

1:30 PM

(GOMD-162-2021) Insights into the contribution of optical spectroscopies (Brillouin, Raman) to the study of v-SiO_2 at high pressure (Invited)

C. Weigel*[1]; M. Foret[1]; B. Hehlen[1]; S. Clément[1]; A. Polian[2]; R. Vacher[1]; B. Ruffle[1]

1. University of Montpellier, France
2. University Paris 6, France

High-pressure Brillouin and Raman spectra of vitreous silica are measured in diamond-anvil cell at room temperature. On one hand, measurements are led in the elastic domain, P<10 GPa. In that pressure range, the equation of state obtained from the Brillouin sound velocities corrected for dispersion agrees with volume measurements. Brillouin frequencies also show that the well-known anomalous maximum in the pressure dependence of the compressibility is suppressed by a surprising sizeable amount of He incorporated into the silica network. This allows relating the elastic anomaly to the collapse of the largest interstitial voids in the structure. Raman experiments, combining the use of either a nonpenetrating pressurizing medium—argon—or penetrating ones—helium and neon, allows assessing density from stress effects on the Raman frequencies on in-situ compacted samples. On the other hand, Brillouin light scattering experiments along loading-unloading paths are used to investigate the compressibility of v-SiO_2 up to 60 GPa. It shows that complete pressure cycles certainly deserve more attention and that precise density and sound velocity measurements, when combined, give fundamental insights into pressure-induced transformations in glasses.

2:00 PM

(GOMD-163-2021) A Multi-faceted Experimental Approach to Studying Structural Modification in Silicate Melts and Glasses under High-Pressure (Invited)

Y. Ryu*[1]; T. Yu[1]; C. Prescher[2]; S. Chariton[1]; E. Greenberg[1]; T. Officer[1]; F. Shi[1]; V. Prakapenka[1]; S. Tkachev[1]; P. Eng[1]; J. Stubbs[1]; P. Dera[3]; H. Watson[4]; M. L. Rivers[1]; Y. Wang[1]

1. The University of Chicago, Center for Advanced Radiation Sources, USA
2. Deutsches Elektronen-Synchrotron, Germany
3. University of Hawaii Manoa, Hawaii Institute of Geophysics and Planetology, USA
4. Union College, Department of Physics and Astronomy, USA

It is believed that the early Earth underwent large-scale melting during the accretion process, thank the bombardment of massive asteroids. As the melts crystallized, chemical differentiation occurred, which was largely responsible for the layered structure of the present-day Earth. In order to understand the origin and significance of deep melt throughout the Earth's history, it is important to obtain laboratory constraints on thermodynamics and physical properties of silicate melt at high-pressure (HP) and high-temperature (HT). However, little is known about the structures of liquid silicates throughout the Earth's pressure regime because the HP and HT environments impose severe experimental difficulties. Thus, there were still many issues that need to be addressed. Here, we report the evolution of structural modifications and physical properties in Mg2Si2O6, Ca2Si2O6, and CaMgSi2O6 glasses using a combination of experimental techniques including Raman spectroscopy, synchrotron X-ray scattering, and Brillouin spectroscopy, conducted in the diamond anvil cell under pressure up to ~70 GPa. Such information is vital in uncovering missing links between the structure and physical properties of silicate glasses. Further studies may help extend such a relationship to silicate melts and help facilitate the synthesis of more advanced technological materials.

2:30 PM

(GOMD-164-2021) Structure and Properties of the Silica glass Pressure-quenched at Liquid phase

M. Ono*[1]; Y. Tanabe[1]; K. Zagarzusem[1]; M. Jeem[1]; M. Fujioka[1]; J. Nishii[1]

1. Hokkaido University, Research Institute for Electronic Science, Japan

Silica glass is the most indispensable material for optical communications and high-power laser applications due to its excellent optical properties. The transmission loss of silica glass has been reduced over the past 30 years by continuous efforts toward decreasing density fluctuations by lowering of fictive temperature, e.g., through improvements in processing or doping. Our recent study has shown that shrinkage of structural voids (=empty spaces) by hot-compression at 2000K is a promising way to further decrease the loss. Other studies have shown that the structural fluctuation of silica glass is reduced when hot-compression is done at 1400K. However, hot-compression of silica glass tends to cause crystallization, particularly within the region where SiO_2 crystalline phase exist on the phase diagram, and/or it contains impurity, or the applied pressure is anisotropic. Thus, in this study, we apply hydrostatic compression to the liquid phase SiO_2 at 2000 K to ensure the vitrification. The structure, optical and thermal properties of the obtained hot-compressed glasses are discussed.

Glass under Extreme Conditions III

Room: Saltspring Island A/B

Session Chair: Akihiro Yamada, University of Shiga Prefecture

3:20 PM

(GOMD-165-2021) Glass Processing Under Electromagnetic Extreme Conditions (Invited)

S. K. Sundaram*[1]

1. Alfred University, Inamori School of Engineering, USA

Powerful radiations produced by gyrotrons and ultrafast laser pulses can deposit high energy into bulk dielectric materials in a short period of time. This can be exploited in materials processing including glasses. Gyrotrons producing intense millimeter wave power in the frequency range of 30-300 GHz originally developed for magnetic confinement fusion research are available commercially in power range of 10 kW to 2 MW. A 95 GHz, 10 kW continuous wave (CW) gyrotron custom is designed and installed at Alfred University for processing of glasses and ceramics. Gyrotron technologies will be reviewed and selected examples of glass processing applications will be presented. Ultrafast femtosecond laser pulses can be used to create surface textures, gratings, and waveguides in glasses. In addition, these pulses can be used to modify local glass structure and thus control glass properties. A femtolaser system with estimated pulse duration of 50 fs and energy of 500 nJ with a control wavelength of 800 nm and a repetition rate of 5.1 MHz installed at Alfred University is used for rapidly modifying glass structures and properties. Ultrafast laser pulse–glass interactions will be reviewed and selected examples of modification glass structures and properties will be presented.

3:50 PM

(GOMD-166-2021) Effect of femtosecond laser exposure on aluminosilicate glass

M. Lesik*[1]; G. Torun[1]; N. Lonnroth[2]; Y. Bellouard[1]

1. Ecole Polytechnique Fédérale de Lausanne, STI/IMT, Switzerland
2. Huawei Technologies, Finland

Over the last two decades, femtosecond laser exposure in the non-ablative regime has emerged as an interesting method not only for patterning glass substrates such as fused silica and borosilicate, but also for modulating the refractive index in a variety of glass systems. Furthermore, thanks to the non-linear nature of the laser-matter interaction, these laser-induced modifications can occur anywhere within the volume of the substrate, opening a new realm of three-dimensional direct-processing with sub-micron resolution capabilities. Here we explore the applicability of these processes in both, pristine and strengthened, aluminosilicate glass substrates. Specifically, we systematically investigate various laser exposure regimes leading to self-organized nanostructures, including polarization-dependent nanogratings and micro-scale periodic gas bubbles networks when thermal accumulation conditions are reached due to high laser-pulse emission rates. Finally, we compare the taxonomy of femtosecond laser-induced non-ablative modifications in aluminosilicate glass with the well-documented case of laser-induced modifications in fused silica and discuss its potential for future applications.

4:10 PM

(GOMD-167-2021) Study of Silica Glass Structural Properties Under Compression Shockwave using Reactive Force Field: Molecular Dynamics Simulation

V. Bihani[1]; A. Yadav*[1]; N. Krishnan[1]

1. Indian Institute of Technology Delhi, Department of Civil Engineering, India

Silica glasses are commonly used in the optical systems. During the manufacturing process, these glasses are subjected shockwaves due to high power lasers. However, the effect of shockwave on the structural and mechanical properties of silicate glasses remain poorly understood. Here, we perform reactive molecular dynamics simulations on silica glass to study the structural changes of silica glass when subjected to compressive shockwave. We show that the Hugoniot curves follow the trends as of experiment trends. In addition, we show that both the short-range order (coordination number, inter-tetrahedral angle, and intra-tetrahedral angle) and the medium-range order (pair distribution function and rings) are significantly affected by the shockwave propagation through the material. The structural changes behind the shock front are due to compaction of tetrahedral units, coordination defects, edge-sharing tetrahedra, and reduction of large rings. Overall, we observe that the propagation of shockwave makes the system increasing disordered, with the degree of disorder increasing with increasing shockwave velocity.

4:30 PM

(GOMD-168-2021) Densification of silica glass after grinding in planetary ball mills: Evidence of shear influence on the atomic structure

J. Struczynski[2]; S. Romeis[2]; J. Schmidt[2]; W. Peukert[2]; D. de Ligny*[1]

1. University Erlangen-Nürnberg, Materials Sciences and Engineering, Germany
2. University of Erlangen-Nuremberg, Department Chemie- und Bioingenieurwesen, Germany

During comminution process, the powder is subjected to a very large number of extreme mechanical solicitation involving crushing, shock impact and attrition. It was observed that in the case of silica glass the resulting powder has a very significantly modified atomic structure suggesting a complex interplay between compressive and shear deformation. These modifications have drastic effects on the mechanical and chemical properties of the obtained powder. Suprasil glass was comminuted at different level by using a stirred media mill. The milling time and grinding media size were modified to obtain a large range of samples at different state of modification. Morphology changes were observed with electron microscopy, density by Helium pycnometry and mass specific surface area by gas sorption measurements. The structure of the glass in these different samples was observed by Raman and IR spectroscopy. The atomic structure was found to present a lot of analogy to the modification observed in hydrostatically densified glass. However a more precise evaluation of the spectra showed that the conventionally used correlation found between density and vibrational band shift and intensity cannot be here applied. It is also shown that further relaxation of the sample does not always allow for the recovery of the pristine glass.

GOMD S2: Glass and Interactions with Its Environment - Fundamentals and Applications

Nuclear Waste Mobilization II

Room: Moresby Island

Session Chairs: Mathieu Bauchy, University of California, Los Angeles; Xiaonan Lu, Pacific Northwest National Lab

1:30 PM

(GOMD-169-2021) Understanding the Effects of Various Anions and Cations on Sulfate Retention and Solubility Within Simulated Low Active Waste Glasses

N. J. Smith-Gray*[2]; N. Stone-Weiss[2]; J. M. Lonergan[1]; E. Nienhuis[1]; J. Marcial[1]; J. McCloy[2]

1. Pacific Northwest National Lab, USA
2. Washington State University, Materials Science and Engineering, USA

In the potential composition space for Hanford low-activity nuclear waste glass (LAW), the waste loading of some formulations is limited by poor incorporation of sulfate. Production of a molten salt causes detrimental effects to a melter and the vitrification process. The aim of the present study is to understand the structural effects that various anions and cations each have upon a simplified sulfate-containing simulated LAW glass and therefore, how sulfate incorporation in the glass is impacted. Both sulfate solubility and retention were studied in carefully-designed glass formulations containing varied levels of the cations and anions of interest. To achieve a comprehensive understanding of these systems, various analytical techniques, including nuclear magnetic resonance, visible absorption and Raman spectroscopy were utilized to investigate changes in the glass structure and phase partitioning. Additionally, other techniques such as x-ray diffraction, x-ray absorption spectroscopy, and electron probe microanalysis were utilized to determine the elemental compositions and redox chemistry within the glass.

1:50 PM

(GOMD-170-2021) Alkali molybdenum iron phosphate glasses for nuclear waste immobilization

J. Bai*[1]; R. Brow[1]; C. Kim[2]

1. Missouri University of Science & Technology, USA
2. MO-SCI Corporation, USA

The chemical durability of partially crystallized iron phosphate glasses loaded with up to 40 wt% Collins-CLT waste depends on the nature of the molybdenum-rich residual glass, which can be optimized to meet DOE standards [Hsu et al.]. In the present study, series of Na- and Cs-molybdenum iron phosphate glasses were prepared to study the effects of composition and raw materials on structure and properties, to better understand the performance of these waste forms. High pressure liquid chromatography (HPLC) shows that the addition of Cs_2O and Na_2O to Mo-Fe-phosphate compositions has little effect on the phosphate anions that constitute the glass structure. The replacement of Fe_2O_3 by MoO_3 creates longer P-anions with the concomitant formation of a molybdate sub-network. Greater fractions of reduced ions (Fe^{2+} and Mo^{5+}) are present in glasses prepared with $NH_4H_2PO_4$ compared to H_3PO_4, and the speciation of Fe and Mo ions was characterized with Mossbauer spectroscopy and Raman spectroscopy, respectively. For the Mo-rich compositions, highly distorted $Mo^{5+}O_5$ sites exist in the reduced glasses, whereas isolated $Mo^{6+}O_6$ sites formed in the oxidized glasses. In general, the aqueous dissolution rates of the glasses increased with increasing alkali content, longer phosphate anions, and greater Mo/(Mo+Fe) ratios. This work was supported by the US Department of Energy, SBIR/STTR phase I&II contract DE-SC0011906.

Nuclear Waste Mobilization III

Room: Moresby Island

Session Chairs: John Vienna, Pacific Northwest National Lab; Yu Song, University of California, Los Angeles

2:50 PM

(GOMD-171-2021) Low Melting Temperature Stannous Phosphate Glass For Efficient Salt Waste Form Solution

M. Tang*[1]

1. Clemson University, Department of Materials Science & Engineering, USA

Current waste form options for salt-based high-level waste include glasses, ceramics, glass-ceramics, or glass-bonded ceramics (alternatively called glass-composite waste forms). One approach to immobilize salt waste stream is direct incorporation of the full salt with Cl and/or F containing waste forms. Due to the low solubilities (< 1.5 mass% for silicate-based glass) of chloride/fluoride ions and the evolution of Cl_2/F_2 gas at elevated temperature, it is not suitable to employ borosilicate as the host for full-salt waste streams. Phosphate glass, such as iron phosphate glass, has been explored to immobilize electrochemical salts at intermediate melting temperatures. In this study, we have successfully fabricated SnF_2-P_2O_5 glasses with low melting temperature (400 C) and loaded up to 50 wt% salt waste streams. Chlorophosphate glasses in the systems Sn–Fe–P–O–Cl and Sn–O–P–Cl have been developed via replacing fluorine by chlorine. The range of melting temperature in these glasses system is between 400-500 C or even lower which could avoid the volatilization of chloride and fluoride and improve salt retention or solubility. Low glass transformation temperatures, acceptable chemical durability, and good water resistance suggest that these glasses may be suitable for immobilizing chloride-based and/or fluoride-based salt streams through vitrification.

3:10 PM

(GOMD-172-2021) Mathematical modeling of cold cap in nuclear waste melter

P. Ferkl*[1]; P. Hrma[2]; J. Klouzek[3]; J. Marcial[1]; A. A. Kruger[4]; R. Pokorny[3]

1. PNNL, USA
2. AttainX, USA
3. UCT Prague, Czechia
4. Department of Energy, Office of River Protection, USA

Radioactive waste stored at the Hanford Site in the US will be vitrified in joule-heated melters. The rate of the glass melting during vitrification is determined by the melter design, melter operating parameters, melt and feed properties, and by the kinetics of the batch-to-glass conversion. We developed a mathematical model of a cold cap for slurry-fed electric melters, which considers a boiling slurry at the top, open porous reaction zone in the middle, and primary foam at the bottom. The cold cap properties and the temperature profiles significantly vary across the cold cap domain. The conversion kinetics model was developed for both high-level waste and low-activity waste feeds based on experimental data obtained by the feed expansion test, evolved gas analysis, thermal gravimetry, and x-ray diffraction, each conducted at several heating rates. Finally, the model was coupled to a computational fluid dynamics model of the melter.

3:30 PM

(GOMD-173-2021) Effect of Irradiation on the Structure and Properties of Borosilicate Glasses

R. Kumar*[1]; A. Jan[1]; N. Krishnan[1]

1. Indian Institute of Technology Delhi, Department of Civil Engineering, India

Borosilicate glasses are used as the base glass for nuclear waste immobilization. However, the effect of radiation on the structure and properties of borosilicate glasses remains poorly understood. Herein, we use molecular dynamics simulations to study the effect of irradiation on the structure and properties of borosilicate glasses. To this extent, we consider a series of borosilicate glasses with varying Si/B molar ratios at fixed network modifier contents. Interestingly, we observe a swelling behaviour in borate rich glasses, while silica-rich glasses exhibit densification. We demonstrate that the swelling and densification behaviour arises from two competing mechanisms—(i) decrease of the Si-O-Si bond angles, (ii) increase in the B^4 content. Finally, we observe that irradiation results in an increase in the fictive temperature of glasses. Overall, the insights from the present study can be used for optimising the compositions of borosilicate glasses to be used for nuclear waste immobilization.

3:50 PM

(GOMD-174-2021) Investigation of physical, optical and corrosion properties of nuclear waste glass simulants and the effects of gamma-ray irradiation

J. A. Jiménez*[1]; C. L. Crawford[2]

1. Augusta University, Chemistry & Physics, USA
2. Savannah River National Lab, USA

Borosilicate glass has been adopted internationally for the treatment of nuclear waste with the object of long-term stabilization through vitrification. Still, the effects of self-radiation on glass properties following radioactive element decay over thousands of years can only be simulated, e.g., exposing simulant glasses to external ionizing radiation. In this work, specimens of the International Simple Glass (ISG) system agreed upon by the international community as representative glass were as a baseline characterized with regards to structural, thermal and optical properties. An aspect of focus in the latter category has been the use of photoluminescence spectroscopy which is currently lacking but considered promising to study radiation-induced effects. Further assessed was the chemical durability through short term, static powder aqueous corrosion tests as part of the comprehensive evaluation. A comparison across all properties evaluated for pristine samples with glasses subjected to gamma-ray irradiation is ultimately pursued.

Surfaces and Coatings II

Room: Cortes Island

Session Chair: Gabriel Agnello, Corning Incorporated

1:30 PM

(GOMD-175-2021) Analyzing the Outer Atomic Layer using Low Energy Ion Scattering (LEIS) (Invited)

T. Grehl*[1]

1. IONTOF GmbH, Germany

The surface of a material, especially the outer atomic layer, is what interacts with the environment. This outer layer is first subject to corrosion, heavily modified in contact to production chemicals and determines the film growth when adding coatings. It is also known that the outer layer of glass differs heavily in composition from the bulk. Various surface analytical techniques like XPS and (ToF-)SIMS are widely applied to glass, but Low Energy Ion Scattering (LEIS) is the only technique that can quantitatively asses the elemental composition of the outer atomic layer. In addition, it non-destructively determines the depth distribution of heavier elements over the first few nm of the sample. This presentation will give a thorough introduction to LEIS and illustrate its key advantages (single atomic layer sensitivity, non-destructive in-depth analysis, quantification) in the context of glass and related materials. Analytical examples will demonstrate how the composition of the outer layer is determined, how it changes with the chemical environment, and how film growth can be optimized especially by controlling the nucleation phase.

2:00 PM

(GOMD-176-2021) Low Energy Ion Scattering for Quantification of Surface Density of Silanols on Fused Silica Glasses

T. Gholian Avval*[1]; S. Prusa[2]; H. Brongersma[3]; M. R. Linford[1]

1. Brigham Young University, Chemistry and Biochemistry, USA
2. Brno University of Technology, Mechanical Engineering, Czechia
3. Calipso, CEO, Netherlands

Glass surface chemistry directly affects device performance in cell phones. Surface hydroxyls (silanols, SiOH) are among the most polar and reactive sites on glass surfaces and play critical role in glass surface modification and coatings. Knowledge of the level of silanol formation following various surface pretreatments is important for the application of thin film coatings. There have been multiple attempts to quantify the surface density of these functional groups on flat surfaces. Sensitivity to the outermost atomic layer is required to do such measurements accurately. Here we present a tag-and-count approach that consists of coupling atomic layer deposition(ALD) with low energy ion scattering(LEIS) to determine the density of surface silanols on chemically and thermally treated fused silica. One ALD cycle with dimethylzinc(DMZ) or diethylzinc(DEZ) and water was used for tagging. The coverage of Zn on the surfaces was then determined using He^+ and Ne^+ as probe ions in LEIS. Surfaces that were treated with DMZ showed slightly higher Zn coverages than for DEZ. The SiOH density reported here for samples treated at 200 °C is in agreement with the accepted value for silica

(4.6 OH/nm^2). The general methodology presented here should be useful for understanding surface composition, surface cleanliness, and industrial cleaning procedures on glass surfaces.

2:20 PM

(GOMD-177-2021) Dispersion-derived Tribological Coatings to Enhance Scratch Resistance of Glasses

S. Sahoo*[1]; O. P. Khatri[2]; N. Krishnan[3]; N. N. Gosvami[1]

1. Indian Institute of Technology Delhi, Materials Science and Engineering, India
2. CSIR - Indian Institute of Petroleum, India
3. Indian Institute of Technology Delhi, Civil Engineering, India

Glasses, being brittle, are constantly subjected to scratch-induced damages during their service life. Such damage can significantly deteriorate the integrity of the glass surface, leading to catastrophic failure of the system. Here, we show that the interfacial friction-induced deformations in glasses during sliding in an aqueous medium can be significantly minimized using nano-dispersions. To this extent, the self-generation of protective tribological coatings during the sliding of a spherical alumina probe over fused silica glass surface in aqueous nano-dispersion environment at ambient temperatures is investigated in this study. Such lubricious tribo-coatings are observed to profoundly reduce the frictional forces at the sliding contact and limit the shear-induced tensile stresses determining the initiation of partial Hertzian cone cracks. Micro-Raman spectroscopy analysis confirms the localized deposition of dense dispersion-derived films within the scratch track. Furthermore, the AFM analysis of the tracks reveals the thin and lubricious disposition of the films coating the scratched region. The optical and SEM micrographs display clear differences in the crack density and severity of sliding with and without the presence of the tribo-coating. The results provide new insights into the scratch-induced surface damage mechanisms of glasses in the presence of aqueous nano-dispersions.

2:40 PM

(GOMD-178-2021) Subsurface Structural Change of Silica upon Nanoscale Physical Contact: Chemical Plasticity beyond Topographic Elasticity

S. H. Kim*[1]

1. Pennsylvania State University, Chemical Engineering & Materials Science, USA

Once manufactured into functional shapes, glass materials will be physically contacted with foreign materials during storage, transport, installation, and in-service. Such contacts can introduce extrinsic defects to the surface. Then, the in-service optical properties and mechanical strengths may be severely deteriorated by the extrinsically-introduced surface defects. Thus, it is of paramount importance to understand what structural changes are made in the glass network by the physical contact with foreign objects. This talk addresses the subsurface structural changes of silica upon nanoindentation and nanoscratch using nanoscale infrared spectroscopy and reactive molecular dynamics simulations. The results reveal an elongation of the Si-O bond length distribution even after the topographically-elastic contact, indicating a "chemical plasticity" at the sub-Angstrom level. In the plastic region with subsurface densification, the Si-O bond is found to be longer than the pristine region, indicating the decrease in molar volume is accompanied with the elongation, not shortening, of the Si-O bond. These findings provide unprecedented insights into the sub-Angstrom level alterations of chemical bonding environments of a glassy material upon physical contact that cannot be delineated based on the topographic deformation of the surface.

Surfaces and Coatings III

Room: Cortes Island

Session Chair: Matthew Linford, Stanford University

3:30 PM

(GOMD-179-2021) A mechanistic approach to understanding the effects of exposure to variable acidic treatments on the intrinsic properties of flat glass surfaces (Invited)

G. Agnello*[1]; D. Cabrera[1]; C. Simons[1]; N. J. Smith[1]; J. Banerjee[1]; C. V. Cushman[1]; A. Antony[1]; H. Park[1]

1. Corning Incorporated, USA

Acids, both by themselves as well as in endless different mixtures, have been used to treat glass surfaces for ages. The purpose(s) of these treatments can vary significantly; from adding a purely decorative aesthetic, to improving the sense of surface touch/feel for a user, to modifying surface property(s) for functionality enhancement. Somewhat surprisingly, detailed studies that examine the relationships between variations in chemical treatment, such as relative concentration(s) of modifying molecule/ion(s), and resulting physical/chemical surface states are sorely lacking. Furthermore, reported investigations of the potential connections between these types of relationships and the ionic/molecular equilibrium concentrations in the as-prepared treatments, which should determine the dominant dissolution mechanism(s) of the solutions themselves, are equally scarce. Here, we explore the relationship between carefully varied HF-based treatment chemistries and resulting glass surface properties through the lens of dominant dissolution mechanism(s). Through a combination of experimental and theoretical methods, we show that shifts in surface properties can be at least qualitatively related to the interplay of different dissolution mechanisms present based on starting treatment chemistry and perhaps may even be directly attributed to them

4:00 PM

(GOMD-180-2021) Structural analysis of sputtered amorphous silica thin films a Raman spectroscopy investigation

S. ben Khemis*[1]; L. Cormier[1]; E. Burov[2]; H. Montigaud[2]; E. Gouillart[2]

1. Sorbonne Université / UPMC, IMPMC, France
2. Saint Gobain Recherche, Laboartoire Mixte Saint-Gobain/CNRS, France

Owing to its chemical and physical properties, silicon dioxide is a well-known thin-film material in the semiconductor industry. Moreover, they have found application in large area of active coating systems on glass substrates, as anti-reflective coating, diffusion masks, hard masks... Silica film performances are governed by the deposition process parameters that may affect its microscopic structural features. Therefore, there is much technological and fundamental interest in gaining a deep understanding of the structural properties of the silica film. However, unlike bulk glass, the structural analysis of amorphous thin films, especially deposited on a glass substrate, is very challenging due to the weak signal of the film disordered materials and its small thickness, causing an overlap of film signal with the one of the vitreous substrate. In this presentation, we report an advanced structural investigation (Raman, PDF) of sputtered silica films deposited on a glass substrate and doped with different amounts of aluminum. This investigation is based on different methodologies to extract characteristic silica film signal. Thanks to these latter, interesting structural properties of the sputtered silica film have been brought to light. Particularly, we will be demonstrated that the PVD-sputtered silica films exhibit different structural properties compared to fused silica glass.

4:20 PM

(GOMD-181-2021) Ab initio modeling of the electronic structure at multicomponent silicate glass surfaces

A. Antony*[1]; S. Goyal[1]; D. Thelen[1]; R. G. Manley[1]

1. Corning Incorporated, USA

Contact triboelectrification occurs when two materials are brought together, charge is transferred from one surface to the other, and some residual charge remains after separation. Based on surface state theory, electronic states present within the apparent insulating band gap are responsible for electronic charge transfer between glass and other materials. This presentation uses atomistic simulations and density functional theory to model electronic structure of glass surfaces and quantify properties, including work function, insulator band gap, and electronic density of states. Silica, TiO_2-SiO_2, and CaO-Al_2O_3-SiO_2 compositions are considered and we discuss how local atomic environments and hydroxylation states influence the electronic structure of these surfaces. Lastly, atomistic models of dehydrated and hydrated calcium aluminosilicate surfaces are developed to gain insight into how electronic surface properties can vary in humid environments.

GOMD S3: Optical and Electronic Materials and Devices - Fundamentals and Applications

Laser Interactions with Glass II

Room: Saturna Island

Session Chair: Maxime Cavillon, University Paris Saclay

1:30 PM

(GOMD-182-2021) Effect of polarization and surface effects on the orientation and growth of Sb_2S_3 crystals on the surface of Sb-S-I glass

C. Au-Yeung*[1]; D. Savytskii[1]; K. J. Veenhuizen[2]; V. Dierolf[1]; H. Jain[1]

1. Lehigh University, USA
2. Lebanon Valley College, Physics, USA

Laser heating of chalcogenide glasses has successfully been used to fabricate rotating lattice single crystals. To increase the utility of these crystals for use in opto-electronic and energy storage applications, the lattice orientation of the initial crystal seed and the subsequent rotation of the lattice must be controlled. To realize this goal, a spatial light modulator is used to configure the beam shape, thermal gradient, and/or the polarization of the laser. Using the spatial light modulator to fabricate Sb_2S_3 seed crystals on the surface of Sb-S-I glass, we observe a preferred orientation as well as an anisotropic crystal lattice rotation rate. It has been found that polarization has the most significant influence on the seed orientation and can be used as an effective tool for controlling the nucleation and therefore the orientation of the Sb_2S_3 crystals. Additionally, this orientation of the crystal can influence the rotation rate of the crystal based on the inclusion of dislocations. This presentation will discuss the implications of the effects of the experimental parameters on the orientation and rotation rate of Sb_2S_3 crystals.

1:50 PM

(GOMD-183-2021) Effect of Glass Composition on the Laser-Induced Nucleation and Growth of Lithium Niobate Crystals in Lithium Niobosilicate Glass

K. J. Veenhuizen*[1]; C. Barker[2]; J. Franklin[1]; S. McAnany[3]; D. Nolan[4]; B. Aitken[4]; V. Dierolf[2]; H. Jain[3]

1. Lebanon Valley College, Physics, USA
2. Lehigh University, Physics, USA
3. Lehigh University, Materials Science and Engineering, USA
4. Corning Incorporated, USA

Spatially selective growth of lithium niobate ($LiNbO_3$) crystals deep within lithium niobosilicate (LNS) glass can be achieved via femtosecond laser irradiation of the glass, wherein these crystals could serve as active optical elements in photonic integrated circuits. In practice, it is challenging to achieve continuous, single-crystal growth of $LiNbO_3$ within LNS glass as would be desired for practical applications. This work reports on progress in overcoming this challenge by exploring how the formation of single crystal $LiNbO_3$ is affected by the incongruent composition of the glass matrix. An investigation of the parameter space of glass composition ((100-x) $LiNbO_3$ - x SiO_2 for x = 22, 26, 30, and 34), laser scanning speed, and laser power leads to the observation that the critical growth rate for the $LiNbO_3$ single crystal, above which only polycrystalline growth occurs, can be controlled by glass composition. At this critical growth rate, the ability to grow cm-long crystal lines is demonstrated. Recording the time of formation of crystals dots in the different glass compositions leads to the conclusion that the crystal nucleation rate can be tailored by the glass composition, with a lower fraction of glass former yielding higher nucleation rates.

2:10 PM

(GOMD-184-2021) Recent advances in Mid-Infrared germano-gallate glasses: Fiber drawing and Direct laser writing

T. Guerineau*[2]; R. Ferreira Falci[2]; J. Lapointe[2]; J. Lafreniere-Greig[2]; L. Loi[3]; L. Canioni[3]; Y. Petit[3]; R. Vallée[2]; T. Cardinal[1]; Y. Messaddeq[2]

1. CNRS - ICMCB, France
2. Université Laval, Center for Optics, Photonics and Lasers, Canada
3. University Bordeaux, CELIA, France

The quest for groundbreaking technologies in integrated optics faces many challenges to achieve efficient and compact photonic systems. To this issue, femtosecond direct laser writing (DLW) in Mid-IR transparent vitreous materials, such as barium-germanium-gallium glasses (BGG), reveals a high-potential approach. By tailoring BGG glass composition and doping, DLW has allowed for controlling the type of photo-structurations embedded inside the glass. In pristine BGG glass, a DLW investigation has been performed in both gallium- and germanium-rich glasses. While no difference in the DLW-induced glass structure modification was reported in both glass families, an extended range of Type I refractive index modification has been found in the gallate-rich BGG. In silver-doped BGG glasses, two laser-induced modification regimes have been highlighted depending on the glass structure, both based on the photochemistry of Ag^+ ions. At low content of non-bridging oxygens (NBO), DLW leads to the formation of micrometric single-track refractive index change with a double-track fluorescence. Meanwhile, at high content of NBOs, DLW leads to a double-track fluorescence, followed by nanometric double-track refractive index change. Finally, fiber drawings of built-in-casting core-clade BGG preforms have been demonstrated, emphasizing the high potential of BGG materials for photonic applications.

2:30 PM

(GOMD-185-2021) Investigating femtosecond laser interaction with tellurite glass family

G. Torun*[1]; T. Kishi[3]; D. Pugliese[2]; D. Milanese[4]; E. Descrovi[2]; Y. Bellouard[1]

1. EPFL, Switzerland
2. Politecnico di Torino, Department of Applied Science and Technology and RU INSTM, Italy
3. Tokyo Institute of Technology, Department of Chemistry and Materials Science, Japan
4. Università di Parma, Department of Engineering and Architecture and RU INSTM, Italy

Focusing ultrafast laser pulses induce localized permanent structural modifications on the surface or in transparent materials, that are of particular interest for photonic applications. Among the materials of interest, the tellurite glass family is attractive for near-infrared and photonics applications due to its broad-transparency window and high optical nonlinearity. Here, we systematically investigate structural changes occurring in various TeO_2-based glasses exposed to femtosecond laser with various laser parameters. Remarkably, in a regime where heat accumulated after successive pulses, we observed the formation of polarization-controlled self-organized patterns expanding well beyond the focal volume, suggesting the presence of an evanescent coupling mechanism enhancing the self-organization. In addition, our results, obtained with compositional elemental analysis coupled with Raman spectra suggest different ion migration mechanisms in the laser affected zone at the surface and inside the glass. The formation of crystalline tellurium (t-Te) from glass structural units due to photo-induced elemental dissociation was observed only at the surface. The formation of ultrathin layer of crystalline tellurium offers the possibility to explore structural transitions in two-dimensional (2D) glasses by observing changes in the short- and medium- range structural orders, induced by spatial confinement.

2:50 PM

(GOMD-186-2021) Sub-micrometric femtosecond laser structuring in silver-free and silver-containing gallo-germanate glasses

R. Zaiter*[1]; T. Guerineau[1]; T. Cardinal[1]; Y. Petit[2]; B. Sapaly[3]; J. Harb[2]; L. Canioni[2]; M. Lancry[3]

1. Institut de Chimie de la Matière Condensée de Bordeaux (ICMCB) - Université de Bordeaux, Centre National de la Recherche Scientifique : UPR9048, France
2. Centre d'Études Lasers Intenses et Applications (CELIA) – Centre National de la Recherche Scientifique : UMR5107, Commissariat à l'énergie atomique et aux énergies alternatives : DAM/CESTA, Université Sciences et Technologies - Bordeaux 1, France
3. Institut de Chimie Moléculaire et des Matériaux d'Orsay (ICMMO) – Université Paris Saclay, France

Femtosecond Direct Laser Writing technique (FDLW) has been widely used during the last two decades to perform 3D photo-structuring in glasses such as for PIC (Photonic Integrated circuits). The controlled formation of the photo-induced modifications allows developing remarkable optical properties such as birefringence, photoluminescence, 2nd and 3rd harmonic generation. The understanding of correlations between the glass structure, the laser induced process and the resulting properties remains a tremendous challenge. To explore new opportunity in terms of applications and wavelength ranges, novel vitreous materials need to be explored. Among the different oxide glasses, heavy metal gallo-germanate oxide glasses have attracted attention and lead to a growing interest since they could combine local functionality, extended transparency window in the mid-IR up to 5.5 μm, chemical durability and robust mechanical properties. We report on the formation and structural evolution of embedded self-organized, polarization-dependent nanogratings in gallo-germanates induced by femtosecond laser pulses. Micro-Raman spectroscopy will be used for establishing local structure property relationship. Micro-fluorescence, micro-absorption, phase-contrast imaging, and confocal fluorescence imaging of the FDLW structure will be discussed.

Laser Interactions with Glass III

Room: Saturna Island

Session Chair: Theo Guerineau, University Laval

3:30 PM

(GOMD-187-2021) Formation of Nanoparticles in Glass Under Femtosecond Laser Irradiation

C. Barker*[1]; S. McAnany[2]; K. J. Veenhuizen[3]; D. Nolan[4]; B. Aitken[4]; V. Dierolf[1]; H. Jain[5]

1. Lehigh University, Physics, USA
2. Lehigh University, Materials Science & Engineering, USA
3. Lebanon Valley College, Physics, USA
4. Corning Incorporated, USA
5. Lehigh University, International Materials Institute for New Functionality in Glass, USA

Formation of metal nanoparticles (NP) such as of gold, copper and silver in silicate glass via dissolution of metal salt at high temperature followed by precipitation at low temperature has been known to produce ruby color uniformly for millennia. This method exploits the strong temperature dependence of the solubility and stability of metal ions under near equilibrium conditions. By contrast, recently NPs have been produced in spatially selective regions deep inside the glass by fs laser irradiation under highly nonequilibrium conditions. We report on the formation of Ge NP in La_2O_3-B_2O_3-GeO_2 glass by this method using Pharos 1026 nm wavelength laser focused 90 μm below the glass sample surface using a 50x, 0.6 NA objective lens with a pulse width of 175 fs, and a repetition rate of 200 kHz. It is shown that such NPs form only within a narrow range of pulse power of 2-3 μJ. Structural changes in the glass are characterized for various laser irradiation conditions by X-ray absorption near edge structure, Raman spectroscopy, and electron microscopy. Likely mechanisms for the formation of NPs in glass by fs laser irradiation will be presented.

3:50 PM

(GOMD-188-2021) Structure-Terahertz Properties-Ultrafast Laser Irradiation Correlations in Borosilicate, Tellurite, and Chalcogenide Glasses

N. Tostanoski*[1]; S. K. Sundaram[1]

1. Alfred University, USA

We have studied Na_2O WO_3 TeO_2, La_2O_3 WO_3 TeO_2, and Na_2O B_2O_3 SiO_2 borosilicate and tellurite glass systems, along with additional chalcogenide glass systems, with their chemistries systematically varied within the glass forming region, changing the glass structure, which affects their terahertz properties. We prepared the glass samples via melting and casting, followed by annealing. The glasses were then characterized using differential scanning calorimetry (DSC), x-ray diffraction (XRD), ultraviolet-visible (UV-Vis), Fourier transformation infrared (FTIR), Raman spectroscopy, and time-domain terahertz spectroscopy (THz-TDS). In addition, ultrafast laser pulses (e.g., ~ 40–100 femtoseconds) were used to irradiate these glass samples, modifying the local glass structure and their optical properties. Glass structural changes made via laser irradiation are correlated to terahertz properties of these glasses. Linear correlation between optical and THz refractive indices will be reported for the glass systems. Any deviation from linearity and its origin will be presented.

4:10 PM

(GOMD-189-2021) Gallium Telluride Phase Change Materials for Nonvolatile Memory and Neuromorphic Computing

E. Bychkov*[1]; A. Tverjanovich[4]; C. J. Benmore[3]; M. Khomenko[2]

1. University of Littoral and IPLIT, Russian Academy of Science, LPCA – EA 4493, France
2. Russian Academy of Science, IPLIT, Russian Federation
3. Argonne National Lab, APS, USA
4. St Petersburg University, Chemistry, Russian Federation

Binary Ge–Te and ternary Ge–Sb–Te systems belong to flagship phase-change materials (PCMs) and are used in nonvolatile memory applications and neuromorphic computing. The working temperatures of these PCMs are limited by low-T glass transition and crystallization phenomena. Promising high-T PCMs may include gallium tellurides; however, the atomic structure and transformation processes for amorphous Ga–Te binaries are simply missing. Using high-energy X-ray diffraction supported by first-principles simulations, we elucidate the short- and intermediate-range order in bulk glassy Ga-Te and thin film Ga_2Te_3, following their thermal, electric, and optical properties, revealing a semiconductor–metal transition above melting. We also show that a phase change in binary Ga–Te is characterized by a very unusual nanotectonic compression with the high internal transition pressure reaching 4–8 GPa, which appears to be beneficial for PCM applications increasing optical and electrical contrast between the SET and RESET states and decreasing power consumption.

4:30 PM

(GOMD-190-2021) A method for mapping localized changes in the oxidation state of elements in glass with sub-micron resolution

S. McAnany*[1]; A. Kiss[2]; J. Thieme[2]; D. Nolan[3]; B. Aitken[3]; V. Dierolf[4]; H. Jain[1]

1. Lehigh University, Materials Science & Engineering, USA
2. Brookhaven National Laboratory, NSLS II, USA
3. Corning Incorporated, USA
4. Lehigh University, Physics, USA

X-ray absorption near edge structure (XANES) spectroscopy is a technique employed for investigating the chemistry and oxidation state of elements irrespective of the glassy or crystalline state of the material. Recent developments in synchrotron radiation instrumentation have opened the potential of performing such measurements with sub-micron spatial resolution. Consequently, it is possible to create maps of a sample to locally discern the oxidation state of the elements. In this work, the benefits of this methods were explored with a case study of femtosecond laser heating of La_2O_3 - B_2O_3 - $2GeO_2$ glass. Using this X-ray absorption spectroscopy mapping technique, we demonstrated unequivocally that the application of femtosecond laser heating to the glass causes a local reduction of Ge^{4+} to Ge^0. Pros and cons of the current method are compared with other X-ray and electron beam based techniques available to obtain similar information about the chemistry of solids.

Tuesday, December 14, 2021

GOMD Posters

Room: Regency A/B

5:30 PM

(GOMD-SP041-2021) Thermal Conductivity of Zeolitic Imidazolate Framework Glasses

S. S. Sørensen*[1]; M. B. Østergaard[1]; M. Stepniewska[1]; H. Johra[2]; Y. Yue[1]; M. M. Smedskjaer[1]

1. Aalborg University, Department of Chemistry and Bioscience, Denmark
2. Aalborg University, Department of the Built Environment, Denmark

Generally, crystals are known to have higher thermal conductivity than their isochemical glasses due to stronger phonon scattering in the latter. In this work, we report the inverse relation for a recently discovered family of zeolitic imidazolate framework (ZIF) glasses by both experiments and reactive molecular dynamics (MD) simulations. That is, we find that the studied ZIF-4 and ZIF-62 glasses possess higher thermal conductivity than their isochemical crystals. We ascribe the effect to the anomalous density increase of the ZIF systems upon vitrification combined with the strong phonon scattering in the crystalline ZIFs. The effect is further verified by simulating a ZIF-8 glass and its corresponding and highly porous crystal. Finally, we probe the phonon characteristics by MD simulations, showing that the low-frequency modes are the main contributors to heat conduction.

(GOMD-SP042-2021) Boron-11 MAS-NMR Study of Superstructural and Ionic Groupings in the Lithium Borate Glass System

R. M. Wilson*[1]; I. Slagle[1]; N. Barrow[2]; A. Ruckman[1]; G. Beckler[1]; W. Guthrie[1]; B. Royle[1]; D. Shephardson[1]; M. Boyd[1]; M. R. Jesuit[1]; C. B. Bragatto[1]; S. Feller[1]

1. Coe College, Physics, USA
2. Johnson Matthey, United Kingdom

Using the program DMFit, we have fit Boron-11 MAS-NMR spectra obtained at 14.1 T for both the three-coordinated and four-coordinated borons. This was done over a wide compositional range of lithium borate glasses ($x = 0.1$ to 0.65, where x is the molar fraction of lithium oxide). Additionally, we have used Monte Carlo error analysis to not only estimate the error of each of our fitting parameters from the quadrupole interaction, but also to further visualize and understand the fits from larger groupings including triborate and diborate units. We have found that the Boron-11 data agrees with previous work with Boron-10 NMR, and shows quantitative evidence for ring and non-ring borons as well as ionic units such as pyroborate and orthoborate triangles. Also, the fraction of four-coordinate borons is in good agreement with the literature. We would like to thank the National Science Foundation for funding under NSF grant DMR-1746230.

(GOMD-SP043-2021) Comparisons of atomic arrangements in binary borate glasses with oxygen and modifier packing fractions

S. Weiss*[1]; I. Slagle[1]; S. Feller[1]

1. Coe College, Physics, USA

Packing Fractions provide an important way of looking at glass density. By calculating the packing efficiency of atoms in a glass, we can infer the atomic structures present at various glass compositions. Our focus has been on connecting the overall packing, the oxygen packing, and the modifier packing for various binary borate glasses to the fraction of four-coordinated borons. We propose that the presence of four-coordinated borons can be shown most directly through the oxygen packing. Also, we looked at the maximal random packing for a collection of two dissimilar spheres as a model for borate glass structure. This research was supported by the NSF under grant DMR-1746230.

(GOMD-SP044-2021) Spectroscopic study of the structure of alkaline alumino-silicate glasses

J. L. Hunt*[1]; D. Möncke[1]; R. Youngman[2]; A. Herrmann[1]

1. Alfred University, USA
2. Corning Incorporated, USA

Understanding the structure of alumino-silicate glasses is important for many applications. Changes in the coordination of aluminate polyhedra, connectivity and the degree of polymerization of the network impact the materials properties and coordination sites of dopants, such as rare earth ions and consequently their photo-luminescing properties. The structure of two alkaline earth alumino-silicate glass series is studied by infrared and Raman spectroscopy. The glasses contain either barium or magnesium oxide as modifier component. Four glasses have been prepared for each series with MO: Al_2O_3 ratio of 15:15, 20:20, 30:10, and 45:5 the third component making the difference to 100 is SiO_2 in mol%. The aluminum coordination was quantified by 27Al-NMR. The high field strength of Mg^{2+} compared to Ba^{2+} has a significant impact on the aluminum speciation as well as on the number of non-bridging oxygen ions in the glasses and therefore also the glass properties.

(GOMD-SP045-2021) Structures and optical properties of silica glass hot-compressed in solid medium

Y. Tanabe*[1]; M. Jeem[1]; M. Fujioka[1]; K. Akatsuka[2]; S. Kohara[3]; M. Ono[1]; J. Nishii[1]

1. Hokkaido University, Research Institute for Electronic Science, Japan
2. AGC CO.LTD, Japan
3. National Institute for Materials Science (NIMS), Japan

Silica glass is an indispensable material for optical communications and high-power laser applications due to its favorable optical properties. Over the last 30 years, the optical transmission loss of silica glass has been minimized by continuous efforts toward decreasing density fluctuations by reducing fictive temperature, e.g., through improvements in processing or doping. However, the rate of loss reduction by conventional methods becomes lower in the past few years. Thus, new methods are highly desired. Recently, it was suggested that shrinkage of structural voids (empty spaces) by hot compression under 0.2 GPa at 2000 K was a promising way to further decrease the loss. However, there are no systematic studies of loss by hot compression higher than 0.2 GPa at various temperatures. In this work, a hot-compression technique was used to study silica glass in high temperature (above 1223 K) and high pressure ($0.2 < P < 4.0$ GPa) conditions. For this purpose, hydrostatic pressure in a solid medium is applied to the glass using a multi-anvil cell and its vitrification behavior is investigated. To reveal the structures of hot-compressed glasses, High-energy X-ray diffraction technique was applied. We calculated glass structural models using initial MD model and X-ray pattern by Reverse Monte Carlo modelling method. The influence of the structure on the optical properties will be finally discussed.

(GOMD-P046-2021) Densification of oxide glasses at different time-temperature-pressure routes

X. Ren*[1]; M. Bockowski[2]; M. M. Smedskjaer[1]; L. R. Jensen[3]; R. Youngman[4]

1. Aalborg University, Denmark, Department of Chemistry and Bioscience, Denmark
2. Institute of High-Pressure Physics, Poland
3. Department of Materials and Production, Denmark
4. Science and Technology Division, USA

Hot compression is an interesting method for changing the structure and properties of oxide glasses, and it can also help to address some of the unanswered questions regarding the nature of the glassy state through temperature-pressure studies. In this work, we prepared sodium aluminoborate (NAB) and sodium aluminoborosilicate (SNAB) glasses and subjected these to various hot compression treatments, i.e., various pressure/temperature/time paths were explored to access different glass states. We then determined the density of the permanently densified samples and characterized the glass structure using Raman and NMR spectroscopy. At the same time, the influence of hot compression under different conditions on the glass mechanical properties was also explored through indentation experiments.

(GOMD-SP047-2021) Mechanical testing and void structure analysis of irradiated quartz using molecular dynamics simulation

Y. Hakozaki*[1]; T. Ohkubo[1]; I. Maruyama[2]; K. Murakami[3]; K. Suzuki[4]

1. Chiba University, Japan
2. Nagoya University, Japan
3. Nagaoka University of Technology, Japan
4. Mitsubishi Research Institute, Inc., Japan

The life-time of concrete materials constructing the nuclear power plant have to be accurately assessed for the long-term operation. The α-quartz in concrete was expanded and amorphized by neutron irradiation, and the expansion governs the life-time. However, the relationship between the void in amorphized α-quartz and its mechanical properties has not been well understood. Therefore, we analyzed the irradiation-induced voids related to the mechanical properties using molecular dynamics (MD) simulations. Void analyses were performed through Voronoi decomposition based on the atomic configurations. Under the assumption that the atoms were spherical, a void was defined as a sphere centered on a Voronoi node. Micro voids defined as a pore diameter of 1.2 Å, which did not exist in α-quartz, were found in the amorphized α-quartz. These micro voids were distributed in the amorphized α-quartz at a volume fraction of 0.005 %, became an origin of extended cracks for further strain. As a result, fracture stress of amorphized α-quartz was lower than that of α-quartz and silica glass. On the other, the failure strain of amorphized α-quartz was higher than that of α-quartz and silica glass with a factor of 1.2 and 1.5. This is due to preventing propagation of crack owing to the micro voids in amorphized α-quartz.

(GOMD-P048-2021) Glassy primer induced fracture-toughness enhancement in polymer-metal system

B. Han[1]; T. Hao[1]; J. Boerio*[2]; Z. Hossain[1]

1. University of Delaware, USA
2. University of Cincinnati, USA

Silica-like layers were deposited using atmospheric pressure plasma polymerized HMDSO onto titanium substrates with film thicknesses between 30 and 300 nm. The glassy primer layer, along with 3-aminopropylsilane coupling agent slowed the cathodic delamination rate of polyurethane on titanium by a factor of 4 compared with the one without the adhesion promotion layer, and also greatly increased the initial fracture toughness. For the unprimed control surface, before CDL exposure, XPS results indicated the DCB specimens had either mixed mode failure at the metal/polyurethane interfaces or cohesive failure in the polyurethane nearthe titanium surface. After CDL exposure, the control samples failed interfacially at the titanium/polyurethane interface. For the primed samples, before CDL exposure, XPS results indicated the failure was interfacial at the PU/silica interface. After the exposure, the failure occurred cohesive in silica near the titanium surface. This work also uses phase field modeling of fracture toughness to simulate the mechanical behavior of the PU/silica/APS/Ti system and correlate it with the experimental result. The goal of the simulation was to establish that the fracture toughness of varying adhesion layer thicknesses could be simulated, allowing for such models to be used in future primer development.

(GOMD-P049-2021) Effects of indenter tip stiffness and sharpness on deformation and cracking in oxide glasses

J. Christensen*[1]; M. M. Smedskjaer[1]

1. Aalborg University, Department of Chemistry and Bioscience, Denmark

Instrumented indentation is a common technique for evaluation of mechanical properties of glasses. In this work, we systematically investigate the effect of the indenter tip properties on the indentation response for three glasses in the sodium borosilicate system. For indentation with pyramidal diamond tips of different angles between the faces, we find that the tip sharpness does not affect the measured hardness at high loads, whereas it affects the extent of the indentation size effect observed at lower loads. Furthermore, the propensity for crack initiation and the amount of permanent densification of the glass are greatly affected by the tip sharpness. Moreover, the propensity for crack initiation and the amount of densification are correlated, supporting the understanding of densification as a mechanism that reduces residual stress. For indentation with tips of the Vickers geometry, we observe a large effect of the tip material (diamond vs. silicon carbide vs. fused quartz) on, e.g., the propensity for crack initiation and ascribe this behavior to differences in tip stiffness. These findings are relevant for understanding the relationship between laboratory indentation testing and the real-life damage and wear of glasses.

(GOMD-P050-2021) Influence of Post-indentation Recovery on the Hardness of Oxide Glasses

K. S*[1]; N. N. Gosvami[1]; N. Krishnan[2]

1. Indian Institute of Technology Delhi, Department of Materials Science and Engineering, India
2. Indian Institute of Technology Delhi, Department of Civil Engineering, India

Oxide glasses are used in various applications like decorative items, laboratory glass wares, and nuclear waste immobilization owing to their transparency, chemical stability, and high hardness. Hardness is one of the fundamental properties that are measured from the indentation response of oxide glasses. However, hardness depends on various factors such as method, measurement conditions, indentation load, tip geometry, etc. Herein, we evaluate the hardness of three oxide glass samples, namely, pure silica (0B), borosilicate (37B), and sodium borate (75B) using Vickers and nanoindentation. We observe that the annealing of the aforesaid glass samples after polishing has least influence on the hardness and modulus values. Further, nanoindentation is not able to capture the full extent of elastic recovery in these glasses and thereby underestimating the hardness values. We present that the combination of nanoindentation and AFM imaging is necessary for capturing the complete elastic recovery and thereby estimating the true hardness values of these glass samples from their respective impression depth profiles. Altogether, we demonstrate that post-indentation elastic recovery may play a significant role in dictating the hardness of glasses.

(GOMD-P051-2021) Assessing Scaled Melters for Nuclear Waste Vitrification

D. Dixon*[1]; M. Hall[1]; T. Jin[1]; C. Stewart[1]; D. Cutforth[1]; J. Lang[1]; W. Eaton[1]

1. Pacific Northwest National Lab, USA

The Hanford Waste Treatment and Immobilization Plant (WTP) will process and stabilize nuclear waste stored in tanks on the Hanford Site. At the WTP, the tank waste will be combined with glass-forming additives to make a slurry called melter feed, which can be vitrified in joule-heated melters. Two small-scale melter systems have been designed by PNNL to investigate specific aspects of the nuclear waste vitrification process. The continuous laboratory-scale melter (CLSM) uses an externally heated, metal vessel designed to operate under radioactive conditions or with non-radioactive simulants. This gives it the ability to vitrify actual tank waste and gain information about melter feed processability, the partitioning of components of interest, and validation of simulants used in testing. The research-scale melter (RSM) may use a joule-heated, ceramic lined vessel to gain glass production information or an externally heated design connected to a prototypically designed offgas system for unit analysis. A series of runs with simulated low-activity waste streams were performed in the CLSM, the results of which were compared to each other and to runs with radioactive waste streams. A high chromium, high-level waste simulant was run in the RSM and offgas was sampled during processing for production rate calculations. The mass balance results from both systems are described to give a full picture of scaled melter capabilities.

(GOMD-P052-2021) Er doped TeO_2-$PbCl_2$-WO_3 glasses for temperature sensing

R. Yatskiv*[1]; P. Kostka[2]; J. Grym[1]; J. Zavadil[1]

1. Institute of Photonics and Electronics, Czech Academy of Sciences, Czechia
2. Laboratory of Inorganic Materials, Institute of Rock Structure and Mechanics AS CR, Czechia

Precise temperature measuring is very important in many industrial and scientific applications. The optical temperature sensor (OTS) offers several advantages over traditional temperature measurement devices, in particular suitability for use in different environments, electromagnetic passivity, and the range of temperatures for which an optical sensor can be used. The Er ions, which possess thermally coupled energy levels of H and S, could be used as suitable activators in OTS. Apart from strong green emission (GE) intensity, Er ions can also provide sufficiently intense up-conversion GE under 980 nm excitation. This fact allows to use a low-cost NIR laser as an excitation source for sensing applications. Tellurite glasses are highly efficient materials for RE ion incorporation and are suitable for the development of efficient up-conversion materials. The sensing properties of tellurite glasses containing for example TiO, ZnO, etc. as modifiers have been recently reported for temperatures higher than 300 K. Preparation of new stable glasses and study of their thermal sensing properties in temperature ranges other than near RT is very important. We will demonstrate that TeO -PbCl -WO glasses doped with Er can be used for OTS also for temperatures starting from 100 K. They show promising temperature sensing properties and can be good candidates for non-contact optical temperature sensor applications.

(GOMD-SP053-2021) Impact of modifier-rich coatings on the ionic transport in silicate glasses

C. Nieves*[1]; M. Yuan[2]; E. Furman[3]; M. Lanagan[3]; P. Clem[1]

1. Sandia National Laboratories, USA
2. Pennsylvania State University, Materials Science and Engineering, USA
3. Materials Research Institute, Pennsylvania State University, USA

Silicate glasses are excellent dielectric materials for power electronic applications since they support exceptionally high electric fields and have high electrostatic energy density. The electrical conduction in silicate glasses is associated with ion migration and is mainly influenced by the concentration and mobility of the glass network modifiers (Ba^{++} and Ca^{++}) and impurities (Na^{+}). Studies have shown that when silicate glass is exposed to a combined effect of electric field and temperature, cations drift away from the sub-anodic region forming a depletion layer, which causes a distortion of the electric field within the glass. However, several uncertainties (i.e., overlapping mechanisms associated with cation migration, proton injection, and diffusion of non-bridging oxygen (NBO)) controlling electrical conduction and breakdown in these glasses are still not fully understood. In this study, a systematic experimental analysis will be conducted on several silicate glass structures coated in modifier-rich solutions. The ionic migration is investigated by absorption current measurements and thermally stimulated polarization/depolarization current measurements on as-received silicate glass and coated-silicate glass samples. The focus is to separate and understand the impact of univalent and divalent cations on electrical performance and reliability.

(GOMD-SP054-2021) Formation of Continuous Lithium Niobate Single Crystals in Lithium Niobosilicate Glass Via Femtosecond Laser Irradiation

J. Franklin*[1]; C. Barker[2]; S. McAnany[3]; D. Nolan[4]; B. Aitken[4]; V. Dierolf[2]; H. Jain[3]; K. J. Veenhuizen[1]

1. Lebanon Valley College, Physics, USA
2. Lehigh University, Physics, USA
3. Lehigh University, Materials Science and Engineering, USA
4. Corning Incorporated, USA

Lithium niobate ($LiNbO_3$) crystals can be grown with spatial selectivity deep inside lithium niobosilicate glass via femtosecond laser irradiation of the glass. $LiNbO_3$ possesses favorable nonlinear optical, electro-optic, and piezoelectric coefficients, making these crystals in glass potentially useful as optically active waveguides in multifunctional photonic circuits. Researchers have demonstrated the formation of $LiNbO_3$ crystals in glass in the form of oriented nanocrystals, polycrystalline lines, and segmented single crystal lines. For the first time, this work demonstrates the creation of continuous $LiNbO_3$ single crystals in glass which can be grown to indefniite length (centimeter or more). These crystals have their c-axis oriented along the laser scanning direction. For each investigated glass composition ((100 – x) $LiNbO_3$ – x SiO_2 where x = 22, 26, 30, and 34), a different optimal laser scanning speed is needed to ensure continuous, single crystal growth.

(GOMD-P055-2021) Phase-selective laser-induced crystallization of lead bismuth gallate glass

O. Magneson*[1]; C. Barker[2]; A. Hearsey[1]; J. Marsh[1]; H. Jain[3]; V. Dierolf[2]; K. J. Veenhuizen[1]

1. Lebanon Valley College, Physics, USA
2. Lehigh University, Physics, USA
3. Lehigh University, Materials Science and Engineering, USA

Spatially selective devitrification of glass can be achieved via laser irradiation, allowing for the creation of single crystal lines, which can serve as passive or active optical interconnects within photonic integrated circuits. It should be possible to create multiphase structures within the inhomogeneous temperature profile, allowing greater flexibility in designing multifunctional crystal elements. We have tested this hypothesis of phase selective crystallization by laser heating the lead bismuth gallate model glass system, particularly the $50PbO–40Bi_2O_3–10Ga_2O_3$ composition. DSC indicated the formation of multiple crystal phases within this glass. This observation was verified by XRD of glass-ceramics prepared via conventional heat treatment at each respective peak crystallization temperature, showing the formation of cubic bismuth lead oxide phase at lower temperature and triclinic phase at higher temperature. Laser fabricated crystals in glass were studied using Raman microscopy, within which the presence of multiple phases was observed. The triclinic structure formed in the center of the laser modification and cubic structure in the periphery, consistent with DSC and XRD measurements. By controlling the temperature profile by variation of laser power and laser scanning speed, phase-selective crystallization of glass was observed, whereby only the cubic phase formed at lower powers.

(GOMD-P056-2021) Synthesis and characterization of phosphate glasses containing luminescent ions: An inorganic chemistry lab practice for undergraduate students

T. I. Rubio*[1]; C. Soterio[1]; D. dos Santos Francisco[1]; G. F. Bigotto[1]; H. Maza e Silva[1]; T. V. Ferracini[1]; R. S. Baltieri[1]; D. Manzani[1]

1. University of Sao Paulo, Chemistry, Brazil

This work presents an undergraduate lab practice that aims to introduce and reinforce concepts of inorganic, physical-chemistry and spectroscopy using glass as the target material. Phosphate glasses are relatively simple to prepare because of their low melting temperatures when compared to silicate. The samples were synthesized using different amounts of Na_2O. Then, doped samples were produced by using trace concentrations of transition metals ions Mn, Ni, and Cu. A sample doped with Eu_2O_3 was also synthesized to introduce basic concepts of f-f transitions. The glass samples were characterized by infrared, UV-visible absorption, Raman scattering, and photoluminescence spectroscopies and DSC thermal analysis. After synthesis and characterization, more than 10 reports were produced from groups of 3 students each. They were able to show the glassy nature of the sample through DSC, verify the influence of Na_2O on the matrix crystallization, and measure the glass transparency window through UV-vis and IR absorption spectra, as well as determine the bands assigned to the phosphate units from Raman spectra. UV-vis was also used to characterize the d-d and f-f electronic transitions of the doped glass samples. PL measurements were used to understand the optical properties of the Mn and Eu-doped samples. This hands-on experience was successful in achieving its goals, relating theory concepts to a practical class, using a simple methodology for undergraduate students.

(GOMD-P057-2021) Energy transfer microparameters between Yb^{3+}/Tm^{3+} in highly IR transparent gallate glasses

R. F. Falci*[1]; T. Guerineau[1]; Y. Messaddeq[1]

1. Université Laval, Physics and Engineering, Canada

Material for photonic devices in the IR spectral region such as eye-safe fiber lasers, remote chemical sensing, optical amplifiers, integrated optical fiber have recently been investigated widely by different researchers. Due to its extended IR optical window ranging from 280 nm to 6 μm, mechanical strength, high glass transition temperature (T_g ~ 700 °C), high laser damage threshold, and good rare-earth oxide solubility, Heavy-metal oxide glasses are a suitable glass material for IR-application optical material. As active ion, Tm^{3+} shows two interesting emissions at 1.4μm and 1.8μm corresponding to $^3H_4 \rightarrow ^3F_4$ and $^3F_4 \rightarrow ^3H_6$ levels. But for commercial purposes, Tm^{3+} has a small absorption on 808 nm and no absorption on 980 nm, where there are several low-cost diode pumping. To overcome this issue co-doping with Yb^{3+} has been shown to be a suitable solution. In this work, we present an extensive spectroscopy characterization like luminescence, optical absorption, lifetime, etc. in the IR region for a set of gallate glasses doped with Tm/Yb. The energy migration between the ions ($Yb^{3+} \rightarrow Tm^{3+}$ and $Yb^{3+} \rightarrow Yb^{3+}$) was evaluated in terms of the energy transfer microparameters as donor-donor and donor-acceptor. We also compare our results with other similar studies in different oxide glasses.

Author Index

*** Denotes Presenter**

D

E

F

*Denotes Presenter

N

O

Author Index

T

*Denotes Presenter

Author Index

The American Ceramic Society
550 Polaris Parkway, Suite 510
Westerville, Ohio 43082

ISBN 978-1-7138-6324-3

CURRAN ASSOCIATES INC.
proceedings.com